AF559741

Agribusiness Management

Agribusiness Management

Virender Kamal Vanshi

RANDOM PUBLICATIONS
NEW DELHI (INDIA)

Agribusiness Management

ISBN 978-93-5111-584-7

Published in 2015 in India by

RANDOM PUBLICATIONS

4376-A/4B, Gali Murari Lal, Ansari Road
New Delhi-110 002
Phone : +9111-43580356, 011-23289044, 011-43142548
e-mail: sales@randompublications.com,
info@randompublications.com, randomexports@gmail.com

Reprinted 2025

Type Setting by : Friends Media, Delhi-110089
Digitally Printed at: Replika Press Pvt. Ltd.

Preface

Agriculture has evolved in to agribusiness and has become a vast and complex system that reaches for beyond the farm to include all those who are involved in bringing food and fiber to consumers. Agribusiness include not only those that farm the land but also the people and firms that provide the inputs, process the output, manufacture the food products, and transport and sell the food products to consumers.

Agribusiness management encompasses many aspects of the economy: agricultural producers, businesses that provide supplies and services to the producers (including cooperatives), businesses that add value to agricultural products, and those that facilitate the marketing of agricultural products to an ever-growing marketplace.

Business management is aimed at developing analytical and cognitive skills of micro and small rural entrepreneurs to cope with decision-making and problem-solving for their enterprises. Support could be for entrepreneurs dealing with farms and agro-industries on an individual or associative basis. Training materials are developed to provide practical advice and information on management aspects to help entrepreneurs or potential investors to run a sustainable business.

Agribusiness is synonymous with corporate farming. It combines the words agriculture and business and it involves a range of activities and methods used involving modern food production. This involves farming, seed supply, agrichemicals, farm machinery, wholesale and distribution of products, processing, marketing, and retail sales. They do not necessarily take into consideration environmental and social best practices when doing business. Their ultimate result for their bottom line is profit.

The Agribusiness Management major is designed to meet the needs of students interested in careers with agricultural input supply, agricultural production, agricultural finance, commodity assembly and processing, and agricultural marketing organizations.

The aims of this book providing students of agribusiness, teachers, researchers, professionals and all those interested in the field of agriculture with a broader understanding of agribusiness as a system and the key concepts needed to successfully manage an agribusiness enterprise.

I would like to thank my team for standing beside me throughout my career and writing this book. My special thanks go to "Random Publications" who have published the book.

– Virender Kamal Vanshi

Contents

1

Introduction to Agribusiness Management

AGRIBUSINESS

Agribusiness is the business of agricultural production. It includes agrichemicals, breeding, crop production (farming and contract farming), distribution, farm machinery, processing, and seed supply, as well as marketing and retail sales.

Ray A. Goldberg coins the term agribusiness together with coauthor John H. Davis. They provided a rigorous economic framework for the field in their book A Concept of Agribusiness (Boston: Division of Research, Graduate School of Business Administration, Harvard University, 1957). That seminal work traces a complex value-added chain that begins with the farmer's purchase of seed and livestock and ends with a product fit for the consumer's table.

The discipline of agribusiness is changing to market centric. All agents of the food and fibre value chain and those institutions that influence it are part of agribusiness system.

Agribusiness boundary expansion is driven by a variety of transaction costs. Nobel Prizes Ronald Coase and Oliver Williamson showed how transaction costs push firms to innovate due to the increased costs of resources used for the creation of goods. The term transaction cost is frequently thought to have been coined by Ronald Coase, who used it to develop a theoretical framework for predicting when certain economic tasks would be performed by firms, and when they would be performed on the market.

According to Williamson, the determinants of transaction costs are frequency, specificity, uncertainty, limited rationality, and opportunistic behaviour.

Argentine economist Manuel Alvarado Ledesma (CEMA University) explains the implications of institutions on agribusiness and writes that institutions are sets of rules, regulations, guidelines, codes and implied and express traditions which prevail in a society, which govern the relations among citizens, and also the relationship between the citizens with the government. He emphasizes that weak institutional environment allows for capricious tax, trade, pricing and investment policies by all governments to the point of creating business uncertainty. He also provides a thorough review of the empirical literature on contract farming, paying attention to broad implications for economic development. Alvarado Ledesma states that the discipline of agribusiness should contribute to the conservation of natural resources and biodiversity.

Within the agriculture industry, "agribusiness" is used simply as a portmanteau of agriculture and business, referring to the range of activities and disciplines encompassed by modern food production. There are academic degrees in and departments of agribusiness, agribusiness trade associations, agribusiness publications, and so forth, worldwide.

The UN's Food and Agriculture Organization (FAO) operates a section devoted to Agribusiness Development which seeks to promote food industry growth in developing nations.

In the context of agribusiness management in academia, each individual element of agriculture production and distribution may be described as agribusinesses. However, the term "agribusiness" most often emphasizes the "interdependence" of these various sectors within the production chain.

Among critics of large-scale, industrialized, vertically integrated food production, the term *agribusiness* is used negatively, synonymous with *corporate farming*. As such, it is often contrasted with smaller family-owned farms.

EXAMPLES

Examples of agribusinesses include seed and agrichemical producers like Dow AgroSciences, DuPont, Monsanto, and Syngenta; AB Agri (part of Associated British Foods) animal feeds, biofuels, and micro-ingredients, ADM, grain transport and processing; John Deere, farm machinery producer; Ocean Spray, farmer's cooperative; and Purina Farms, agritourism farm.

As concern over global warming intensifies, biofuels derived from crops are gaining increased public and scientific attention. This is driven by factors such as oil price spikes, the need for increased energy security, concern over greenhouse gas emissions from fossil fuels, and support from government subsidies. In Europe and in the US, increased research and production of biofuels has been mandated by law.

STUDIES AND REPORTS

Studies of agribusiness often come from the academic fields of agricultural economics and management studies, sometimes called agribusiness management. To promote more development of food economies, many government agencies support the research and publication of economic studies and reports exploring agribusiness and agribusiness practices. Some of these studies are on foods produced for export and are derived from agencies focused on food exports. These agencies include the Foreign Agricultural Service (FAS) of the U.S. Department of Agriculture, Agriculture and Agri-Food Canada (AAFC), Austrade, and New Zealand Trade and Enterprise (NZTE). TheFederation of International Trade Associations publishes studies and reports by FAS and AAFC, as well as other non-governmental organizations on its web site.

AGRIBUSINESS MANAGEMENT, ITS MEANING, NATURE AND SCOPE

Literally speaking business means bushes. In simple words "business means the state of being busy". Broadly, business involves activities connected with the production of wealth. It is an organized and systematized human activity involving and purchase of goods and service with the object of selling them at a profit. Business concerns with buying and selling goods, manufacturing goods or providing services in order to earn profit.

WHAT IS AGRIBUSINESS

The word agriculture indicate plowing a field, planting seed, harvesting a crop, milking cows, or feeding livestock. Until recently, this was a fairly accurate picture. But to days' agriculture is radically different.

Agriculture has evolved in to agribusiness and has become a vast and complex system that reaches for beyond the farm to include all those who are involved in bringing food and fibre to consumers. Agribusiness include not only those that farm the land but also the people and firms that provide the inputs (for ex. Seed, chemicals, credit etc.), process the output (for ex. Milk, grain, meat etc.), manufacture the food products (for ex. icc cream, bread, breakfast cereals etc.), and transport and sell the food products to consumers (for ex. restaurants, supermarkets).

Agribusiness system has undergone a rapid transformation as new industries have evolve and traditional farming operations have grown larger and more specialized. The transformation did not happen over night, but came slowly as a response to a variety of forces. Knowing something about how agribusiness came about makes it easier to understand how this system operates today and how it is likely to change in the future.

Initially agriculture being the major venture it was easy to become a farmer, but productivity was low. Average farmer produced enough food to feed just four people. As a consequence most farmers were nearly totally self-sufficient. They produced most of the inputs they needed for production, such as seed, draft animals, feed and simple farm equipment. Farm families processed the commodities they grew to make their own food and clothing. They consumed or used just about everything they produced. The small amount of output not consumed on the farm was sold for cash. These items were used to feed and cloth the minor portion of the country's population that lived in villages and cities. A few agricultural products made their way into the export market and were sold to buyers is other countries.

Farmers found it increasingly profitable to concentrate on production and began to purchase inputs they formerly made themselves. This trend enabled others to build business that focused on meeting the need for inputs used in production agriculture such as seed, fencing, machinery and so on. These farms involved into the industries that make up the "agricultural inputs sector". Input farms are major part of agribusiness and produce variety of technologically based products that account for approximately 75 per cent of all the inputs used in production agriculture.

At the same time the agriculture input sector was evolving, a similar evaluation was taking place a commodity processing and food manufacturing moved off the farm. The form of most commodities (wheat, rice, milk, livestock and so on) must be changed to make them more useful and convenient for consumers. For ex. consumers would rather buy flour than grind the wheat themselves before backing a cake. They are willing to pay extra for the convenience of buying the processed commodity (flour) instead of the raw agriculture commodity (wheat).

During the same period technological advance were being made in food preservation method. Up until this time the perishable nature of most agriculture commodities meant that they were available only at harvest. Advance in food processing have made it possible to get those commodities all throughout the year. Today even most farm families use purchased food and fibre products rather than doing the processing themselves. The farms that meet the consumers demand for greater processing and convenience also constitute a major part of agribusiness and are referred to as the processing manufacturing sector.

It is apparent that the definition of agriculture had to be expanded to include more than production. Farmers rely on the input industries to provide the products and service they need to produce agricultural commodities. They also rely on commodity processors, food manufactures, and ultimately food distributors and retailers to purchase their raw agricultural commodities and to process and deliver them to the consumer for final sale. The result is the food and fibre system. The food and fibre system is increasingly being referred to as "agribusiness". The term agribusiness was first introduced by Davis and Goldberg in 1957. it represents three part system made up of (1) the agricultural input sector (2) the production sector and (3) the processing-manufacturing sector. The capture the full meaning of the term "agribusiness" it is important to visualizes these there sectors as interrelated parts of a system in which the success of each part depends heavily on the proper functioning of the other two.

WHAT IS MANAGEMENT?

There is no single definition of management. Henry Fayol who is considered as the father of principles of management, - "To manage is to forecast, to plan, to organize, to command coordinate and to control".

Freederick Winslow Taylor, - "Management is knowing exactly what you want men to do and then seeing that they do it in the best and cheapest way".

Mary parker, - "Management is the art of getting things done through people".

Per Drucker, - "Management is a multi-purpose organ that manages a business, manages manager and manages worker and work".

George Terry, - "Management is a distinct process consisting of planning, organizing actuating and controlling performance to determine and accomplish the objectives by the use of people and resources". Having gone through the above definitions of management, now it can be defines as getting things done through others/subordinates. In other words, it is a process of various functions like planning, organizing, leading and controlling the business operations in such a manner as to achieve the objectives set by the business firm. It consists of all activities beginning from business planning to its actual survival.

NATURE OF SUCCESSFUL AGRIBUSINESS

Today the business has become very competitive and complex. This is mainly due to changing taste and fashion of the consumers on the one hand, and introduction of substitute and cheaper and better competitive goods, on the other. The old dictum "produce and sells has changed overtime into "produce only what customers want". In fact, knowing what customers want in never simple. Nevertheless, a farmer operator/farmer manager has to give proper thought to this consideration in order to make his business a successful one. The important requisites for success in a modern business are:

1. *Clean objectives:* Determination of objectives is one of the most essential pre requisite for the success of business. The objectives set forth should be realistic and clearly defined. Then, all the business efforts should be geared to achieve the set objectives. In a way, objectives are destination points for an agribusiness. As a traveler must know here he/she has to reach, *i.e.* destination similarly business also must know what objectives.
2. *Planning:* In simple words, planning is a pre-determined line of action. The accomplishment of objectives set, to a great extent, depends upon planning itself. It is said that it does not take time to do thing but it takes time to decide what and how to do. Planning is a proposal based on part experience and present trends for future actions. In other words, it is an analysis of a problem and finding out the solutions to solve them with reference to the objective of the farm.

3. *Sound organization:* An organization is the art or science of building up systematical whole by a number of but related parts. Just as human frame is build up by various parts like heart, lever, brain, legs etc. similarly, organization of business is a harmonies combination of men, machine material, money management etc. so that all these could work jointly as one unit, *i.e.* "business" "the agribusiness". Organization is, thus such a systematic combination of various related parts for achieving a defined objective in an effective manner.
4. *Research:* As indicated earlier, today the agricultural production philosophy "produce what the consumer want". "Consumers" behavuiour is influenced by variety of factors like cultural, social, personal and psychological factors. The business needs to know and appreciate these factors and then function accordingly. The knowledge of these factors is acquired through market research. Research is a systematic search for new knowledge. Market research enable a business in finding out new methods of production, improving the quality of product and developing new products as per the changing tastes and wants if the consumers.
5. *Finance:* Finance is said to be the life-blood of business enterprise. It brings together the land, labour, machine and raw materials into production. Agribusiness should estimate its financial requirements adequately so that it may keep the business wheel on moving. Therefore, proper arrangements should be made for securing the required finance for the enterprise.
6. *Proper plant location, layout and size:* The success of agribusiness depends to a great extent on the location. Where it is set up. Location of the business should be convenient from various points of view such as availability of required infrastructure facilities, availability of inputs like raw materials, skill labour, nearer to the market etc. Hence the business men must take sufficient care in the initial stages to selected suitable location for his business.

   The sine of the business is also important because the requirement for infrastructural facilities and inputs varies as per the size of the business. The requirement for raw materials, for example, will be les in a smaller sized firm than a larger size firm.
7. *Efficient management:* One of the reasons for failure of business often attributed to as their poor management or inefficient management. The one man, *i.e.* the proprietor may not be equally good in all areas of the business. Efficient businessman can make proper use of available resources for achieving the objectives set for the business.
8. *Harmonious relations with the workers:* In an agribusiness organization, the farmer operator occupies a distinct place because he/she is the

main living factor among all factors of production. In fact, it is the human factor who makes the use of other non-human factors like land, machine, money etc. Therefore, for successful operation of business, there should be cordial and harmonious relations maintained with the workers/labours to get their full cooperation in achieving business activities.

SCOPE OF AGRIBUSINESS

It was already indicated that agribusiness is a complex, system of input sector, production sector, processing manufacturing sector and transport and marketing sector. Therefore, it is directly related to industry, commence and trade, Industry is concerned with the production of commodities and materials while commerce and trade are concerned with their distribution.

INDUSTRY:

Industry refers to the processes of extraction and production of goods meant for final consumption or use buy individual or buy another industry for its production. Thus goods used by the final or ultimate consumers are called "consumer goods" such as edible oils, fruit jams, papaya, pickles etc.

TYPES OF INDUSTRIES:

According to nature, the industries are broadly classified into following types.

1. *Extractive industries:* These industries are concerned with the extraction;and utilization of natural resources. Example – fishing, fruit gathering, agro-based industries, forestation.
2. *Genetic Industries:* These industries include breeding of plants, seeds, cattle breeding farm, fish hatcheries, poultry farms. Of course, factors like nature, climate and environment play a dominant role in these industries, yet human skill involved in their production cannot be ignored. For example intensive agriculture is possible with greater amount of capital and larger number of workers.
3. *Manufacturing Industries:* These industries are engaged in the conversion of raw material or semi finished goods produced in the extractive industries. Some prominent examples are – cotton textile industry, spinning and weaving mills etc. Manufacturing industries can further be classified into five types: (i) Analytical industry (ii) Processing industry (iii) Synthetic industry, (iv) Service industry (v) Assembly industry.

COMMERCE

Commerce is the another major component of agribusiness. It includes all those activities which are necessary to bring goods and services from the place of their production to the place of their consumption. Thus, sit includes the buying and selling of goods and service and all those activities which facilitate trade such as storing, grading, packaging, financing, insurance and transportation.

In simple words, commerce includes trade and aid to trade. The principal function of trade (commerce) are to remove the hindrance of person, place, time exchange, knowledge etc. and ensure a free and smooth flow of goods from the producers to the consumers.

Trade in fact is a branch of commerce itself. In a way, it is the final state of business activity involving sale and purchase of commodities or goods. It does not include and to trade like transportation, insurance, banking, finance etc. On the basis of its coverage and volume, trade is normally classified into the following types:

1. On the basis of volume:
   (i) Wholesale trade
   (ii) Retail trade
2. On the basis of coverage:
   (i) Regional trade
   (ii) National trade

MANAGEMENT – NATURE, TYPES, TASKS AND RESPONSIBILITIES

Traditional concept of management restricted management to getting things done by others. According to modern view, management covers wide range of business related activities. It is considered as a process, an activity, a discipline and effort to coordinate control and direct individual and group effort towards attaining the cherished goal of the business. Management may also play the role as science, as an art, as a profession and as a social process.

As a participant in any management programme, one may either be a practicing manager or aspiring to be one. Responsibility and performance are really the key words in defining a manager's role. Performance implies action, and action necessitates taking specific steps and doing the following tasks to produce desired results.

1. Providing purposeful direction to the firm.
2. Managing survival and growth.
3. Maintaining farm's efficiency.
4. Meeting the challenge of increasing competition.
5. Managing for innovation
6. Coping with growing technological sophistication.
7. Maintaining relation with various society segments etc.

An agribusiness is a social institution. Its very existence is dependent upon its harmonious relationship with various segments of the society. This harmonies relationship originates from the farm's positive responsiveness to the various segments and is closely associated with the tasks a manager is expected to perform. The process of evolving this mutual relationship between agribusiness farms and various interest groups begins by acknowledging the existence of the responsibilities of manager. These responsibilities are towards consumers, suppliers, distributors, workers, financiers, government and the society.

2

Agribusiness Institutions and Management System

AGRIBUSINESS INSTITUTIONS

India is one of the very few countries in the world, endowed with rich and diverse agroclimatic zones, vast land resources for cultivation (about 160 million hectares), huge chunk of human resources in agriculture (nearly two-thirds of the population), wide variety of crops grown, and strong R&D facilities. Indian agriculture has performed at the most satisfactorily (Deshpande and Indira, 2004) and have comparative advantage in much farm production but does not enjoy competitive advantage. However, its share in world trade is insignificant though it has 20 per cent of world's irrigable land under foodgrains, is second largest grower of vegetables, and produces nearly 17 per cent of world's cotton. Many paradoxes still exist, for instance, inspite of good surplus of food-grains there are millions of poor dying of hunger, many farmers are poor and in debt in spite of having reasonable marketable surplus, despite good R&D facilities, farmers still practice subsistence agriculture.

One of the underlying reasons for such paradoxes is the existence of a large number of small farmers.Through successive family generations and the resulting fragmentation of land, small farmers in India are poor, undernourished, poverty stricken, depend on monsoons and. practice subsistence agriculture. As their risk-taking abilities are low, productivity declines. Poor market orientation and low value addition capacity, gives them low margins leading to low risk-taking abilities. Thus, small farmers enter a vicious circle and find it difficult to break. Little bargaining power and heavy dependence on their farm produce for basic needs, add to their woes and they end up becoming victims to the exploitative market intermediaries who deny them a fair share of the consumer rupee.

Small farmers in India in a bid to hedge their miseries embraced diversification of agriculture. During the past two decades, there has been a perceptible increase of incomes from livestock, fisheries and forestry. Food crops have been replaced by cash crops in many areas. However, the experience of agricultural development in India has shown that benefits will not accrue to the small farmers unless they are integrated into the markets as recognisable players through innovative institutional arrangements. Institutions, especially cooperative and corporate, have attempted to integrate small farmers into the market by removing bottlenecks in marketing of agricultural produce and regulating the market in favour of small farmers. A few institutional arrangements have claimed to have reduced the transaction costs by steadily transforming from manpower driven to technology driven, from disintegrated supply chains to integrated supply chains, from finance as a source of control to information as a source of control.

Two institutional arrangements, namely, Anand pattern of cooperatives and e-choupal have received wide attention for the benefits they claim to have passed on to the small farmers. There have been various studies conducted to evaluate and verify such claims, but most of the studies evaluated these two institutions exclusively. Hardly any study has been conducted comparing and contrasting both. We attempt to do the job. The purpose of comparison is to bring out the salient features embedded into the arrangements that benefit small farmers and look for the institutional arrangement that offers better services to the small farmers. The outcome of the analysis would be helpful for policy makers to take appropriate initiatives.

In analysing the institutional arrangements we first look at how these institutional arrangements have reduced transaction costs in terms of opportunism, information asymmetry and asset specificity at various levels. Our premise is that institutional arrangement that has been able to reduce transaction costs for the small farmers and for itself should be preferred. Our premise also configures design of the institution as a key determinant of transaction costs. Thus, in our analysis we go beyond organisational structure, physical layout of facilities and consider creative process of problem solving of the small farmers, services offered and procedures and policies of the institutions that expand the opportunities for the small farmers. The chapter is based on an exploratory study and does not empirically measure transaction costs or design elements but analyses theoretically. Our analysis is based on the available literature and field visits to a few select villages where the institutions are in operation.

COOPERATIVE INSTITUTIONS

Voluntary and democratic in nature, participatory in approach, commitment towards ethical standards, socially responsive, are some of the distinguishing and differentiating features of cooperatives. Unlike corporates which have profit as major motive, cooperatives attempt to reduce income disparities, improve the social conditions, combat exploitation and strive for a better society through sustainable developmental activities. Cooperatives offer a good alternative for small farmers when there are wider market imperfections. Cooperatives can procure supplies at lower transaction cost through suitable backward linkages and aggregate them appropriately to harness the benefits of economies of scale. They can be helpful to small farmers in providing credit, inputs and extension services. More often they are also involved in the processing of the produce so that a higher value can be generated. Noted among the agribusiness cooperatives in India and world across is the Anand pattern of cooperatives.

E-CHOUPAL INDIAN TOBACCO COMPANY'S (ITC) PIONEERING AGRIBUSINESS INITIATIVE

Initiative 'e-choupal' is unique in many ways. The e-choupal: is a complete end-to-end solution as it delivers real-time information and customised knowledge to improve farmers' decisionmaking ability to align farm output with market demands, and to improve productivity; aggregates demand like a virtual producers' cooperative and provides access to high-quality farm inputs at lower cost; acts as a direct marketing channel with more efficient price discovery and lower transaction costs in output marketing; is scalable as it is built on market principles; is replicable across different crops and geography in India because of its conceptual strength; and integrates farmers to the market place and simultaneously, meets the needs and challenges of consumers at the bottom of the economic pyramid. Currently, there are 5,500 choupals in 26,000 villages covering 2.6 million farmers spread across six states of India (personal communication, 2006). E-choupal, in addition to forward integrating farmers into the markets, sells agricultural inputs and other products of the organisations through backward supply chain movements.

The extension services offered by e-choupal at the village level include e-governance, health and education services. A second layer of infrastructural facility 'choupal sagar' is also offered with facilities like warehousing, weighment of farmers' produce, soil-testing laboratory, training centre for farmers, life and general insurance, supermarket, e-health facility, pharmacy, diesel station and so on.

In the next years, e-choupal intends to be present in 15 states, one lakh villages, and deal with wider range of commodities like grains, oilseeds, coffee, spices, cotton, horticulture and aquaculture. At present, there are four types of choupals tailored specifically for four different commodities, *viz.*, soybean, wheat, coffee, and shrimps.

We discuss the soy-choupal model for two reasons; firstly soya in India is grown largely by small farmers and secondly, it is the first commodity for the model becoming the benchmark for other choupal models. Internet-enabled multimedia computers with solar power backup and VSAT connectivity is set up at company's cost for every 5 or 6 villages.

This set up is called e-choupal and it is installed in a selected farmer's place called sanchalak who is chosen on the basis of high social status, entrepreneurial and development orientation. Sanchalaks once chosen, take an oath in public to serve the village and are given training by ITC on basics of computers and internet, design of echoupal and the role they need to play in villages.

They receive 0.5 per cent of commission on the procured crop and bear the operational costs of the installed equipment. The company has reintermediated rather than disintermediating the role of the traditional commission agents by making them sanyojaks (coordinators) in the network. They are the source of liquidity as they manage the physical flows of the soya, collect price information from the local mandis and maintain records.

COMPARING TRANSACTION COSTS

In transaction cost analysis, the extent of information available between the parties of exchange, opportunistic behaviour of the members in the exchange and the level of asset specificity to the transaction outline the cost. Further, as exchange relations are not always cooperative in nature, the perspective of the implications would differ from the sellers to buyers. While comparing the transaction costs of two institutional arrangements, we consider these aspects.

As mentioned earlier, the evaluation of institutions' performance and the choice of better institution emerging thereof is based on their ability to reduce transaction costs on various fronts and especially, for the small farmers. Farmers have been able to reduce their transaction costs considerably, after becoming members of the cooperative society. As cooperatives did not act as any typical middlemen who hide market information, the costs associated with information asymmetry were minimised to a great extent.

As each member has equal voting rights, the opportunistic tendencies within the cooperative society are negligible provided the practices are institutionalised and rules and regulations enforced properly. The Anand pattern of cooperatives has been able to cut transaction costs for itself on three fronts—information asymmetry, opportunism and asset specificity—by evolving governance structures which harmonise exchange relations across parties. Before the farmer sells his milk to the society, s/he knows the base price of procurement of milk offered by Union. In general practice, prices are same for a given district.

Any change in the prices would be informed to the societies immediately and societies in turn inform the farmers. Such a mechanism removes any information asymmetry. As cooperatives at village level have continuing relationship with the farmer members, milk can be collected regularly and hedged to the uncertainties of markets and opportunistic behaviour of the trading partners/ middlemen.

It has also avoided a major chunk of implied cost of agency by eliminating possible alienation across membership and conflict with management among the members. By training the society staff on specialised skills like artificial insemination and animal health care, it has reduced transaction costs considerably. Staff members of the cooperatives, including chairman are equal share holders in the society. They are equally benefited as any poor or low caste member of the society.

Such an arrangement would also ensure little opportunity for transaction costs to arise due to opportunistic behaviour. Further, as membership of the cooperatives is long term, asset specificity-related transaction costs also are reduced to a considerable extent. Most importantly, if the cooperative society has been able to reduce transaction costs substantially, the benefits would be trickled down to each of its members.

This kind of savings isunique to cooperative system which unfortunately is not possible in any other institutional arrangement, more so in a corporate. By generating enough economies of scale, in terms of collection of milk and by routing appropriately so that maximum milk is collected within minimum distance of transportation, the union has been able to reduce transaction costs to a good extent. Costs in the way of asset specificity have been reduced for the union through outsourcing the transportation activities. Supply of inputs and cattle feed is done through the same transport arrangement further reducing transaction costs.

The arrangement of APC is such that all unions do not produce same milk products, thus nullifying costs arising out of conflict of interests and competition among unions. With ERP packages, federations have been able to network through the dealers and able to reduce transaction costs on all three aspects, *viz.*, information asymmetry, opportunism and asset specificity in the forward supply chain.

E-choupal has also been able to reduce transaction costs for the farmers and in its own operations. Farmers have saved transaction costs approximating to ₹ 270 per MT which would have been incurred in mandi (trolley freight to mandi = 100, filling and weighing labour = 70, labour Khadi Karai = 50, Handling Loss = 50), (as given to us in presentation by the company officials). Another study stated that e-choupal has enabled the organisation to reduce transaction costs of procuring soya from 8 to 2 per cent. About half of these savings are shared by the farmers.

As an institution, e-choupal has been able to considerably reduce transaction costs in terms of information asymmetry and asset specificity. However, we suspect that the transaction costs which might arise out of opportunism still need to be reduced. For example, one of the choice criteria for considering a farmer as sanchalak is his entrepreneurial aptitude. An entrepreneur in any system would always attempt to maximise his returns. This attitude would most likely make him more opportunistic, thus raising the scope for increase in the transaction costs. Therefore there is need to improve the accountability of Sanchalaks.

COMPARING DESIGN

Broadly, both the institutional arrangements have been successful in reducing transaction costs to a considerable extent and increase financial returns to the small farmers. However, Anand pattern of cooperatives' design scores as a better alternative compared to e-choupal in terms of the producers, control over the agribusiness, consistency in the share of consumer rupee, response to the farmer's needs, removal of intermediaries, transparency of operations from the small farmers' perspective and purchase guarantees. E-choupal fares equal to Amul design on fronts like inputs and technical assistance, and response to farmer requirements and fares better on response to market requirements.

CHALLENGES OF AGRIBUSINESS INSTITUTION

Challenges before any agribusiness institution is to integrate small farmers into the market and benefit them in an efficient, equitable, sustainable and transparent manner. Unless they are in control of the business, they will be in control of the business. Cooperatives have been successful in benefiting small farmers for a long time as they are owned and controlled by the farmers.

Corporate on the other hand, is known for its shrewd profit motives and poor social concern. Their inclination towards sustainable agricultural practices is obviously less as they are neither owned nor controlled by the farmers. These arguments have received support from our analysis of the two institutional arrangements; Anand pattern of cooperatives and echoupal. In most of the criteria for evaluation of the institutional offers to the benefit of the small farmers, Anand pattern of cooperatives stand out. A World Bank report on Anand pattern of cooperatives and its replication across India has concluded "...there should be no further lending to the dairy sector in states which have not yet adopted the full Anand Pattern of cooperatives or which do not treat these cooperatives equally with private cooperatives".

In another recent development, World Bank is attempting to replicate Anand pattern of cooperatives in South Africa. In a statement, World Bank states, The World Bank agrees that the model used in Operation Flood is a viable business model for farmers as it provides high returns. World Bank will now identify the African countries where Operation Flood could be replicated so that the poor farmers there could become self-sufficient.

These statements are the testimony of the success of Anand pattern. Hence, we strongly feel that policy formulators should encourage relatively more of cooperative form, more so, Anand pattern of institutions than that of corporate institutions. However, we do not insist that the role of corporate institutions should be undermined and not encouraged. The comparison that we have done here is in the present context and only relative in nature, not in absolute terms of benefits offered. Further, Anand pattern of cooperatives is more than half a century old which has withstood the test of time and has evolved over years. E-choupal is a recent emerging institutional arrangement and might take some time to settle down and benefit the small farmers. Whether or not small farmers would benefit in same manner as the large farmers though e-choupal of ITC, would largely depend upon benevolence of top management and concern for small farmers. There is no inbuilt mechanism in e-choupal to ensure that the small farm households would not be ignored once their initial euphoria of social concern dies down and efficiency aspects begin to count more than equity and equal participation of all farm households.

DEVELOPING AN AGRIBUSINESS STATISTICS AND INFORMATION SYSTEM

One strategy to access global markets was to ensure the availability of agribusiness/market information. The agribusiness/market information system, therefore, would need to be developed to link and match supply in rural areas and demand in the cities. The advantage was that the farmers would be given access to the market and would be able to use agribusiness/market information for better decision-making. In countries with a fairly developed ASIS, three major types of information, and a fourth (other information) constituted the dominant scope of the system.

- Price information would always be the centerpiece of any ASIS as it was necessary to make informed marketing decisions at all levels of an agribusiness system.
- The next most important sets of information for ASIS were those that relate to public and private supplies of commodities at various levels of the agricultural marketing system, along with regular information on the volume of arrivals and outflows (origin/destination) of marketed commodities.
- Information on acreage as well as production would be the third most important type of information for ASIS, a necessary component of the agricultural situation and outlook service.
- Other information on commodity losses, commodity quality specifications, new marketing or post-harvest technologies that improve drying and storage, and alternative costs of transportation from production to consumption areas would also be important, although they might not collected as frequently as the other three.

The development of agribusiness statistics and information system in Indonesia began in 1995 and lasted until 1997. This was through an FAO Technical Cooperation Project, which developed the agribusiness sector by setting up a user-oriented, demand-driven agribusiness statistics and information service. The TCP set up the Agribusiness Statistics and Information System (ASIS) to fulfill four specific functions: to inform, to understand, to infer, and to decide. Each function was translated into clearly defined service that ASIS provided: Agribusiness News Service; Price and Volume Monitoring Service; Agribusiness Analysis Service, and Agribusiness Advisory Service.

The launching areas were West Java and the grocery market Kramat Jati in Jakarta, with regular monitoring of market arrivals done in the latter, and the pilot demonstration of the statistical index area approach in the former. To facilitate data capture, the Project installed a computer unit in the Pasar's administrative office, two staff members assigned to undertake data entry, and CADI sent a disk file of the records of market arrivals. The information available from special reporting forms for vegetables and fruits (called SPII and SPIII) was computer-processed.

The first few months of the project's implementation of the system proved encouraging. CADI thus, requested the Project to expand the application of the statistical index area approach to include Central and East Java. The data accumulated from these operations, along with the price information obtained directly from the Directorate-General for Food Crops and Horticulture, provide the statistical inputs to the Agribusiness Statistical Indicators bulletins initiated by the Project.

The data collection frame was based on the concentration of the area of vegetables and fruits. Vegetables included shallot, garlic, potato, cabbage, chilli, and tomato; and fruits included mango, papaya, banana, pineapple, and orange. The sampling domain was the province, with the object of data collection at sub-districts within the sampled district. Data collection was done by agricultural officials in each sub-district.

The data collection methodology used for acreage and production of vegetables and fruits was complete reporting, using special forms within the concentration area, through interviews with farmers or farmers groups, and head of villages, and sometimes through eye estimates.

Grocery prices and supply volume data including point of origin were collected directly from grocery market, while retail prices were collected daily at retail markets. After one month, the data was processed to get the average grocery price as well as retail price, and the total volume for each commodity. The results were analyzed and published in the monthly bulletin, which were distributed to all related agencies. In addition, the publications were posted in the web site of the Ministry of Agriculture so that more users could read them.

THE AGRIBUSINESS SYSTEM APPROACH TO AGRICULTURE

The true meaning of the agribusiness system paradigm remains widely misperceived; it is confused with large farm enterprises. The agribusiness approach is then misinterpreted as a strategy which focuses merely on promoting large agricultural corporations to achieve high economic growth and high export figures, ignoring the small family farms and hence, is no help to the rural poor. While the approach may be good for high growth it would be at the cost of the loss of livelihoods of many apart from environmental destruction.

The agribusiness system approach is misperceived as part of the neo-liberal economic policy which blindly advocates the free market, ignores agrarian institutions and hence will fail to address issues such as fairness in division of surplus, participation and income equity.

The agribusiness system approach is a misfit with the present context of Indonesian agriculture. The agribusiness system is identical with the Dutch colonial policy that promoted large plantation enclaves, created a dual agrarian structure, pushed out small family farms and created a situation wherein the laborers were exploited by greedy capitalists.

This argument further says that as long as small family farms are still predominant, readoption of the agribusiness system approach can only repeat the above colonial history.

The basic tenet of the agribusiness paradigm may be summarized as follows. First, farms, small and large, are profit-oriented business enterprises. This is why Davies and Goldberg introduced a new word “”agribusiness””, agriculture-related business, to replace the old term “”farming””.

The agribusiness system approach assumes that even a very small family farm is actually a profit-oriented business, hence all government policies must be based on this basic premise.

This premise is now accepted as a universal truth. Policy wise, it has two important implications. The prime objective of agricultural policy must be to increase farming profit and thus farmers’ income. Here this implies a policy paradigm shift from production orientation, adopted during the Soeharto administration, to farmers’ income orientation.

If increasing farm profits, including and especially of the small ones, is a key to poverty alleviation in rural areas, then the agribusiness approach to agricultural development is consistent with this objective. By focusing on increasing farmers’ income, the agribusiness system approach puts the farmers, rather than officials, at the center of agricultural policies. The agribusiness system approach is a farmers’, or people-driven approach to agricultural development.

Second, farming is a key link in the chain of the commodity system, thus farming performance is determined by the commodity system’s performance. A commodity system can be divided into input suppliers, on-farm, output processors and distributors, as well as supporting infrastructure and services. The performance of the farming (on-farm) component depends on the other factors. Most farming constraints are off-farm, characterized by an inefficient input supply system (insufficient availability, low quality, high price of agricultural inputs), underdeveloped agroprocessing industries and inefficient marketing systems (low farm gate price), and insufficient supporting infrastructure (lacking irrigation system, credit availability, agrarian institutions).

The agribusiness system approach is an integrated commodity system based on development: “”How to get the commodity system right””, rather than “”how to get the farming right””.

Access to a sufficient size of productive land is indeed a key to uplift the marginal farmers out of the poverty trap. A seemingly plausible policy option for this is agrarian reform. But it is misleading if agrarian reform is said to be inconsistent with the agribusiness system approach.

Agrarian law is part of the supporting infrastructure or institutional policies in the above commodity system. Mistakes would not have been caused by the agribusiness system approach but in the implementation of the strategy.

3

Financial Management for Agribusiness

FINANCE IN AGRICULTUREFINANCE IN AGRICULTURE

Finance in agriculture is as important as development of technologies. Technical inputs can be purchased and used by farmer only if he has money (funds). But his own money is always inadequate and he needs outside finance or credit. Professional money lenders were the only source of credit to agriculture till 1935. They use to charge unduly high rates of interest and follow serious practices while giving loans and recovering them. As a result, farmers were heavily burdened with debts and many of them perpetuated debts. There were widespread discontents among farmers against these practices and there were instances of riots also. With the passing of Reserve Bank of India Act 1934, District Central Co-operative Banks Act and Land Development Banks Act, agricultural credit received impetons and there were improvements in agricultural credit.

A powerful alternative agency came into being. Large-scale credit became available with reasonable rates of interest at easy terms, both in terms of granting loans and recovery of them. Although the co-operative banks started financing agriculture with their establishments in 1930's real impetons was received only after Independence when suitable legislation were passed and policies were formulated. Thereafter, bank credit to agriculture made phenomenal progress by opening branches in rural areas and attracting deposits.

Till 14 major commercial banks were nationalized in 1969, co-operative banks were the main institutional agencies providing finance to agriculture. After nationalization, it was made mandatory for these banks to provide finance to agriculture as a priority sector. These banks undertook special programmes of branch expansion and created a network of banking services throughout the country and started financing agriculture on large scale. Thus agriculture credit acquired multi-agency dimension. Development and adoption of new technologies and availability of finance go hand in hand. In bringing "Green Revolution", "White Revolution" and now "Yellow Revolution" finance has played a crucial role. Now the agriculture credit, through multi agency approach has come to stay. The procedures and amount of loans for various purposes have been standardized. Among the various purposes "Crop loans" (Short-term loan) has the major share. In addition, farmers get loans for purchase of electric motor with pump, tractor and other machinery, digging wells or boring wells, installation of pipe lines, drip irrigation, planting fruit orchards, purchase of dairy animals and feeds/fodder for them, poultry, sheep/goat keeping and for many other allied enterprises.

AGRICULTURE GROWTH RATE IN INDIA

Agriculture Growth Rate in India GDP had been growing earlier but in the last few years it is constantly declining. Still, the Growth Rate of Agriculture in India GDP in the share of the country's GDP remains the biggest economic sector in the country.

India GDP means the total value of all the services and goods that are produced within the territory of the nation within the specified time period. The country has the GDP of around US$ 1.09 trillion in 2007 and this makes the Indian economy the twelfth biggest in the whole world.

The growth rate of India GDP is 9.4 per cent in 2006-2007. The agricultural sector has always been an important contributor to the India GDP. This is due to the fact that the country is mainly based on the agriculture sector and employs around 60 per cent of the total work-force in India. The agricultural sector contributed around 18.6 per cent to India GDP in 2005. Agriculture Growth Rate in India GDP in spite of its decline in the share of the country's GDP plays a very important role in the all round economic and social development of the country.

The Growth Rate of the Agriculture Sector in India GDP grew after independence for the government of India placed special emphasis on the sector in its five-year plans. Further the Green revolution took place in India and this gave a major boost to the agricultural sector for irrigation facilities, provision of agriculture subsidies and credits, and improved technology. This in turn helped to increase the Agriculture Growth Rate in India GDP. The agricultural yield increased in India after independence but in the last few years it has decreased.

This in its turn has declined the Growth Rate of the Agricultural Sector in India GDP. The total production of food grain was 212 million tonnes in 2001-2002 and the next year it declined to 174.2 million tonnes. Agriculture Growth Rate in India GDP declined by 5.2 per cent in 2002-2003. The Growth Rate of the Agriculture Sector in India GDP grew at the rate of 1.7 per cent each year between 2001-2002 and 2003-2004. This shows that Agriculture Growth Rate in India GDP has grown very slowly in the last few years. Agriculture Growth Rate in India GDP has slowed down for the production in this sector has reduced over the years.

The agricultural sector has had low production due to a number of factors such as illiteracy, insufficient finance, and inadequate marketing of agricultural products. Further the reasons for the decline in Agriculture Growth Rate in India GDP are that in the sector the average size of the farms is very small which in turn has resulted in low productivity. Also the Growth Rate of the Agricultural Sector in India GDP has declined due to the fact that the sector has not adopted modern technology and agricultural practices. Agriculture Growth Rate in India GDP has also decreased due to the fact that the sector has insufficient irrigation facilities. As a result of this the farmers are dependent on rainfall, which is however very unpredictable.

Agriculture Growth Rate in India GDP has declined over the years. The Indian government must take steps to boost the agricultural sector for this in its turn will lead to the growth of Agriculture Growth Rate in India GDP.

FINANCE FOR AGRICULTURE AND FOOD PROCESSING

Banks: The Food Processing sector has access to credit from Commercial Banks-Indian and Foreign, Cooperative Banks and the Regional Rural Banks. The formal institutional sector assists the food processing companies through long term loans for capital investments and short term loans for working capital. The total lending to the food processing sector is clubbed with lending to the agriculture sector and separate details for financing to the food processing sector are not available.

- *National Bank for Agricultural and Rural Development (NABARD):* NABARD offers refinance facilities for food processing, agri infrastructure, developmental assistance for RRBs, DCCBs, SSI and SHG linkages, and assists in infrastructure development and research.
- *Small Industries Development Bank of India (SIDBI):* SIDBI has been involved in assisting the entire SSI Sector including tiny, village and cottage industries through suitable schemes tailored to address the funding requirements for expansion, diversification, modernisation and rehabilitation.
- *Export Import Bank (EXIM Bank):* Exim Bank assists in financing and facilitation of foreign trade. The Bank also offers financial support to companies engaged in exports.
- *National Cooperative Development Corporation (NCDC):* NCDC assists in promotion, planning and financing the agricultural supply chain from production, processing, storage and trade of agricultural produce and food products. NCDC also provides assistance for marketing of certain notified commodities *e.g.* fertilizers, pesticides, agricultural machinery etc.
- Ministries/Government bodies
- *Ministry of Food Processing Industries (MFPI):* MFPI is the Nodal

agency for development of the processed food sector in the country MFPI's financial schemes include schemes for technology upgradation, Human resource development, Quality testing, R&D,TQM, backward and forward integration, development of infrastructure including food parks.

- *Agricultural and Processed Foods Products Export Development Authority (APEDA):* APEDA facilitates market linkages between Indian producers, manufacturers and the international market. APEDA provides financial assistance for market development, infrastructure development and development of quality enhancing facilities.
- *Ministry of Agriculture (MoA), Government of India:* The Ministry of Agriculture under various schemes, provides financial assistance for development of specific crops for investment in seeds, irrigation, farm implements, inputs, infrastructure and training.
- *National Horticultural Board (NHB):* NHB promotes integrated development in horticulture, assists in development of post harvest management infrastructure, promotes production and processing of fruits and vegetables, strengthening of market information systems and assists in R&D programmes in cultivation and processing. NHB's financial schemes are directed towards commercial horticulture and infrastructure related to post harvest techniques.

Financial assistance from these organizations are in the form of grants, back-ended subsidies, soft loans, refinance etc., with most of the schemes directed to specific sub-sectors of the agri/food processing industry. In addition, some State Governments offer financial schemes to companies in the sector.

The key issues with Government schemes are highlighted below:

- *Overlap among schemes:* Various Government agencies have separate schemes which aim to achieve the same objectives. Thus, an applicant can potentially approach different agencies and obtain assistance from more than one agency. Some of the examples of overlapping schemes are as follows:
- *Several areas addressed inadequately:* There are a host of areas which are not addressed by these schemes such as financing of freezer cabinets for retail outlets, financing of bulkcoolers for milk, financing of alternate markets/mandis, financing of vending machines for tea/ coffee/beverages etc.
- *Creation of excess capacity:* There are instances where assistance has been provided to set up cold storages in locations which have excess storage capacity. This has lead to an unviable situation for all cold storage operators in the region. It is essential for the nodal agencies to achieve development of the food processing sector in a sustainable manner.

- Inadequate financial assistance^ most instances, there is a cap on total financial assistance provided at INR 5-5.5 mn. This leads to fragmentation in capacity creation, and often, a situation of overcapacity. At the same time, it does not allow players to scale up their operations.
- *Inadequate Monitoring of progress of schemes:* It is important to track progress of the projects funded by the Government. Periodic appraisal of the progress will ensure that funds are appropriately utilized. This will also enable assessment of success and lacunae of these schemes and record the multiplier effect of the financial support provided.
- Most of the schemes are back-ended which leads to a funds crunch during implementation.
- Institutions providing subsidies do not undertake their own assessment and rely on the appraisal report of the bank which provides loans to the project.
- There are significant time lags from the date of application for financial assistance, to release of funds, and may affect the project schedule, and lead to cost overruns
- These schemes do not address working capital requirements, which due to the nature of the industry are significant and often higher in terms of quantum than capital investment requirements
- Applicants claim that they are often unaware that their applications have been rejected on account of lack of communication from the funding institution. Further, applicants are unaware of the reasons for rejection.

SOURCES OF FINANCE FOR AGRICULTURE

At present nearly 60 per cent of rural credit is used for agriculture and allied activities, 10 per cent for non-farm activities and the balance 30 per cent for funding household consumption expenditure. Nearly 40-45 per cent of rural credit needs are catered to by formal credit institutions and the balance by the informal sector including commission agents, input suppliers, traders, Self Help Groups (Joint Liability Groups), processing industries and professional money lenders.

The Indian commercial banks are mandated by the Reserve Bank of India (RBI), to undertake directed lending for agriculture and rural development. Commercial banks are required to achieve priority sector lending of 40 per cent of net bank credit of which 18 per cent should be for agriculture.

Further, sub-targets are also specified *i.e.*, Direct Agriculture (credit for direct farmer benefit, Target: Minimum of 13.5 per cent of the net bank credit) and Indirect Agriculture (Target: Maximum of 4.5 per cent of the net bank credit).

In FY 03, total agricultural advances outstanding under priority sector lending, by Public Sector Banks stood at INR 735 billion, on a total net bank credit portfolio of INR 4779 billion. As compared to this, agricultural advances by Private Sector Banks stood at INR 119 billion, on a total net bank credit portfolio of INR 718 billion. The following Exhibit provides details of direct and indirect priority sector credit amongst Public and Private sector banks during FY03.

In case of a shortfall in lending to agriculture, banks are required to invest in Rural Infrastructure Development Fund (RIDF) deposits, which offers interest rates linked to the bank's performance in lending to agriculture. These rates are inversely proportional to the shortfall in agricultural lending. The interest rate on RIDF is provided in slabs, based on the achievement of the priority sector lending target, and can be as low as 4 per cent to 5 per cent.Therefore, it is imperative to achieve the agriculture target and more specifically the Direct Agriculture target as it accounts for 75 per cent of the total.

Established in 1996, RIDF is being utilized for making investments in rural infrastructure projects. Nine tranches of RIDF have been established with an aggregate corpus of INR 340 billion, with the ninth tranche of INR 55 billion. Cumulative sanctions and disbursements under various tranches of RIDF stood at INR 334 billion and INR 193 billion respectively, at the end of February 2004.The RIDF funds are used to finance minor irrigation projects, shallow wells, drinking water facilities, roads, bridges and storage infrastructure.

AGRIBUSINESS ECONOMICS

Agribusiness economics is concerned with understanding how institutions, organizations, and markets affect vertical and horizontal coor-dination within the food system. This chapter describes five significant contributions that members of our profession have made to the design and analysis of institutions, organiza-tions, and markets.

The importance of under-standing basic economic principles in farmer decision making when initiating joint verti-cal integration. Emphasizing scale economies, asymmetric information and other market fail-ure elements as rational economic reasons for forming cooperatives, agricultural economists often confronted the advocacy frenzy of politi-cians and farm leaders rushing to organize producers into agricultural cooperatives in the post–World War I depression. They also published warnings of the dangers of moving too quickly without understanding market structure forces or the debilitating implications of inadequate capital and human resources.

During these early years, agricultural economists advanced understanding of mar-ket coordination through detailed descriptions of market functions and the costs of partic-ipating privately or collectively in specific supply chains.

Their emphasis on analyzing the performance and welfare role of mar-keting cooperatives relative to multipurpose cooperatives continues today.

Transforming agricultural commodities into food products typically requires conversion of large amounts of lower-value materials into more valuable products and transport (of agri-cultural inputs and food product outputs) over considerable distances. To address this eco-nomic challenge, managers need to be able to assess both cost of production alternatives within a single production facility as well as the total cost-effectiveness of locating several facilities across a region. A similar challenge exists in terms of designing the most effec-tive production and transportation system to provide inputs to farm production operations. Greater operational efficiency results in higher performance levels for agribusiness firms and enhances social welfare through lower food costs and higher-quality food products.

Over the last 100 years, application of microeconomic principles along with use of evolving quantitative analytical tools has pro-vided significant opportunity for innovation throughout the agribusiness sector. Especially in the middle decades of the 1900s, agricul-tural economists led both intellectually and in application of these capabilities to enhance operational efficiency of supply, processing, and distribution in the sector.

In the 1940s, the economic engineering approach to estimation of plant cost rela-tionships began to be employed to address the more managerially relevant question of optimal plant size for a specific commodity and setting. The work of R. G. Bressler and numerous colleagues made a particularly pro-found contribution to application of the eco-nomic engineering approach in the food sector. Early work focused on Connecticut and the dairy industry, while numerous later efforts were conducted at the Univer-sity of California. The economic engineering approach focused on synthesizing cost func-tions from engineering, biological, and other sources of information, including accounting data, based upon process level input-output relationships.

Although an individual food manufactur-ing facility might achieve exceptional internal efficiency, the overall economic performance of that unit can be materially affected by the costs of obtaining agricultural inputs and of distributing the factory's output.

AGRIBUSINESS STRATEGY

Business strategy focuses on optimizing the linkages between the firm and its surround-ing business environment. This managerial function inherently incorporates a longer-run perspective, striving to ensure that the firm not only is well aligned with current conditions but also is being positioned to adapt to the dynamic and uncertain business environ-ment of tomorrow. A key element of strategy can be paraphrased as a "how will we compete" question. Business management concepts such as competitive advantage, dis-tinctive competencies, and the resource-based view of strategy have been intertwined in the scholarship addressing this question, especially during the last three decades.

Understanding how to compete within the context of a dynamic agricultural sector and an evolving global economy is a key challenge for agribusiness decision makers. Throughout the last 100 years, the scholarship of agricul-tural economists has contributed to improving decision-maker understanding of the business environment of agriculture.

In addition to his seminal work with Davis, Goldberg pioneered in employing the case study method within agribusiness. He utilized this form of scholarship to uncover relation-ships and to communicate the dynamics of change to legions of decision makers. The more than 300 Harvard Business School case stud-ies he has coauthored address a vast array of factors affecting global agriculture. The Agribusiness Seminar he has led also has been a powerfully effective means of educating agribusiness managers.

4

Agricultural Marketing and Cooperation

AGRICULTURAL MARKETING

Agricultural marketing involves in its simplest form the buying and selling of agricultural produce. In olden days when the village economy was more or less self-sufficient the marketing of agricultural produce presented no difficulty as the farmer sold his produce to the consumer on a cash or barter basis.

MARKETING FUNCTIONS

In modern marketing the agricultural produce has to undergo a series of transfers or exchanges from one hand to another before it finally reaches the consumer.

This is achieved through three marketing functions:

1. Assembling,
2. Preparation for consumption
3. Distribution.

Concentration pertains to the operations concerned with the assembly and transport from the field to a common assembly field or market. The produce may be taken direct to the market or it may be stored on the farm or in the village for varying periods before its transport. It may be sold as obtained from the field or may be cleaned, graded, processed and packed either by the farmer or village merchant before it is taken to the market. Some of the processing is done not because consumers desire it, but because it is necessary for the conservation of quality.

At the market the produce may be sold by the farmer direct to the consumer or more usually through a commission agent or a broker. It may also be purchased by traders, wholesalers or retailers. The transactions may be carried out by direct negotiation or through middlemen, by barter or cash, by open or under cover auction, on the spot or in future markets. The transactions take place at one or more levels in the primary, secondary or terminal markets or all three. Distribution (dispersion) involves the operations of wholesaling and retailing at various points. By a series of indispensable adjustments and equalising functions, it is the task of the distribution system to match the available supplies with the existing demand.

MARKETING FUNCTIONARIES (AGENCIES)

The transfer of produce or goods takes place through a chain of middlemen or agencies. In the primary market the main functionaries are the producer, the village or itinerary merchant, pre-harvest contractors, commission agents, transport agents etc. In the secondary market the processing and manufacturing agents are the additional functionaries. Financing agents such as shroffs, banks and co-operatives may also take part. In the terminal or export market the commercial analyst and shipping agent also gets involved in the transfer of goods.

The functionaries have their own set up. They may be individuals, partners or co-operatives who may buy and sell on ready and future basis at a price determined by forces of supply and demand. Each functionary renders some service in the process of marketing and also earns a varying margin of profit for himself. This procedure makes marketing rather complicated and inflates the price of the produce. The nature of some of the agricultural products for example their bulk form and perishability and their seasonal availability further add to the complexity of agricultural marketing.

MARKETING IMPROVEMENTS

India being a primary producing country, agriculture plays a vital role, both as an essential infrastructure and a development component in generating and sustaining a higher national income. Out of a national income of about ₹38,921 crores in 1972-73 as much as ₹17,500 crores or about 44.9 per cent is contributed by agriculture and allied sectors. It is estimated that about 50 per cent of the agricultural produce is available as marketable surplus. The marketing system in India provides sustenance for about 3 million persons who are engaged in performing various marketing function. In the field of exports too, the agricultural sector accounts for about 50 per cent of the total value.

The process involved in the disposal of such a substantial produce of great economic importance are significant not only for the farmer but also for the country as a whole. The unreasonably low return that the farmer gets for his produce and the excessive margin of profit retained by the intermediaries attracted the Government's attention and it was felt that the economic condition of the agriculturists could not be improved unless determined steps were taken to establish an orderly system of marketing in the country.

GOVERNMENT REGULATORY PROGRAMMES

With this object in view, a number of marketing surveys were conducted by the Directorate of Marketing and Inspection which revealed the shortcomings in the country's marketing system. A rectification of these deficiencies was sought to be achieved by rationalizing various activities and standardizing various practices in the markets through legislation or otherwise. The primary objective of improving the system of agricultural marketing was not only to remove the handicaps from which the producer-seller was suffering but also to increase his income by ensuring him a fair price.

REGULATION OF MARKET

Prevailing market practices and market charges made a deep cut in the share of the producer in the price paid by the consumer. Some of the market charges were authorized whereas others were more than what the service rendered warranted. It was felt that a remunerative price to the producer could only be ensured if the market practices and market charges were regulated and rationalised. And thus the regulation of the markets has been given a high priority in the various Five Year Plans. The markets are sought to be regulated through an Act of each legislature. The Act is generally known as the Agricultural Produce Markets Act and it is provided for the removal of various malpractices widely prevalent in the markets for the settlement of disputes between sellers and buyers and for the promoting of orderly marketing of farm produce in general.

Various state Governments have made considerable progress in this field by bringing in the necessary legislation. The Acts enabling the respective states to regulate the markets generally provide for the notification of market areas and the commodities to be covered in the act in different areas. A 'Marketing Committee' consisting of the representatives of growers, traders, merchants, local bodies and Government nominees administers the working of each market. The functions of the market committee are to frame bye-laws, define local market prices, fix market prices payable to various functionaries, license the functionaries, settle disputes, supervise weightment and promote the development of orderly marketing in general. The committee is generally empowered to raise funds for its working by levying a small fee on the produce bought and sold in the market in addition to the license fees received from the functionaries.

Market Surveys

A survey conducted in 496 markets in 1961-62 has shown that after the regulation the market charges have been reduced by 48 per cent. Another survey of selected commodities revealed that for some commodities market charges have been reduced by as much as 98 per cent.

Although the first market to be regulated was Karanja in 1886 the regulation of markets did not make headway till the first Five Year plan. It got a fill up in the second and third five year plan.

The states and union territories which have regulated the markets are: Andhra Pradesh(335), Bihar(62), Chandigarh(1), Delhi(3), Gujarat(212), Mysore(102), Madhya Pradesh(233), Haryana(80), Kerala(6), Maharashtra(212), Orissa(34), Punjab(95), Rajasthan(90), Tamil Nadu(136), Tripura(1), Uttar Pradesh(264), Goa, Daman and Diu(1), West Bengal(15), and Himachal Pradesh(5). The states and union territories which are yet to regulate markets are Assam, Andaman and Nicobar islands, Arunachal Pradesh, Dadra and Nagar Haveli, Jammu and Kashmir, Laccadive and Minicoy islands, Meghalaya, Mizoram, Nagaland, and Pondicherry. Necessary measures to regulate markets in these states and union territories are at various stages of progress. It is expected that these states and union territories will regulate the markets by the end of the Fifth plan.

All the states where necessary legislation has since been passed, have formulated phased programmes for the regulation of markets. By the end of the fifth plan it is expected that all the wholesale assembling markets would be brought within the regulatory orbit.

As a result of this scheme, excessive commissions and other market charges have been substantially rationalized. Unauthorized and arbitrary deductions have been prohibited and malpractices stopped. The issue of sale slips by licensed commission agents to the sellers, indicating the details of sale proceeds, deductions effected etc. has been made obligatory. Weightment is also done by licensed weightmen of the market committee.

The dissemination of marketing information and news is one of the functions of the market committees. This is done through the displaying of the prices prevailing in the market and also in the neighbouring markets on the notice boards and announcements through loud speakers at regular intervals. This information is also supplied to the Central Government and the State Government and also to other market committees. Arrangements have been made for the maintenance of reliable statistics arrivals, sales, stocks, prices etc. which are maintained by the market committees.

It is also obligatory on the part of the market committee to provide the market yards with the necessary amenities. In some of the markets that have been regulated amenities like rest houses, cattle sheds and water troughs have been provided by the market committees for the convenience of the producer sellers. Facilities for grading before sale and storage have also been provided.

A survey of 500 regulated markets was undertaken by the Directorate of Marketing and Inspection in the Ministry of Agriculture in 1970-71 and 1971-72, with a view to assessing the adequacy and efficiency of the existing regulated markets and highlighting their drawbacks and deficiencies and suggesting measures to develop them. One of the most important drawbacks has been the inadequate financial resources of some of the market committees. During the fourth plan, a central sector scheme was drawn up by the Ministry of Agriculture to provide a grant at 20 per cent of the cost of development of market, subject to a maximum of ₹ 2 lakhs. The balance will have to be provided by the commercial banks.

An important development in the field of regulated markets is the keen interest taken by the International Development Agency (IDA) in the development of the infrastructure in regulated markets. The IDA is financing the development of infrastructure in 50 markets of Bihar.

The World Bank has approved a loan assistance of 6.5 crores to Karnataka also for the development of markets.

CONTRACT TERMS

Under the existing trade practices, the sale of produce in a primary market takes place on the basis of the visual inspection of the goods, and in the secondary and terminal markets on the inspection of the samples. Thereafter the buyer and the seller decide upon the terms either orally or through written contracts. The contract terms specify the quality and quantity of the produce, the time and place of delivery, the price and terms of payment, handling and incidental charges, the procedure for settlement of disputes and penalties. The terms of contract were not standardized and thus varied for every individual transaction, and were more favourable to the buyer.

With a view to improving trade practices, All-India standard contract terms have been drawn up for a number of commodities. In standard contract terms the definition of quality and allowances in respect of refraction, damaged goods have been specifically standardized though the adoption of these standard contract terms by traders is voluntary they have to a large extent strengthened the position of the producer-seller and have improved the quality of the product marketed.

STANDARDIZATION AND GRADING

In order to gain the confidence and establish a rational relationship between the quality of a produce and its price, it is necessary to devote some attention to the proper preparation sifting and sorting of a material according to certain attributes before it is taken to the market. This is sought to be achieved by grading the produce in conformity with certain accepted quality standards *viz.* shape, size, form, weight, and other physical and technical characteristics. The produce brought to the market is very often contaminated with dust, stones and other foreign matter added either deliberately or by accident. Sometimes the produce is immature or not properly dried or contains shrivelled grains or damaged and rotten material. Such a produce brings a lower price to the farmers. Care should be exercised while assembling the produce of different farmers so that the good material is not mixed with the inferior material brought in by some farmers.

The Government of India had recognized the need to introduce the standardization of agricultural produce which would enable the farmers to derive the benefits of grading in terms of fair practices according to the prescribed standards and enacted the Agricultural Produce Grading and Marking Act in 1937. The Act empowers the central Government to prescribe grade standards indicating the quality of articles included in the schedule and specify grade designation marks to represent particular grades or qualities.

The Act provides for the grading and marketing of agricultural produce. The grade standards prescribed under this act are based on both physical and chemical characteristics and are formulated after analysing representative samples of each commodity collected from different regions and different seasons. Besides the international standards and special requirements of overseas consumers are also taken into account while formulating these standards for the commodities which are exported. The grade standards are reviewed and amended from time to time in the light of the shift of the pattern of production and trade and changes in the consumer's preferences. The grades are designated as the 'Agmark' grades.

A central Agmark Laboratory at Nagpur with sixteen regional laboratories at Guntur, Madras, Bombay, Kanpur, Cochin, Rajkot, Calcutta, Sahidabad, Jamnagar, Bangalore, Patna, Tuticorin, Virudhunagar, Mangalore, Alleppey and Kozhikode are assisting to provide adequate laboratory facilities for fixing grade standards for new commodities, for revising old grade standards and for routine quality control work.

Grade for Export

Grading of agricultural produce under the A.P. (G and M) Act is voluntary. Exports of certain agricultural commodities have however been prohibited unless duly graded and marked in accordance with the grade standards laid down under the A.P. Act 1937. The power to so prohibit exports was derived under the provisions of the sea-customs act.

The Directorate of Marketing and Inspection under the Ministry of Agriculture exercises a three tier control on the quality of agricultural commodities that are graded under 'Agmark' before they are exported. This is done through inspection.

Grading for Internal Trade

Commodities such as cotton, ghee, butter, rice, wheat, atta, gur, eggs, arecanut, potatoes, fruits, bura, pulses, vegetable oils and ground spices are being presently graded under 'Agmark' on voluntary basis.

Grading at Farmers Level

The grading of agricultural commodities under 'Agmark' has been consumer oriented. Generally the grading was done at the level of the traders. At this stage the producer was not a direct beneficiary of the grading scheme. It was felt the need to introduce grading at the producers' level. Thus the Directorate of marketing and inspection introduced a scheme for setting up commercial grading units.

Grading of Fruits and Vegetable Products

With a view to exercising quality control over fruits and vegetables the Government promulgated the Fruits Product Order under the essential Commodities Act. The preservatives and colours to be used are also clearly laid down. The order also stipulates the hygienic and sanitary methods which must be adopted by the manufacturers. The license for all this is issued by the executive Director, food and Nutrition board.

The total number of licensed factories was 1194. The total average production of fruits and vegetable products in the country during 1970 has been estimated as 156 thousand tonnes valued at ₹32.66 crores. Some of the products are very popular in foreign markets and are a good source of foreign exchange. In 1970 alone fruit products worth ₹2.92 crores were exported from India.

Regulation of Cold Stores

Most of the problems relating to the marketing of fruits and vegetables can be traced to their perishability. Perishability is responsible for high marketing costs, market gluts, price fluctuations and other similar problems. At low temperature, perishability is considerably reduced and the shelf life is increased and thus the importance of cold storage or refrigeration.

The first cold store in India was reported to have been established in Calcutta in 1892. However significant progress in the expansion of the cold storage industry in the country has been made only after independence.

An ad-hoc survey of the cold stores carried out by the directorate of marketing and inspection in 1955 showed that the total available cold storage in the country was only 77 thousand tonnes. The survey also highlighted the need to regulate the cold storage industry in a planned manner.

With a view to ensuring the observance of proper conditions in the cold stores and to providing for development of the industry in a scientific manner, the Government of India and the ministry of agriculture promulgated an order known as "cold storage order, 1964" under Section 3 of the Essential Commodities Act, 1955. The order is applicable in respect of every cold store with a capacity more than 8.4 cu m. The jurisdiction of this Order extends to the whole of India except West Bengal. Under this order, it is obligatory on every operator of the cold store to obtain a license from the Licensing Officer before using the installation for storing any food stuffs *e.g.* fruit, vegetables, meat, fish, dairy products. The Agricultural Marketing Advisor to the

Government of India is the Licensing Officer. The directorate of marketing and inspection is enforcing the cold storage order. The field staff posted in the regional and sub offices located in different states regularly inspects the cold stores and offers necessary guidance for better and scientific preservation of foodstuffs. Besides the directorate of marketing and inspection gives general guidance on all technical matters concerning the setting up of cold stores to the intending entrepreneurs.

The Government of India constituted a Central Cold Storage Advisory Committee consisting of official and non- official members, representing the growers, owners, machinery manufacturers, research organizations etc. The Committee advises the Government on all matters pertaining to the enforcement of Cold Storage Order and the future development of the industry. At the end of 1973-74 there were 1503 cold stores in the country with a capacity of 18.70 lakh tonnes. Uttar Pradesh had the maximum number of cold stores of 9.37 lakh tonnes followed by Bihar with 169 cold stores with 2.10 lakh tonnes capacity and West Bengal with 133 cold stores with 3.15 lakh tonnes capacity. Out of 1217 cold stores licensed during 1970, 1021 representing 84 per cent of the total were owned by the private sector, whereas 121 and 75 were owned by the public and co-operative sector respectively. The respective capacity is 13.88 lakh tonnes, 0.25 lakh tonnes and 0.73 lakh tonnes. Though numerically the cold stores in the co-operative sector were less than those in the public sector the capacity was more. The National Co-operative Development Corporation (NCDC) has formulated a scheme for financial assistance for setting up new cold stores in the co-operative sector.

Potato is the most important commodity which is presently placed in the cold stores accounting for as much as 92 per cent of the total capacity in the country. The remaining 8 per cent is being utilized for other perishables *viz.* fruits, vegetables, meat, sea-foods, dairy and poultry products.

With the attainment of self-sufficiency in the production of food grains, greater attention is being paid to the increased production of protective foods such as fruits, vegetables, fish, poultry and dairy products. These products being highly perishable are required to be kept in cold stores. Owing to the inadequate cold storage facilities at present considerable losses occur in the case of these commodities. There is enough scope for expanding the cold-storage industry with a view to providing facilities for preserving and prolonging the shelf life of these protective foods which are essential for human health.

CONSUMER PROTECTION

In the marketing process the producers and consumers are the two weak ends of the chain. It is incumbent on the part of the Government to protect the interests of both of them. Producers are sought to be protected through the regulation of the markets, grading at the producers level and other similar measures. Consumer's interests on the other hand are safeguarded by grading under 'Agmark' at the level of traders. The progress towards making the consumer quality conscious is slow. With the growing popularity of semi-processed foods the danger of sub-standard food articles being marketed has become manifold. With this in view steps have been taken in many directions, *e.g.* Acts relating to grading and standardization of agricultural commodities, certification marks in respect of manufactured goods, pure food laws, laws relating to weights and measures, and laws relating to the manufacture of fruits and vegetable products have been passed and enforced. Commodities such as ghee, vegetable oils, butter, honey and powdered spices are being graded and marked under 'Agmark' under the provisions of the Agricultural Produce (Grading and Marking) Act 1937. The Agmark attempts to provide a third party guarantee for the consumer. It not only certifies the purity of the product but also gives an indication of their quality by the grade-designating mark. The economic incentive to the producer-manufacturer is reflected in the premium the Agmarked product fetches in the market over the ungraded products.

Running parallel to the enforcement of laws, certain measures to complement the effort of achieving the overall objective of consumer protection are also being adopted. Monopoly procurement of food grains being operated by the Food corporation of India and other state departments achieves the two objectives of ensuring a remunerative price to the farmers and a uniform and reasonable price commensurate with the quality of the produce to the consumers.

The task of consumer education and protection is formidable and the Government machinery alone will be inadequate to accomplish this task. There is a need of active consumers' movement as in other countries. The Consumers' Guidance Society has been established with the objective of educating the consumers about their rights and responsibilities and advising them on the quality of products available in the market. Similarly it has advised the producers and manufacturers to abide by such standards as are necessary for the health and safety of the users.

CO-OPERATIVE MARKETING

The existing institutional structure of co-operative marketing is such that the co-operatives are functioning at the primary level, at the secondary level (taluka or district) and at the state level. In pursuance of a recommendation of the Dantwala Committee, efforts have been made to persuade the state Governments to divert the middle tier *viz.* the district marketing federations, of the functions legitimately falling within the purview of the state or primary marketing societies, so that a two-tier system can be brought into operation.

The co-operatives in different states have been federated into a central level federation *viz.* the National Agricultural Co-operative Marketing

Federation (NAFED). At the end of 1960-70 as many as 3335 primary cooperative marketing societies were operating. Of these more than 500 were specialized commodity marketing societies for special crops, cotton, fruits and vegetables.

In certain states intermediate organizations at district and state levels have also been established. At the end of 1960-70, 232 such societies were functioning. They included some commodity federations also. At the state level 28 apex co-operative marketing federations are functioning. They are normally handling all the commodities. There are a few apex cooperative societies which are handling exclusively a particular commodity *e.g.* two apex societies in Gujarat are handling cotton, one is handling fruits and vegetables and one in Uttar Pradesh is handling sugarcane only. In order to strengthen and develop co-operative marketing to an extent where it may have an impact on the marketing of agricultural produce, several measures have been initiated. Steps are also being taken to see that there is an effective co-ordination between the state cooperative departments and their counterparts dealing with agricultural marketing. Consequently there has been a significant expansion in the operations of marketing co-operatives. This is exclusive of the value of agricultural requisites and consumer articles handled by the marketing cooperatives. The main commodities marketed by the cooperatives were sugarcane, cotton, oilseeds, fruits, vegetables and plantation crops.

In pursuance of the decision of the Government, a scheme of outright purchases of agricultural produce by the co-operative marketing societies was launched in 1964-65 in 200 selected marketing societies. The basic theme of this scheme has been to bring the small producers within the fold of co-operative marketing together for such farmers. To provide the necessary financial backing the scheme envisaged the creation of a price-fluctuation fund. The fund envisages meeting losses if suffered by the co-operative marketing societies at different levels as a result of outright purchases of agricultural produce.

The scheme has been operative in several states. During 1969-70 the value of the agricultural produce purchased under the outright purchase scheme was about 34 crores. As a result of this scheme there has been a greater involvement of cooperative societies in the marketing of agricultural produce.

Besides this scheme has given impetus to inter-state trade by the cooperatives. They transacted about ₹ 66.55 crores worth of agricultural produce during 1969-70.

INTERSTATE TRADE

The co-operative marketing societies are devoting increasing attention to interstate trade in agricultural produce. The main commodities are wheat, pulses, plantation crops, copra, spices, fruits and vegetables. The bulk of the transactions were made by the Punjab Apex Federation. During 1970-71 the

NAFED acted as the agency of the co-operatives of Jammu and Kashmir for marketing their apples outside the state.

CO-OPERATIVE EXPORT OF AGRICULTURAL PRODUCE

The export of agricultural produce by the co-operative sector continues to be a growing activity. The bulk of the exports are made by National Agricultural Cooperative Marketing Federation Ltd., which accounted for exports worth ₹5.64 crores. Besides, the Gujarat State Cooperative Marketing Federation Ltd., the Khanna Cooperative Marketing Federation Ltd., the Kerala State Cooperative Marketing Federation Ltd., the Jalgaon district fruit sales societies and the coconut oil millers society also cooperated to exports to countries like Kuwait, Malaysia, Ceylon, Singapore, Bahrain, Doha, Dubai, Iran, Muscat.

The Khanna Solvent Extraction Plant in the Punjab State exported de-oiled cake worth ₹ 35 lakhs during 1969-70. The main commodities exported by the cooperatives were pulses, chillies, onions, pepper, de-oiled cake, potatoes and kardi extraction meal. The exports were mainly made to Ceylon and other important markets were Mauritius, Kuwait, Doha, Bahrain, Hong Kong, the USSR, the UK, Iran, Czechoslovakia and France. Some of the traditional items of export have been marketed in non- traditional areas. Pulses were exported to Cuba and onions to Malaysia and Singapore. The cooperatives also assisted various agencies to export agricultural commodities. In this connection exports of coffee and raw sugar were made. During 1970-71, NAFED exported agricultural produce worth ₹5.26 crores. The Jalgaon District Fruit Sale Societies Cooperative Marketing Federation Ltd., directly exported bananas worth ₹34 lakhs to Kuwait and Bahrain Islands.

CO-OPERATIVE COLD STORES

Co-operatives are also to facilitate the storage and marketing of perishable commodities especially seed potatoes. By the end of the Third plan the cooperatives had established 87 cold stores and the target for the fourth plan was set at 45 more. At the end of December 1971 there were as many as 96 cooperative cold stores with a capacity of 1.42 lakhs tonnes.

AGRICULTURAL MARKETING AND AGRI-BUSINESS DEPARTMENT

STATE SCHEME

Own building for Agmark Laboratories, Strengthening of Agmark Laboratories and Provision of Computers (₹ 3.07 crores): There are 30 State Agmark Grading Laboratories, one Principal Laboratory and 15 Agricultural marketing centres functioning in the State. It is proposed to construct building for Agmark Labs as well as modernize the equipments during the Tenth Five Year Plan for which a sum of ₹ 3.07 crores is provided.

SCHEMES PROPOSED TO BE IMPLEMENTED AVAILING FINANCIAL ASSISTANCE FROM THE OTHER FINANCIAL INSTITUTIONS

Provision of Infrastructure Facilities (₹ 100 crores)

i. *Creation of infrastructure facilities:* There is a need to provide infrastructure facilities like transaction shed, drying yard, farmers rest sheds, sanitation facilities, drinking water supply electronic weighing scales, cleaning and washing facility, moisture meters, scientific instruments for grading, rural godowns etc., for which an amount of ₹ 100 crores is provided.
ii. *Revamping Tamil Nadu Agricultural Marketing Board:* There is an imperative need to revitalize the Board activities by making this body more responsive and independent decision- making body. It should be given administrative and financial powers.
iii. *Revitalising the market committees:* At present the Act provides for nomination of members to the Market Committees. To make it more accountable and democratic, it is necessary that members be elected so that there is a sense of belonging. Secondly, there is lack of professionalism in the staff of the Market Committees. Their mind set is more of regulation than market oriented approach. Training to the existing staff and inclusion of professionals is a must.
iv. *Provision of infrastructure facilities in Post Harvest Centres:* At present the post harvest centres in the Regulated markets conduct training programmes to the farmers and exhibitions on various post harvest practices. The regulated markets should expand their activities through provision of whole storage CA, MA, Retardation and ripening facilities for the agricultural produce.
v. *Agricultural Marketing Extension:* Strong network of marketing extension is very much necessary at block level to effectively advise farmers on various aspects of marketing, advise on product planning, marketing information, and securing market for farmers, and advise on improved market practices and advise on post harvest management practices.

To strengthen the regulated market it is proposed to provide infrastructure facilities like transaction shed, drying yard, farmers resting shed, sanitation facilities, drinking water supply, electronic weighing scales, cleaning and washing facility, moisture meters, scientific instruments for grading, rural godowns etc. at a cost of ₹ 100 crores with financial assistance from Marketing Committee, NABARD, GOI.

Provision of infrastructure facilities in Post Harvest Centres-Cold Storage (₹ 200 crores- National Horticultural Board/Private Sector)

It is estimated that around 30 per cent of the horticulture produce is wasted due to inadequate cold chain facility and appropriate technology for the preservation of horticultural produce. At present the combined cold storage capacity of 133 units in Tamil Nadu is around 1 LMT. The existing capacity is not sufficient to store horticultural, dairy, and marine products. In order to meet the demand it is proposed to establish an additional 1 LMT capacity of cold storage facility, at an estimated cost of ₹ 200 crore. Taking into consideration the initial capital investment, high recurring cost and low capacity utilisation and project failure, it has been proposed to attract investment by private sector through measures like financial incentives in addition to NHB's subsidy and power tariff concession.

Agricultural Marketing Extension/ Packaging Training (₹ 42.95 crores)- Tamil Nadu State Agricultural Marketing Board (TNSAMB)

Agricultural farmers require advice on various aspects of marketing like selection of the crops to be grown with marketability in mind, current price, market arrival and forecasting of market trends, and on post harvest management practices.

Agro-processing Food Park (₹ 10 crores)- Private Sector

It is estimated that about 10-20 per cent of the food grain production and 30-35 per cent of fruits and vegetables are wasted at various stages from picking to consumption. This indicates the need to establish food-processing industries to preserve and minimize the wastage. To encourage private sector to establish agro-processing industries, the Government of India, Ministry of Food Processing Industries has formulated two models of Food Park for Food Processing Industries apart from other incentives. It is proposed to set up a Food Park at a cost of ₹ 10 crores either in private sector or by Market Committees.

Setting up of Terminal Markets and Collection Centres (₹ 160 crores)

Private sector or growers association in partnership with private sector can organise Terminal Markets for specified products with backward integration with collection centres (value addition centres). Fifteen to twenty collection centres established nearer to the production area, can feed one Terminal Market. Two such a Terminal Markets will be established one each at Chennai and at Coimbatore with NDDB assistance.

Mega Markets (₹ 10 crores)

A mega market centre can cater to the needs of wholesale dealers, exporters and food processing industries. The availability of horticultural products in large scale in one place would promote exports and provide single sourcing to agri processing industries. It is proposed to establish a mega market for vegetables at Oddanchatram in Dindigul District at a cost of ₹ 2.59 crore. Depending on the success of this project, mega markets could be developed at Mettupalayam in Coimbatore District and Athur in Salam District either through Market Committees or private sector.

Agro Processing Industries (₹ 10 crores)

Food Processing Industries provide the crucial farm - industry linkages, which help to add value to the produce, generate employment opportunities and increase the net income to the farmers. The Market Committees will establish these through private sector in 10 places with an incentive of 20 per cent of the capital cost. The project cost is estimated to be ₹ 10 crores. Packaging fresh fruits, vegetables and other farm produce is an important process as it reduces post harvest losses, increases the income of the farmers and ensures clean and hygienic farm produce to the consumers. It is proposed to create awareness and impart training on farm produce packaging to the technical officers of Agriculture, Horticulture, Agricultural Marketing and Agri Business departments during the Tenth Plan period. It is proposed to train 600 officials and 400 farmers at a total estimated cost of ₹ 25 lakhs. It is also proposed to conduct studies for developing low cost packaging techniques and material for some specific commodities at a cost of ₹ 0.35 crore.

Agri Export Zone (₹ 150 crores) –GOI

The most critical factor to meet the challenges of export will be enhancing exporting capability of our State in a highly competitive environment. To promote export of agri products, Government of India, Ministry of Commerce has announced in 2001 - Exim Policy, a scheme of establishing Agri Export Zone (AEZ). In AEZ, institutional and physical infrastructure would be created as per the needs of the specific commodity. Steps have been taken to establish AEZ for cut flowers, mangoes, bananas, medicinal plants and vegetables. To promote agricultural exports, it is proposed to educate and train the growers of identified crops in producing, grading and packing for international market, and in establishment of analytical laboratories, setting up of Export promotion cell in Agri Business department to disseminate information and harmonization of standards of Indian products with international standards (Codex).

MARKETING EXTENSION

Network is proposed to be formed integrating with the extension network already available with Agriculture Department. Officers of Agriculture, Horticulture and Agricultural Marketing departments will be given training on various aspects of Agricultural Production and Marketing for the purpose of carrying out extension works effectively and efficiently.

A sum of ₹ 65 lakhs is proposed during the Tenth Five Year Plan. Further, the Government will also facilitate private sector to carryout extension work for the quick reach of farm information to the farmers.

To improve market practices and also sale of specific agricultural products, two products' specific market Complexes *viz*- Turmeric Market Complex at Erode at a total cost of ₹ 32.30 crores, and Jaggery Market Complex at Trichy at a total cost of ₹ 10 crores will be established. These market complexes will be established incurring expenditure initially from Market Committees and financial institutions like NABARD and then recovered from the traders in installments.

ALTERNATIVE MARKETING FORMS

Role of Government in managing markets is on decline worldwide. It is not easy to bring major changes in the traditional marketing system. The only way to modernize marketing is to promote alternative marketing system and that may operate parallel to and in addition to present marketing system. The purpose of the proposed alternative marketing is to promote modern trade practices, which in turn will pave way for transperancy and effiecncy in market.

Even though, the various forms of alternate marketing like:

- Direct marketing,
- Marketing through farmers interest group,
- Setting up of terminal markets,
- Forward and future market,
- e-commerce,
- Setting up of mega markets,
- Negotiable warehouse receipt system etc. have been suggested by Expert Committee on Agricultural Marketing headed by Shankarlal Guru, three important marketing methods could be considered in the State *viz.*, Terminal Market, Mega Market and Direct Market.

AGRO PROCESSING INDUSTRIES

Food Processing Industries provide the crucial farm - industry linkages which helps to accelerate overall agricultural development, adding value to the produce, generating employment opportunities and increasing the net income to the farmers.

In India, only 2 per cent of the total horticultural produce is processed. In countries like Brazil, it is in the range of 70 per cent. Considering the rising demand for good quality products, there is an urgent need to enhance capacities for value added and processed products. At present, value addition is estimated at 7 per cent of the total production within next 5 years.

There is a need to increase value addition to 20 per cent and processing at 7 per cent. To increase our country's share in the world trade of agri products which stands at less than 1 per cent at present, the most critical factor in the highly competitive market environment is quality processed products. In Tamil Nadu, food processing in the form of drying, vegetable oil, grain processing, sugar breweries are in existence for a quite long period. Lack of adequate infrastructure facilities like storage, processing, marketing besides technical know-how have been the major constraints affecting the growth of the industry. Tamil Nadu with a coastline of 922 km and surface boundary of 1200 km with tropical and sub-tropical climate is ideally suited for the production of a host of agricultural, horticultural, acquacultural and animal husbandry produces. Thus there is vast scope for setting up of Food Processing Industries in Tamil Nadu.

The growth of food processing industry, which is included in the priority lending sector, will bring immense benefit to the economy. Economic liberalization and raising consumer prosperity is opening up new vistas in food processing sector. Tamil Nadu produces 103 lakh metric tonnes of food grains, 22 lakh metric tonnes of oil seeds, 37 lakh metric tonnes of sugarcane, 53.89 lakh metric tonnes of fruits, 54.65 lakh metric tonnes of vegetables, 2.95 lakh metric tonnes of spices and 6.94 lakh metric tonnes of plantation crops. 10 to 20 per cent of food grain production is lost every year both at pre and post harvest stages. Similarly 30 to 35 per cent of fruits and vegetables produced are also wasted at various stages from picking to final consumption. This only indicates the immense potential and scope that exists for setting up of food processing industries so that the available raw materials may be put into the most judicious use. This also indicates the scope that exists for adoption of post harvest modern techniques ensuring preservation and minimizing wastages of agricultural produce.

CONSTRAINTS

- Very high difference in price between the farmers' realisation and consumer even for the fresh produce. In processed food the high price of raw materials, excessive spoilage, inefficient and costly transportation, high cost of finance due to high taxes and duties leads to low demand of processed foods.
- Lack of linkage with R&D institutions with the users like farmers and industry.
- Impediment in the flow of credit from financial institutions to the food processing industry due to the improper understanding of this sector to attain the required level of imparting skill.
- Low margins, seasonality and high perishability being the distinct features of this industry the access to seed capital and working capital is not easy.

- Indian brands of processed food are yet to be established in the international Market. Competition with imported goods in the wake of liberalization of world trade.
- Week database and lack of market intelligence.
- Backward linkage between the farmers and the processor is yet to take proper shape to tide over the impediments.
- Multiplicity of laws and regulatory authorities affect the growth of industry.
- Prevailing packaging system lacks requisite quality and shelf life.
- Lack of knowledge of quality parameters and standards.
- Lack of participation by people, local bodies, NGOs farmers' organisation and industrial association.

Food Processing Policy will therefore, have to address itself to:

- Promoting innovative measures for fostering group co-operation in adoption of pre-harvest and post-harvest technologies.
- Development of cropping pattern as per food processing units' requirements.
- Speedy development in infrastructure to promote food-processing industry.
- Removing legal/statutory hurdles affecting growth of food processing industry.
- Facilitating liberalized financial assistance to the industry considering the high risk and capital-intensive nature.
- Processing standards may be upgraded by introduction of mechanised cleaning, sorting/grading of agricultural produces
- The policy will seek to create an appropriate environment for entrepreneurs to set up food processing industry through Fiscal initiatives/ intervention like rationalization of tax structure on fresh foods as well as processed foods and machinery for production of processed foods.
- Enactment of food laws to enable the industry for effective implementation.
- Creation of infrastructure development connected with fruits, vegetables, meat, fish, poultry etc.
- Land ceiling exemption for captive farming/ contract farming duly taking care to protect the rights of farmers.
- Creation of a venture capital fund for soft loan assistance to the food processing industry
- Incentive like power tariff concessions and sales tax.
- Creation of technology development centres in private sector for the dissemination of modern technology
- Providing cold chain, cold storage system, tissue culture plant for

the production of planting materials. To encourage private sector to establish agro-processing industries, the Government of India, Ministry of Food Processing Industries has formulated *two models of Food Park for Food Processing Industries* apart from other incentives.

INDUSTRIAL ESTATE MODEL

In this model, the entrepreneurs are encouraged to set up agro processing industries in well-developed, well laid out industrial plots. The developer of the food park also provide common facilities, like water supply, storage, laboratories, warehousing facilities, common effluent treatment plant and other facilities.

Another model which is being tried out in Punjab State, where all common facilities are located in a hub with agro processing units function in a radius of 50 kms. In the second model, the initial capital outlay is less and small agro processing units and farmers can utilise the services at a reasonable cost. This can be either in the private sector or set up by the Market Committees.

Agri Export Tamil Nadu endowed with cheap labour, diverse agro climatic conditions and soil resource with favourable Government policies is poised for accelerated growth in agricultural/ horticultural commodities export. With trade barriers falling apart, we should take advantage of new international trade environment. The most critical factor to meet the challenges will be enhancing exporting capability of our State in a highly competitive environment. To take advantage over other States in India, Tamil Nadu should create necessary infrastructure for export like post harvest facilities, food-processing capability, handling facilities at port, food analytical laboratories to certify the quality aspects etc. To promote export of agri products, the Government of India, Ministry of Commerce has announced in 2001 - Exim Policy, a scheme of establishing Agri Export Zone (AEZ). In AEZ institutional and physical infrastructure would be created as per the needs of the specific commodity. AEZ may be established for cut flowers, mangoes, bananas, medicinal plants and vegetables.

The following steps may help to boost agri exports:

- Vigorous extension work to educate and train the growers of identified crops in producing, grading and packing for international markets.
- Establishment of analytical laboratories.
- Setting up of Export promotion cell in Agri Business department to disseminate information on export opportunities and destinations, obtain sanitary and phyto sanitary standards of various countries and make it known to the prospective exporters.
- Harmonisation of standards of Indian products with the international standards (Codex).
- Provision of common facilities/ infrastructure in AEZ.

There is a great demand for organic farm products in developed countries. In the recent report of FAO, it has been stated that domestic production of organic products in developed countries is expected to rise within the next few years. But it is unlikely to meet the demand.

TYPES OF MARKETING INTERVENTION

There are many different ways in which NGOs or CBOs may intervene to improve access to agricultural markets.

In this chapter interventions are discussed in eight non-exclusive categories that describe aspects of the intervention strategy:

1. Intended beneficiaries
2. Skills and training
3. Access to agricultural inputs
4. Agro-processing technologies
5. Marketing linkages
6. Credit programmes
7. Marketing information
8. Holistic approaches

Any particular marketing intervention may comprise elements from several categories (*e.g.* inputs and training, or technology, training and finance). The concepts and experiences associated with each category are reviewed, permitting some preliminary conclusions on the more promising strategies.

INTENDED BENEFICIARIES

Individuals, Groups or Communities

Different NGO and CBO marketing initiatives operate with different beneficiary or client structures: some work with whole communities, some with groups and others with households/individuals. The choice of appropriate structure will depend on a number of factors. Research carried out by the Plunkett Foundation and experience of CARE's Development through Conservation project in Uganda suggests that working with village associations or whole communities is more difficult than working with smaller groups. The latter are more focused, more specialized and more likely to have a common goal. Many NGOs and CBOs implement their marketing interventions with groups or associations. Not only are there advantages to the NGOs and CBOs of working with groups, but there may be advantages for the farmers themselves of marketing collectively. The groups can be existing groups (*e.g.* women's groups, savings and credit groups, social groups and so on) or newly formed groups.

The potential advantages of farmers collectively addressing marketing constraints include:

- Economies of scale, through joint purchasing of inputs and joint marketing of products;

- Improved access to finance, where credit organizations favour group loans, or where pooled resources provide the necessary down payment; this can overcome problems of larger investment needed in, for example, processing technologies, storage facilities or transport;
- Collective bargaining power;
- Lower transaction costs (for producers and traders).

Yet there is much evidence that, in general, small enterprises owned and managed by individuals are more successful than group enterprises. Technology adoption in particular is felt to be better amongst individuals than groups, although groups can be an appropriate vehicle for technology transfer if the technologies are subsequently employed by individuals. This is echoed by String fellow *et al.* who argue that group enterprises are more likely to succeed when based on joint marketing rather than joint management/ownership of assets, because the latter requires more complex skills and experience. It has also been found that external organization and management of groups can prevent the development of entrepreneurial skills. There are no definitive rules on which structure is appropriate. It will depend on the type of intervention and the objectives of the individuals and NGOs/CBOs.

On the one hand, NGOs need to remain open to working with groups where appropriate, recognizing the potential to build capacity, reduce transaction costs, and introduce activities with a higher investment threshold. However, an understanding of the reasons behind group formation is important and care should be taken not to over-burden groups formed for social rather than economic reasons. Sometimes individuals, often women and particularly in rural areas, prefer to work together. Furthermore, commonly cited problems attributed to groups per se have been found to be less important when there is strong group leadership and cohesion, and when there is good group organization before any external intervention takes place.

String fellow *et al.* point out that a non-interventionist approach – letting producers decide for themselves whether they operate as individuals or groups – allows individuals to develop appropriate structures to build necessary skills and solve their own problems. People will not work well within an imposed structure. As Gibson succinctly states: "Enterprises working in a market environment (whether collectively owned or individually owned) have ultimately to make a surplus in order to survive. The 'bottom line' for NGOs is to support the structures which will work best according to this criterion."

RURAL POOR

The majority of NGOs and CBOs deliberately target their activities at poor households and poor communities. Those that focus on agricultural production, processing and marketing are often found in remote rural areas. This focus has implications for the approach adopted by the NGOs/ CBOs and the marketing initiatives they develop. Whereas the trend in marketing approach adopted by NGOs has moved towards a more business-oriented, facilitative approach in recent years, some authors argue that this is less likely to succeed in remote rural areas than in less remote, higher-potential areas.

Moreover, within rural communities, some poor individuals and groups are considered too remote and disadvantaged to be able to benefit from marketing interventions. The most vulnerable households and women-headed households in rural areas are particularly disadvantaged; they tend to have more restricted access to information and services and tend to be more risk-averse than other households. Some households may only benefit indirectly from a marketing intervention through any impact it has on the labour market. Consideration should also be given to the types of intervention most appropriate to a particularly disadvantaged target group, particularly where the transfer of hardware and entrepreneurial skills are envisaged. For example, food processing technologies are often predominantly adopted by the rural elite, due to their more developed entrepreneurial skills and access to finance. This may nonetheless reduce poverty, through improved markets and employment generation, but the strength of these linkages varies, making it difficult to generalize.

Women In common with other organizations, NGOs and CBOs often find it very difficult to effectively target benefits to women. Yet women are frequently amongst the poorest members of rural communities and face particular constraints in improving their livelihoods. Moreover, their traditional role in food crop production and (sometimes) domestic marketing, as well as the influence they have over child welfare and nutrition, makes them an obvious target for poverty-focused agricultural marketing interventions. In their review of women's role in post-harvest operations, Gordon *et al*. conclude that:

"Where interventions are intended to benefit the poorest women, attention should be focused on particular issues:

- The needs of female-headed households, which feature disproportionately amongst the poor;
- Crops and processes used in marginal areas;
- Carrying fuel and water, because so many women are affected;
- How poor women earn income – so that new technology really does benefit them;
- Understanding the broader processes which determine how benefits are distributed;
- Household level and informal sector activities, where the poorest people earn their living."

Goodland *et al*. make a number of suggestions on how women's participation in rural finance programmes can be increased, the spirit of which is equally relevant here:

- Out-of-hours opening or mobile services, or locating services in places women frequent, for example, marketplaces;
- Relaxing literacy requirements;
- Flexible collateral requirements, for example, accepting jewellery rather than land;
- Allowing loans of a suitable size (usually small).

SKILLS AND TRAINING

The types of training offered by NGOs in support of marketing initiatives can include group strengthening, general extension, marketing and specialized training.

GROUP FORMATION AND STRENGTHENING

The advantages of working with groups have already been highlighted. Despite these, group establishment and operation have generally proven more difficult in practice than expected. Some organizations undertake group strengthening and training (*e.g.* in recruitment and management, structure and leadership, financial and business management skills and so on) before marketing activities are initiated, or as part of the marketing intervention. The case of oil palm processing with the NGO TechnoServe is a good example.

The danger of not assessing the performance of existing groups and providing any necessary training in group strengthening is that marketing activities implemented with these groups can fail due to general group weaknesses, rather than problems with the marketing activities per se. Stringfellow *et al.* studied farmer co-operative enterprises and their findings highlight the importance of not over-estimating group capacities, and the need for long-term involvement in building group capacities. Often groups fail because they have been formed too quickly and too much is expected of them.

They also found that group enterprises are more likely to succeed when based on joint marketing rather than joint management/ ownership of assets which requires more complex skills and experience. It is also important to consider the most appropriate group size and whether there is a culture of working as a group. The Co-operative League of the USA (CLUSA) working with CARE in Mozambique has developed a thorough approach to the development of farmers' associations. They have attempted to make this approach more sustainable by developing the capacity of a local NGO to take-up and continue these activities, and by promoting a structure within the farmers' associations that permits ongoing development. The approach has been criticized for being too costly, but the results to date, albeit over a short duration, are nonetheless very impressive. There is now a need to critically evaluate this approach, to identify the direct and indirect subsidies provided by the NGOs, and assess which components have the most prospects of sustainability.

TECHNOSERVE AND OIL PALM PROCESSING IN GHANA

TechnoServe is an NGO whose activities focus on the provision of food processing technologies for rural communities. The initial focus of its work was small-scale oil palm processing and extraction in Ghana, but it has subsequently expanded to include processing service centres, inventory credit schemes and production and processing of non-traditional export crops. The purpose of the oil palm scheme was to build on traditional processing (which was laborious and slow) and to exploit the potential for expansion in a strong market, by introducing small-scale mechanized oil mills to community-based groups.

The approach adopted by TechnoServe went beyond providing the technology and technical assistance. It was recognized that entrepreneurial skills in the communities were weak, and that group development activities and financial and business training (including linkages to formal credit and extension services) were necessary. This integrated training and support package clearly contributed to the sustainable adoption and management of the oil processing enterprises. However, the Techno Serve approach has also been criticized.

The training and support provided to groups are heavily subsidized. Case studies of the oil palm enterprises indicate that the skills are held by only a small number of active group members, and that because of social norms, there is a reluctance to pass these on. A more serious criticism of the approach is that although it has been successful in transferring skills specific to the oil palm enterprise, it has not managed to create more general entrepreneurial skills and characteristics amongst the group members.

TRAINING AND EXTENSION

The starting point of a number of CBO/NGO marketing initiatives is production. This is because to market successfully, farmers need to produce and sell what is in demand, at a profit. Often existing markets could be accessible to farmers (either on their own or through linkages with traders), but marketing is constrained by the low volumes or poor quality of farmers' crops. Improved production practices are important for increasing yields of existing crops, new varieties and new crops. Yet government extension services are often lacking or extremely under resourced. NGOs, therefore, often assume a role in providing, or facilitating the provision of, relevant extension information. This may be through strengthening existing extension services or through the establishment of alternative services, such as farmer-extension schemes.

When the CARE Egypt Agricultural Reform Programme began, it focused on improving crop and livestock production. Now, however, the project has become more market oriented and requires farmers to examine the market for their products prior to improving production. Other NGOs and CBOs provide marketing training to groups and individuals. This includes training in production and marketing systems, constraints and opportunities, market demands (products and service) and how to assess whether products can be supplied profitably. The benefits of providing producers with this type of training are that it develops their capacity to analyse markets for themselves and, therefore, allows them to respond to changing market opportunities and threats. They are not dependent on the NGO/CBO to identify marketing opportunities for them in the longer term. It is also important for NGOs and CBOs not to create parallel services and for different development organizations and governmental bodies to co-ordinate the training and services they provide. In Uganda, for example, a task force was established to prepare guidelines on how to integrate NGO activities into district agricultural programmes.

Specialized or vocational training is usually provided to individuals or groups when the marketing intervention relates to a new product or new technology, and requires new skills. Examples include production targeted to a new niche market or particularly quality-conscious export market, or training in the use of an agro-processing technology, such as an oil press. Increasingly organizations which provide skills and training for enterprise development are charging trainees a fee. This helps NGOs and CBOs with limited funding to reach a wider audience. More significantly, it has been found that charging a fee increases the proportion of trainees who actually make effective use of their training. The entrepreneur, making the decision to invest money and time in training, is in effect making a risk assessment.

IMPROVING ACCESS TO AGRICULTURAL INPUTS

Poor access to inputs directly influences the level and quality of production. Even in the poorest parts of Africa, there is still demand for farm implements and good quality seed. Less-poor farmers may make selective use also of fertilizer and pesticides. Input subsidies are a particularly vexed issue. Some argue that they are needed to provide a short-run boost to production and incomes.

Yet they are also disruptive and undermining of sustainable commercial development. At a workshop in Uganda, participants highlighted the negative effect of farm input relief programmes in neighbouring countries on the development of commercial input supply networks in Uganda. In Malawi, the starter-pack scheme (distribution of free seed and fertilizer) implemented in 1999 and 2000 has boosted production there, but it has also deprived other low-income farmers in neighbouring countries (notably Mozambique) of a traditional outlet for their surplus production. Gordon discusses five sets of issues affecting access to inputs: affordability; availability; information; risk and uncertainty; and the overall commercial context. Although credit is often assumed to hold the key to improved access, other ways to improve affordability are also identified: timing input sales to coincide with times when farmers have cash; selling inputs (*e.g.* seed) in small pack sizes suited to small producers; and lowering prices, by making cost reductions in distribution and marketing (*e.g.* through bulk purchases, transport sharing arrangements, and farmers' groups taking on more responsibilities).

Whilst NGOs may play a role in providing credit or promoting some of these other strategies, a role for CBOs (*i.e.* farmers associations) is much more apparent. In countries that have succeeded in significantly improving access to inputs, farmers' associations have acted as a key vehicle for input distribution. Many consider the physical availability of inputs to be a more important constraint to access, with thin and unreliable rural distribution networks in most African countries. A recent development involves innovative approaches to the promotion of input stockist networks by NGOs, illustrating what can be achieved through constructive partnerships between the commercial, private non-profit, farming community and government sectors. These initiatives typically involve training stockists and may involve loans or loan guarantees.

There is also a growing interest in ways to improve and build on traditional informal seed systems. Information constraints are also important – be they in terms of information gaps (basic research on fertilizer response, for instance) or information flows. Although farming is inherently risky, better information reduces uncertainty, enabling farmers to make more informed production decisions. Gordon concludes with some general observations, which also have relevance to the activities of NGOs: "In addition to policies aimed towards the general development of rural economies, a number of more specific policy recommendations are made: avoid actions which undermine the development of sustainable commercial input supply networks; support input markets by setting standards and regulations, and providing information and training; promote synergistic partnerships between commercial, private non-profit, farming community and government sectors; and fill critical research and information gaps."

AGRO-PROCESSING TECHNOLOGY

Small-scale farming households in remote rural communities generally find that they operate in markets comprising many producers of undifferentiated products, which leads to price competition and low profit margins. Access to processing technology can provide new market opportunities by reducing perishability or adding value in other ways. Processing technologies can range in scale from household-level 'lowtech' processing to fully mechanized factories. Household level processing has two main functions. It can add value and it can preserve the product, thereby increasing the time available for marketing. Other advantages of small-scale agro-processing enterprises are that they can create employment at low levels of investment that make effective use of local resources. Enterprises owned and managed by individuals or households are often more successful than group enterprises, so technology development organizations need to be aware of this need for small-scale technologies. However, many NGOs and CBOs work with farmer groups either because it is the structure that the farmers prefer (there may be a culture of group activity) or because the NGO/CBO prefers this approach (due to financial and coverage reasons). Yet group approaches to the adoption of agroprocessing technology are often weak and entrepreneurial skills are less evident than when working with active individuals.

In these circumstances, careful group selection is required, as well as consideration of ways in which the need for entrepreneurial skills can be reduced by introducing a third party, such as a private company offering marketing services to small-scale processors. It is necessary to ensure that there is market demand for the technologies and/or their end-products and that the technologies are appropriate (*e.g.* taking into account gender issues), rather than developing technologies for their own sake. It is also important that the wider enterprise environment is considered and that institutional arrangements enable smallholder access to markets. Also, as will be examined further, many marketing constraints faced by farmers are interrelated. In terms of technology development, other marketing factors such as access to credit, adequate managerial and technical skills, and market information all influence technology choice decisions.

THE RAM PRESS – OILSEED PROCESSING TECHNOLOGY INTRODUCED BY ATI

In the 1980s, ATI (now Enterprise Works Worldwide) started developing a manual ram press (also called the Bielenberg press after the engineer who developed the prototype) to produce edible oil in remote parts of Africa without electricity and with poor access to markets. The first presses were intended for soft-shelled sunflower seed in Tanzania, but the model has since been adapted for use with other oilseeds, including coconut and sesame. The first presses were arduous to use but later models improved on this and could be operated easily by one person.

It was hoped that they would improve nutrition by improving access to energy-rich food, whilst also increasing farmer incomes and creating enterprise and employment in rural areas. Indeed, local entrepreneurs involved in small-scale oilseed processing can earn two to three times more gross income, compared with selling the seed to processing factories. In Tanzania, costs could be recovered in just one 3-month season. ATI worked up the promotion methodology over many years in many African countries. The ram press is quite simple and is supplied with a filtration device and tools for maintenance. Training, information and support on proper use and socio-economic and nutritional benefits of an oilseed processing operation are important elements of the extension programme adopted by ATI and its partners. In Tanzania, ATI sold the presses on credit for many years. Elsewhere, whilst varying in their systems and repayment rigour, credit has been an important factor affecting uptake. Depending on the model and manufacturer, the presses cost between US$ 150 and US$ 300. ATI has been innovative in promoting the commercialization of small-scale oil processing in Africa with its Regional Oils Programme.

Its focus shifted from providing technical assistance to NGOs involved in small-scale oilseed processing to a private sector approach that emphasizes commercialization and mass manufacturing of ram presses. As a result, private enterprises have been developed, given assistance with market surveys and business plans, and are becoming responsible for the manufacture and wholesale of ram presses. ATI also stress the importance of the participatory processes used that involve economic actors in the community – owners and workers, press manufacturers, sales agents and oil-consuming families.

MARKETING LINKAGES

Marketing problems identified by producers are often attributed to the commercial sector and the capacity to access it: lack of buyers, unreasonably low farm-gate prices, inflexible requirements, and so on. Links between NGOs/CBOs, the governmental sector and commercial agents are usually weak. Yet working together with the private sector is an important way for farmers to access relevant market information, technologies and new market opportunities.

It is also a way for NGOs and CBOs to reach a wider audience with limited funds. As Kleih has stated, although NGOs are making a very positive contribution to rural development, they generally only reach around 1–2 per cent of farming households in any one country. Private sector agents are sometimes willing to collaborate with NGOs and farmer groups to share the costs of providing training and information, if they see it as an investment through which they can increase their own revenue. In these circumstances, both the producers and traders can benefit.

Developing and strengthening relationships between farmers and traders can reduce transaction costs, transport costs and risk on both sides. It was noted above that approaches that use and strengthen existing private sector production and marketing channels, rather than seeking to override them or invent new ones, have been particularly effective in addressing marketing constraints and in sustaining linkages beyond the life of the NGO marketing project. The Smallholder Agribusiness Development Promgramme (SADP) in Malawi, established by the American Co-operative Development Initiative and Volunteers in Co-operative Assistance (ACDI/VOCA), links farmers to existing private-sector markets, rather than establishing new market channels. This has contributed to its success in working with large numbers of farmers and achieving sustainability. CLUSA and CARE have also focused on developing linkages between farmers and traders in Mozambique.

CREDIT PROGRAMMES

Credit merits particular attention in the context of agricultural marketing and processing. Farmers groups and NGOs often recognize a lack of credit as a critical constraint to the development of new initiatives and many seek to remedy this through credit interventions. The aspects discussed here are drawn from a recent synopsis by Gordon.

Rural Finance

Rural finance comprises credit, savings and insurance (or insurance substitutes) in rural areas, whether provided through formal or informal mechanisms. The word 'credit' tends to be associated with enterprise development, whereas rural finance also includes savings and insurance mechanisms used by the poor to protect and stabilize their families and livelihoods (not just their businesses). Rural finance comprises informal and formal sectors.

Examples of formal sources of credit include: banks, projects and contract farmer schemes. Reference is often made to micro-credit. Micro underlines the small loan size normally associated with the borrowing requirements of poor rural populations, and micro-credit schemes use specially developed pro-poor lending methodologies. Rural populations, however, are much more dependent on informal sources of finance (including loans from family or friends, moneylenders, and rotating or accumulating savings and credit associations). There is an enormous literature on rural finance and micro-credit, much of which relates to small-scale enterprise.

(Many micro-credit schemes specifically exclude agricultural production activities, even in rural areas, because they are considered high risk.) The discussion here focuses on particular aspects of rural finance that are relevant to NGO and CBO agricultural marketing initiatives. First, typical features of micro-credit schemes, which tend to be run by NGOs and often work with community groups, are reviewed.

Attention is then focused on two types of credit scheme with particular relevance to agricultural marketing interventions: inventory credit – a relatively new departure in smallholder agriculture, which is attracting increasing attention from NGOs and private banks; and outgrower schemes, which tend to be supported by the private sector but often work through producer groups. Some NGOs support other 'hybrid' rural finance initiatives that may affect agricultural marketing. These include stamp-based savings groups and rotating savings and credit associations (ROSCAs), where pooled savings may facilitate access to additional loans, and permit crop purchase/assembly/marketing activities or improved access to agricultural inputs.

Micro-credit and the Rural Poor: The Issues

Rural credit would not be the focus of so much development effort were it not for widespread market failure (*i.e.* failure to provide the services people want) in rural financial services in developing countries.

Reasons for market failure include:

- The lender does not know the default risk of each potential borrower and to collect this information is costly;
- It is costly to ensure that the potential borrowers take those actions which make loan repayment more likely;
- It is difficult and costly to enforce repayment;
- The cost of providing services to the rural poor is high because they are located in remote areas, want to borrow small amounts, and illiteracy, lack of experience of banks, and lack of collateral necessitate the development of tailored approaches.

What are the implications of this for agricultural activities? Firstly, all these types of market failure apply to agricultural lending in developing countries. Lack of information on the risk of default is particularly germane to agricultural enterprise. Farmers do not keep records, so it is difficult for them to produce the information that might convince a bank of their creditworthiness.

Rural market transactions are largely informal, so it is difficult for the bank to collect independent information on prices. Farming is clearly a risky business because of weather, pests and market fluctuations and it is difficult for a bank to assess the degree of risk associated with particular activities. The rural poor do not have a track record, or referees who will vouch for their competence and reliability. Making sure that farmers keep to their business plan, using loans as intended, and carrying out tasks to schedule, is also costly – although this might make loan repayment more likely. Enforcing repayment is also difficult.

This requires monitors who know when crops are sold, or agreements with merchants to pay the farmer net of what she/he owes the bank, or effective penalties such as seizure of assets or prosecution. Farmers rarely have collateral acceptable to banks. They may not have clear title to the land they farm, or even if they do, rural land markets may not function well enough for land to be considered a 'bankable' asset. Poor farmers, moreover, rarely have other bankable assets.

They might own a bicycle, and have a store half full of grain, but were a bank to seize such assets the cost of doing so would probably exceed their sale value. The poorer the farmer, the fewer are his/her chances of borrowing from the formal sector. Women, particularly poor women, face even more problems in obtaining credit. Land title, where it exists, is usually held by men. Women often have little control over other factors of production, particularly for the 'bankable' cash-cropping activities. In some countries women may only borrow in the names of their husbands, if at all, and literacy rates for poor women are almost always lower than those for men. The irony is that numerous studies show that women tend to repay loans more reliably than men. Numerous projects, government schemes and NGOs engage in loan programmes targeted at the poor. Some of these work well, but many are unsustainable because of high and subsidized costs, and high rates of default. Moreover, many miss their target, with the benefits captured by the less poor. The poor depend overwhelmingly on the informal sector. Micro-finance is the response to market failure in 'conventional' banking services for the rural poor. It responds to the needs of low-income households.

Sound schemes tend to be characterized by:

- Small, short-term loans, and savings mechanisms;
- Simplified loan appraisal procedures;
- Innovative approaches to collateral;
- Rapid approval/disbursement of repeat loans after repayment;
- High transaction costs;
- High repayment rates;
- Savings and loan services provided at a location and time convenient to the poor.

Thus, micro-credit schemes are often associated with: group-lending (where peer pressure effectively substitutes for collateral, and other group members may take action to prevent one member defaulting, for instance, by providing labour to assure timely harvest); extension inputs arranged by the micro-finance institute (MFI); and mobile banking arrangements. Cashflow analysis may concentrate on overall ability to repay the loan rather than a particular investment project. In some respects, MFIs try to imitate the strengths of the informal sector (using local information to ensure repayment, for instance) and some MFIs are experimenting with ways to link their operations with some of the informal sector financial agents. Agricultural marketing loans are not common in MFI loan portfolios, but some micro-credit schemes do provide marketing finance or input loans.

MFIs can include government and commercial banks, NGOs and savings and loan co-operatives. Some of the largest MFIs are Asian (for instance, Grameen Bank in Bangladesh has 2 million borrowers and savers, Bank Rakyat Indonesia/Unit Desa system has 2 million borrowers and 12 million savers, and the Bank for Agriculture and Agricultural Co-operatives (BAAC) in Thailand has 2.5 million clients). Coverage is less in Africa, but still growing. K-Rep in Kenya has 15000 clients and the Fe´de´ ration des Caisses d´ e´pargne et de Cre´ dit Agricole Mutuel (FECECAM) credit movement in Benin has 200 000 clients.

Whilst these figures are impressive, they indicate a major difficulty: there are too few services to go round – and still fewer likely to become sustainable. Many schemes find it difficult to graduate from subsidized operations to full cost-recovery. Uncosted but critical inputs by NGO staff are commonplace. Interest rates are often set below market rates (even at negative real rates, where high inflation prevails). Too little attention may be focused on repayment and follow-up; and farmer experience of subsidies, free inputs and loan amnesties may all contribute to low repayment rates.

CONTRACT FARMER SCHEMES

Contract farmer or outgrower schemes focus on very specific needs. They usually operate in situations where a processor or trader faces a supply constraint and, therefore, wishes to promote production of crop x, and has access to sufficient resources (own or loaned) to do so. Inputs are usually provided in-kind (to reduce diversion to other activities), accompanied by extension, and the cost of these services is recouped out of the price paid to the farmer when the crop is harvested. The degree of supervision and type of inputs provided varies greatly. High value horticultural products produced for export may be closely supervised, for instance, whilst supervision of a cotton crop (which still has to undergo considerable processing before it reaches the consumer) may be minimal. Viable schemes must include mechanisms to minimize on-farm consumption or side-selling (farmers selling their crop to an alternative buyer and, therefore, avoiding loan repayment), such as:

- Sharing of information and co-operation amongst crop buyers;
- Effective penalties against farmers who default (exclusion from future schemes, seizure of assets, prosecution);
- Peer pressure achieved through group lending.

Some of the longer-established cotton schemes in West Africa are wide ranging in scope. Considerable long-term effort has gone into capacity building of farmers' groups. This enables farmers to take on a greater share of tasks in input supply and crop assembly, which both reduces cotton company costs and increases cash revenues to farmers. Inputs and extension are available for other crops in the farming system (not just for cotton) effectively reinforcing and expanding the benefits accruing from cotton production.

These types of schemes are common for smallholder annuals grown for export such as tobacco, cotton and horticultural production, where the on-farm investment costs are relatively low, and the pay-back period short. (Thus, greenhouse production of flowers for export in Zimbabwe is not a smallholder activity.) In Africa, the cotton schemes are undoubtedly the largest. About 300000 Ugandan farmers benefit from an input scheme organized by the cotton ginners. In Zimbabwe, around 60000 communal farmers take input loans from cotton companies (and many more participate in other schemes intended to increase cotton output). In Mali, around 100000 rural households participate in the cotton input schemes. The schemes in Mali and Zimbabwe both depend on a network of strong farmers groups, and the Ugandan scheme is likely to develop in that direction as it matures, seeking less costly ways to reach its target group.

CROP MARKET CHARACTERISTICS

If there is limited on-farm consumption and local marketing, it may be possible to collect loan repayments at the point of sale. However, unless there is a crop purchase monopoly, or agreement to share information between different buyers, the farmer may be able to avoid repayment by 'side-selling' (selling to another buyer).

Input qualities

Inputs are often provided in-kind to reduce 'diversion' of the input away from the targeted crop. Diversion is lessened if there is a limited alternative use or market for the inputs, or if (unusually) returns to use of the input are greatest for the crop in question.

Commercial/Credit Context

Prospects for viable operation of the scheme are greater if farmers treat farming as a business and are integrated into markets, and if there are supportive legal/political/ contract enforcement institutions. A recent history of loan amnesties, default without penalty, and subsidized inputs may undermine the operation of viable schemes.

Modus operandi of scheme – best practice:

1. Group schemes for peer pressure.
2. Group or individual schemes backed up by monitoring/good information, support staff, and ability to act.
3. Incentives for repayment and penalties for non-repayment.
4. Appropriate incentives for field monitors/co-ordinators.
5. Training provided to farmers extension and business management.
6. Developing relationship/trust/loyalty through field presence/contact.
7. Accessibility of scheme – minimize red tape and transaction costs; organize so that the location and timing of contact is convenient to farmers.
8. Effective and timely monitoring of input use and crop marketing.

Inventory Credit

Another mechanism for financing agricultural trade exists in inventory credit or warehouse receipt systems. This involves a tripartite agreement between a bank, a borrower (usually a trader), and a warehouse operator. Essentially, a trader is able to use an existing stock of, say, grain as collateral for a bank loan providing certain conditions are met. The grain must meet certain verifiable specifications and be stored in a warehouse operated by a third party. The bank will usually lend up to a certain percentage (perhaps 80 per cent) of the value of the grain at that time (at harvest time when prices are low). The trader can then use the loan to acquire additional stocks of grain, and is usually required to repay the loan during the 'lean' season before grain prices start to fall once more with the next season's harvest. The trader must settle all the warehousing costs before she/he removes the grain, and is generally required to repay the loan at this stage also. In the event that she/he is unable to repay the loan, the bank can seize the grain and sell it.

There is a lot of interest in inventory credit at the present time because of the scope it offers to provide traders with capital to fuel agricultural marketing. However, its use is limited by a number of factors: it can only be used for crops that are relatively non-perishable, with reasonably predictable pricing scenarios; by definition it is targeted to larger, more sophisticated traders, able to purchase initial stocks and fairly confident in their negotiations with banks and warehouse operators; and it is critically dependent on the necessary institutional infrastructure (bank willingness to lend against inventories, warehouse systems which can operate to the standards and within the necessary legal framework, and supportive and enforceable legal institutions). NGO involvement in these schemes can take a number of forms. It may include a role as an MFI providing the credit, or it may include working with farmers' groups who effectively become the 'trader' in the tripartite agreement, buying crops from other farmers (or their membership), for storage and sale later in the season when prices are higher.

AGRICULTURAL MARKETING IN DEVELOPING COUNTRIES

Economic reforms have had sweeping impacts on agricultural markets in developing countries.

In general, state intervention has been reduced, notably with respect to:

- The abolition or sharp curtailing of parasitical marketing boards;
- Depreciation of formerly over-valued currencies rendering developing country exports more competitive and imports more expensive;
- A reduced public role in agricultural services, especially in subsidized credit, input and extension networks;
- A shift away from pan-territorial and pan-seasonal crop pricing strategies and pre-announced prices.

Reviewing agricultural markets research in sub-Saharan Africa and Asia, Jones concludes:

- "In newly liberalized [food] markets in eastern and southern Africa... barriers of entry to trade are low, but the marketing system has little

> capacity to channel credit or spread risk. There are strong theoretical reasons for expecting the impact of and response to reforms to vary between different classes of producers. The absence of key markets, risk aversion, high transaction costs and the dual role of agricultural households as producers and consumers are critical features. The marketing system depends on both physical and institutional infrastructure... Collective action by market participants may address this but it may also lead to collusion over prices. Evidence from South Asia shows that food markets exhibit social barriers to entry, massive asset polarization, debt relationships between large and small traders and traders and farmers, diverse institutional and contractual arrangements, and collusive behaviour, enforced in part by manipulation of the state regulatory system."

This conclusion gives some clue to the reasons why NGOs and CBOs intervene in agricultural markets. When extension agents, researchers and development organizations working in rural areas ask farmers to prioritize their problems, agricultural marketing is repeatedly raised as one of the most important problems faced. It may arise in the context of the promotion of new crops or productivity-enhancing technology, or it may be felt particularly acutely in remoter areas poorly served by commercial traders, where parastatals no longer operate.

NGO marketing interventions typically aim to fill critical gaps in the marketing system or address the power imbalances to which Jones refers. Nowhere are those marketing problems felt more acutely than in the areas for which it is most difficult to identify sustainable strategies to improve market access. Farmers in remote areas (either remote because of physical distance from markets or because of poor roads) are almost always poorly served by agricultural traders and are often obliged to accept seemingly unattractive prices for their produce. Distance from markets rules out the production of higher value more perishable crops, and reduces the linkages between these producers and other more specialized markets.

By the same token, CBOs and NGOs seeking to promote alternative strategies for these disadvantaged communities face high costs and tangible obstacles that make their task particularly difficult. Poor access to markets is mirrored by poor access to all kinds of rural services. The poverty that results makes such communities particularly risk-averse. Where rainfall is uncertain, the situation is even worse, whilst the relative absence of trade does nothing to relieve the covariance in production. These are the challenging circumstances that make an examination of marketing interventions worthwhile. There is wide-ranging experience amongst the development NGO community. Some of these initiatives have taken-off and developed into self-sustaining activities, whilst others, although not conceived as such, have effectively become subsidy dependent welfare programmes. This review identifies best practice and the conditions required for such programmes to work.

NGOS AND CBOS - SOME DEFINITIONS

NGOs are part of the development landscape. Increasing amounts of development aid are channelled through NGOs.

The term gives little clue as to their real characteristics but most people associate NGOs with the following:

- A formal and officially recognized organization that is not linked to government;
- Having a purpose that is altruistic rather than commercial;
- Attracting staff who are value-driven rather than financially motivated.

Fowler highlights some features of the voluntary sector by making comparisons with government and business- organizations.

Table. Comparison of Organisations in Different Sectors

	Sector		
Characteristics	**Government**	**Business**	**Voluntary**
Relationship to those served based on:	Mutual obligation	Financial transaction	Personal commitment
Duration of relationship to those served:	Permanent	Momentary	Temporary
Approach to external environment:	Control and authority	Conditioning and isolation	Negotiation and integration
Resources from:	Citizens	Customers	Donors
Feedback on performance	(in) direct politics	Direct from market indicators	"constructed" from multiple users

The term 'community-based organization' or CBO may be associated with similar values but is generally more focused on issues particular relevant to the community from which its membership is drawn. CBOs may be quite formally structured, but can equally be loosely structured, informal organizations; farmers' associations are an example of a CBO.

Stocker and Barbor-Might discuss CBOs in relation to civil society organizations (CSOs): "CSOs are, simply, organizations operating in civil society somewhere between the informal associational world of family, kin, neighbours, friends and the more formal or market-oriented world of business organizations, state – and NGOs. Although in practice there is diversity and complexity, organizationally the idea is a simple one; NGOs act as intermediaries for CSOs...

CSOs sometimes evolve into NGOs. Many of them have nothing to do with development or are not poorer people's organizations. With this in mind, many authors and authorities prefer to use alternative terms, for example, 'GRO' (grass roots organization) or 'CBO' (community-based organizations) to designate CSOs that exist to serve their members, these members being poorer people. CBO is the usage followed by the World Bank."

NGO AND CBO INVOLVEMENT IN AGRICULTURAL MARKETING

DIFFERENT TYPES OF ORGANIZATION

These were not especially helpful in distinguishing between the plethora of organizations present in many developing countries.

A four-way categorization is proposed here, based largely on origins and capacity:

1. Northern NGOs with offices in developing countries, usually obtaining funds from donors (including private individuals); this group is quite broad since it encompasses large NGOs such as Oxfam or CARE, with activities in many countries, as well as small NGOs whose activities may be quite focused on a few countries and issues.
2. Indigenous NGOs who have become relatively large, well-organized and able to attract significant funds from international donors and Northern NGOs; sometimes these NGOs were originally created or strengthened by Northern NGOs.
3. Indigenous NGOs that are small, usually focused on a particular geographical area or issue, that obtain small amounts of funding from donors or government, but who struggle to grow or stay afloat.
4. CBOs, membership organizations serving particular interest groups usually in rural communities, whose focus may be broad or quite narrow; these organizations may be formally structured or quite informal; farmers associations, credit groups, and joint marketing societies could all be considered CBOs in the context of this review.

Many countries have laws governing the registration of different types of organizations that may confer a certain tax status or legal standing. Some developing countries have umbrella associations for NGOs. In any particular country it is useful to find out whether an umbrella organization exists, and if so, the types of CBOs and NGOs that tend to be officially registered – recognizing the potential divergence between official requirements and practice.

THE EVOLVING ROLE OF NGOS AND CBOS IN DEVELOPMENT ASSISTANCE

In the last 20 years, NGOs have become progressively more involved in development assistance, at every level. The shift from a relief and welfare focus has come about partly in an attempt to address the underlying causes of some of those man-made disasters or to limit the negative consequences of the natural disasters at which they assisted. It has been helped by the increased funding they found they were able to attract. Many Northern NGOs now have policy and research departments, and are a legitimate channel for large amounts of donor funding. At the same time, the role of the state has been redrawn, and in developed and developing countries, there is now a much greater focus on civil society as a way to improve democratic processes and bring about greater accountability in government. Governments are also seeking ways to be smaller and to sub-contract functions where feasible. Furthermore, funding developing country organizations to carry out development work is considered a way to build indigenous capacity. NGOs working in developing countries have benefited from this trend – either because they are considered part of civil society or because they work closely with many civil society organizations, including CBOs. As Stocker and Barbor-Might state:

- "From the point of view of the donors, civil society was the 'place' where something could be done and, often enough, NGOs were the intermediary institutions or midwives of such remedial programmes [relating to structural adjustment], spanning the gap between donors and CSOs. The funding channels varied, sometimes being directed through Northern NGOs (which might provide 'aid' directly or channel it to one or more partner Southern NGOs or CSOs) and sometimes going as direct funding to Southern NGOs and in some instances even to CSOs. When governments were irredeemably corrupt or oppressive (as, for instance, in Haiti during the Duvalier regime), these programmes seemed to offer virtually the only hope of channelling assistance to the people who most needed it."

This growth in the funding, remit, competencies and responsibilities of NGOs has meant that they have been closely involved in (if not the instigators of) much of the experimentation with practical solutions to pressing problems in rural areas. This is the context in which NGO experiences with agricultural marketing interventions provide a valid and rich focus for this review.

NGOS AND CBOS

Although many NGOs share similar altruistic goals, their approaches vary enormously. This is particularly evident in the extent to which they embrace and harness commercial activities to promote broader objectives, or reject this as a legitimate means by which to achieve social objectives. Moreover, amongst those NGOs prepared to use commercial activities as a means to an end, there can be considerable variability in the role these activities are accorded within the development strategy and the competence with which they are planned and undertaken.

Organizations that are Primarily Welfare-oriented

A large number of NGOs and CBOs become involved in agricultural marketing activities, but this is rarely their core business. (There are some notable exceptions amongst some of the international NGOs who have become very experienced in agricultural enterprise, agro-processing and marketing. These include, for instance, TechnoServe, the Intermediate Technology Development Group, Enterprise Works Worldwide and the Cooperative League of the USA.) Many NGOs start with welfare (or social or altruistic) objectives, in areas such as education, health, water, infrastructure and agriculture and gradually shift towards a longer-term development focus. With this shift, small business and income-generation activities take on a greater role. Gibson describes this gradual transformation in terms of a continuum of different actions and attitudes.

Table. NGOs' Evolutionary Path in the Development of small Businesses and Income Generation

From ←	→ To
Relief and welfare	Development
Short-term	Long-term
Ideological	Pragmatic
Community-focused	Individual-focused
Targeted	Self-selecting
Grants	Market interest rates
Amateurish	Professional
Income generation	Small business
Social/ Technical	Economic/ Business
Instinctive	Strategic
Beneficiaries	Clients

Often NGOs and CBOs deliberately work in remote and disadvantaged communities and target the poorest households or individuals. These conditions, in combination with a general relief and welfare orientation, influence the strategies they adopt to achieve their objectives. For instance, direct or indirect subsidies may be used to improve access to markets (*e.g.* through the provision of transport, credit or inputs). An example of a direct subsidy is free or below cost use of transport (calculated on the basis of costs of fuel and driver and perhaps some portion of the vehicle costs). An indirect subsidy might involve charging a commercial (or break-even) rate on the vehicle hire but taking no account of the staff costs of implementing and managing the scheme. Whilst few people would suggest that the subsidy could continue indefinitely, there may be little consideration of how these activities can eventually be shifted to a more sustainable basis. The result is often that the programme attracts participation because of the subsidies, and once it ends there is little enduring impact.

Yet in the short-run these types of activities are attractive to NGOs because they have fairly immediate and visible (if not enduring) impacts and can (with varying degrees of success) be targeted to particularly disadvantaged groups (such as the poorest households, women, refugees, the handicapped or other marginalized social groups). An approach that seeks to use commercial channels may take much longer to develop and may place the intended target group at a disadvantage relative to other members of the community. Even when NGOs and CBOs do not intentionally adopt a non-commercial approach, market-oriented interventions are often subordinate to their core business.

This affects the way they are developed and managed, as well as the way they are perceived within and outside the organization.

Stanton gives four reasons for this lower status:

- Ideological – "...the core work assists those who most need it, income-generating work assists those who will exploit it to the greatest economic advantage";
- Resource allocation – reflecting the ideological perspective;
- Management – "the most willing volunteer for the job rather than external recruitment of people qualified and experienced in business management";
- Finance – these activities may be quite costly to implement and effectively monitor.

Furthermore, marketing activities are often managed and evaluated in the same way as other development activities, with insufficient attention to budgeting and profitability. A survey of income-generating programmes carried out by a range of indigenous and international NGOs in Welaita, south-west Ethiopia, found that little attention was given to the income generated and the profitability of different schemes. However, NGOs using subsidies to target disadvantaged groups would argue that this is a legitimate way to improve the livelihoods of poor individuals, households and communities, particularly in remote areas. Yet even with these subsidies, it may be difficult to have much impact on livelihoods in the most geographically and socio-economically disadvantaged communities. Stanton points out the disadvantages of this type of approach: the frequent failure to make a significant profit, high costs that prevent the NGOs/CBOs from reaching a wide audience, and problems concerning long-term sustainability.

BUSINESS-LIKE NGOS AND CBOS

- "Organizations which themselves resemble small businesses – in terms of their people, culture, systems, structure and behaviour – are most likely to be successful in encouraging the growth of small businesses."

In recent years, private sector development has increasingly been seen as a viable and important approach to sustainable development. Thus many governments, NGOs and CBOs have focused on the promotion of marketing and small-scale enterprise to encourage greater participation in the commercial sector, as a route to higher incomes, employment generation and growth. Small enterprise development work, which grew considerably in the 1980s, has contributed to a realization that it is possible to make much greater use of market mechanisms in pursuit of development objectives. This has been shown particularly through the success of micro-credit initiatives, where even very poor individuals are able to repay not only loans but also interest which sometimes covers the costs of providing credit. Also, social objectives and commercial objectives are not mutually exclusive and many NGOs and CBOs pursue both. The fair-trade movement is a good example of this. Fair-trade organizations use commercial methods to generate social development benefits through improved terms of trade. The important thing is to balance potential marketing success with the social benefit needs of the beneficiaries. Furthermore, selective use of subsidies can still lead to sustainable and successful marketing initiatives, depending on the circumstances, as demonstrated by the CARE Egypt Agricultural Reform Programme. The programme provides information services to smallholder farmers and facilitates linkages to help increase farmer income.

The service is highly subsidized but has proven successful for a number of reasons:

- It helps to link farmers to sources of information outside the programme, thereby fostering the long-term sustainability of relationships and networks;
- Farmers contribute financially, *i.e.* they pay fees for the services;
- The demand for services is farmer-driven and project staff work with farmers to identify production and marketing opportunities.

A similar approach is used by Intermediate Technology in Zimbabwe. Assistance in product development is offered but the initiative must come from an existing business, which must be willing to contribute to the costs of product development (*e.g.* through materials, labour or workshop facilities). The two key advantages of greater commercial orientation and awareness are that cost-recovery enables more people to be reached by such programmes, and sustainability becomes a realistic goal. Gibson argues that NGOs appear to have distinct advantages in pursuing income-generation programmes ("smaller, more flexible, innovative organizations"). His conclusions are firmly rooted in the belief that commercial strategies can serve development objectives:

- "The continuing challenge is to progress from this base so that the economic growth of other developing countries is enhanced, is driven by indigenously owned and indigenously managed enterprises, and reaches the poor and disadvantaged sections of the population."

DIRECT INTERVENTION OR FACILITATION

The marketing role that NGOs and CBOs take on lies somewhere along a continuum between being directly responsible for marketing activities to facilitating beneficiaries/clients to market for themselves.

DIRECT MARKETING ROLE

The term 'income-generating programme' (IGP) is used to describe a variety of programmes. These range from enterprises owned and managed by the beneficiaries to enterprises owned and managed by the organization which employs the beneficiaries. A number of NGOs/CBOs have established this latter type of small business to generate income to finance their other programmes and reduce donor dependence. CBOs and NGOs can also be more directly responsible for marketing activities. One way of doing this is through out-grower schemes (sometimes referred to as contract farming or satellite production). Such schemes involve smallholder producers providing agricultural raw materials to trading or processing businesses.

Often growers work as a group, characterized by String fellow *et al.* as linkage-dependent groups. Generally the marketing arrangements are predetermined: prices or a pricing formula are agreed. They help markets function to the benefit of both producers and the companies or organizations involved.

This type of relationship is beneficial for farmers because they have a secure market for their produce at a predetermined price and the buyer benefits from having a guaranteed source of raw materials and lower transaction costs, which reduces his/her risk and costs. Within the fair-trade arena, the role of, and the marketing channels used by NGOs and CBOs (or alternative trading organizations – ATOs), also varies. Some organizations (such as Oxfam Trading and Traidcraft) take a direct marketing role by acting as wholesalers, with the producers acting as subcontractors producing to order. An advantage of this type of arrangement for producers is that they are guaranteed a volume of sales, thereby minimizing their risk. A disadvantage of this, and of out-grower schemes, is that the producers can be dependent on the trader, and may not have access to alternative buyers or markets if for any reason the trader is no longer able to market their produce.

ADC AND BEAN OUTGROWER SCHEMES IN UGANDA

The Agri-business Development Centre (ADC) in Uganda helped to establish bean outgrower schemes in Kasese and Kibaale districts together with the Uganda National Farmers Association (UNFA) and Bugangaizi Export Commodities Limited (BEC). The purpose of the schemes was to integrate poor rural farmers into the bean market, through marketing agencies, to provide them with an additional income source. ADC's role was facilitating the supply of bean seed and training some of the producers as 'farmer-extensionists' to provide follow-up extension advice to other producers. UNFA and BEC promoted and implemented the programmes. They were responsible for managing seed supply, training and extension and organized marketing (procurement and collection) of the crop.

The outgrower schemes have been successful from the producers' perspective in several ways:

- The numbers of farmers reached has steadily increased;
- Output levels have increased (participation, acreage and yields have all increased);
- An effective farmer-extension system has been created, and the adoption of new varieties and improved production methods has been good;
- Planting materials have been maintained and expanded through seed multiplication;
- Household income from bean sales has increased.

From the trader perspective, a strong linkage was formed with a private buyer in Kasese, who procured beans through UNFA and sold to a Kampala-based exporter. In Kibaale the linkage was weaker, due to the inability of BEC to raise finance. This lack of capital was a key constraint, and undermined the efforts that had gone into developing the scheme and building effective outgrower loyalty. Also, competition from other traders emerged for the beans, which affected potential profit. This highlighted the importance of product selection within outgrower schemes and the need to consider diversion factors (whether the product can be used or sold outside the outgrower scheme) and the buyer's (financial) exposure ratio (the likely cost of obtaining the crop against anticipated sales value).

FACILITATIVE ROLE

Other NGOs and CBOs play a more facilitative role. They assist individuals, groups and communities to market for themselves. This includes both improving access to, and benefits generated from, existing products and existing markets as well as creating new products and new markets (*e.g.* through technology development and processing). There are a variety of ways in which organizations facilitate marketing, including: strengthening the capacity of individuals, groups or communities (through group strengthening and training); developing linkages to traders and other stakeholders in the marketing chain (*e.g.* input suppliers, credit sources and transport agents); and linking farmers to relevant market information.

This type of facilitative role is beneficial for a number of reasons: being less interventionist, it is likely to generate more sustainable marketing activities and linkages; it is likely to be achieved at lower cost than if the NGO was more directly responsible for marketing activities; and, therefore, it facilitates reaching a wider audience.

DEVELOPMENT OF AGRICULTURAL MARKETING

Agricultural marketing covers the services involved in moving an agricultural product from the farm to the consumer. Numerous interconnected activities are involved in doing this, such as planning production, growing and harvesting, grading, packing, transport, storage, agro- and food processing, distribution, advertising and sale. Some definitions would even include "the acts of buying supplies, renting equipment, (and) paying labour", arguing that marketing is everything a business does. Such activities cannot take place without the exchange of information and are often heavily dependent on the availability of suitable finance.

Marketing systems are dynamic; they are competitive and involve continuous change and improvement. Businesses that have lower costs, are more efficient, and can deliver quality products, are those that prosper. Those that have high costs, fail to adapt to changes in market demand and provide poorer quality are often forced out of business. Marketing has to be customer-oriented and has to provide the farmer, transporter, trader, processor, etc. with a profit. This requires those involved in marketing chains to understand buyer requirements, both in terms of product and business conditions.

In Western countries considerable agricultural marketing support to farmers is often provided. In the USA, for example, theUSDA operates the Agricultural Marketing Service. Support to developing countries with agricultural marketing development is carried out by various donor organizations and there is a trend for countries to develop their own Agricultural Marketing or Agribusiness units, often attached to ministries of agriculture. Activities include market information development, marketing extension, training in marketing and infrastructure development. Since the 1990s trends have seen the growing importance of supermarkets and a growing interest in contract farming, both of which impact significantly on the way in which marketing takes place.

AGRICULTURAL MARKETING SUPPORT

In the United States the Agricultural Marketing Service (AMS) is a division of USDA and has programmes for cotton, dairy, fruit and vegetable, livestock and seed, poultry, and tobacco. These programmes provide testing, standardization, grading and market news services and oversee marketing agreements and orders, administer research and promotion programmes, and purchase commodities for federal food programmes. The AMS also enforces certain federal laws. USDA also provides support to the Agricultural Marketing Resource Center at Iowa State University and to Penn State University.

In the United Kingdom support for marketing of some commodities was provided before and after the Second World War by boards such as the Milk Marketing Board and theEgg Marketing Board, but these were closed down in the 1970s. As a colonial power Britain established marketing boards in many countries, particularly in Africa. Some continue to exist although many were closed down at the time of the introduction of structural adjustment measures in the 1990s.

In recent years several developing countries have established government-sponsored marketing or agribusiness units. South Africa, for example, started the National Agricultural Marketing Council (NAMC) as a response to the deregulation of the agriculture industry and closure of marketing boards in the country. India has the long-established National Institute of Agricultural Marketing (NIAM). These are primarily research and policy organizations, but other agencies provide facilitating services for marketing channels, such as the provision of infrastructure, market information and documentation support. Examples include the National Agricultural Marketing Development Corporation (NAMDEVCO) in Trinidad and Tobago and the New Guyana Marketing Corporation.

Several organizations provide support to developing countries to develop their agricultural marketing systems, including FAO's agricultural marketing unit and various donor organizations. There has also recently been considerable interest by NGOs to carry out activities to link farmers to markets.

AGRICULTURAL MARKETING DEVELOPMENT

Well-functioning marketing systems necessitate a strong private sector backed up by appropriate policy and legislative frameworks and effective government support services. Such services can include provision of market infrastructure, supply of market information (as done by USDA, for example), and agricultural extension services able to advise farmers on marketing. Training in marketing at all levels is also needed. One of many problems faced in agricultural marketing in developing countries is the latent hostility to the private sector and the lack of understanding of the role of the intermediary. For this reason "middleman" has become very much a pejorative word.

AGRICULTURAL ADVISORY SERVICES AND THE MARKET

Promoting market orientation in agricultural advisory services aims to provide for the sustainable enhancement of the capabilities of the rural poor to enable them to benefit from agricultural markets and help them to adapt to factors which impact upon these. As a study by the Overseas Development Institute demonstrates, a value chain approach to advisory services indicates that the range of clients serviced should go beyond farmers to include input providers, producers, producer organizations and processors and traders.

MARKET INFRASTRUCTURE

Efficient marketing infrastructure such as wholesale, retail and assembly markets and storage facilities is essential for cost-effective marketing, to minimize post-harvest losses and to reduce health risks. Markets play an important role in rural development, income generation, food security, developing rural-market linkages and gender issues. Planners need to be aware of how to design markets that meet a community's social and economic needs and how to choose a suitable site for a new market. In many cases sites are chosen that are inappropriate and result in under-use or even no use of the infrastructure constructed. It is also not sufficient just to build a market: attention needs to be paid to how that market will be managed, operated and maintained. In most cases, where market improvements were only aimed at infrastructure upgrading and did not guarantee maintenance and management, most failed within a few years.

Rural assembly markets are located in production areas and primarily serve as places where farmers can meet with traders to sell their products. These may be occasional (perhaps weekly) markets, such as haat bazaars in India and Nepal, or permanent. Terminal wholesale markets are located in major metropolitan areas, where produce is finally channelled to consumers through trade between wholesalers and retailers, caterers, etc. The characteristics of wholesale markets have changed considerably as retailing changes in response to urban growth, the increasing role of supermarkets and increased consumer spending capacity. These changes require responses in the way in which traditional wholesale markets are organized and managed.

Retail marketing systems in western countries have broadly evolved from traditional street markets through to the modern hypermarket or out-of-town shopping center. In developing countries, there remains considerable scope to improve agricultural marketing by constructing new retail markets, despite the growth of supermarkets, although municipalities often view markets as sources of revenue rather than infrastructure requiring development. Effective regulation of markets is essential. Inside the market, both hygiene rules and revenue collection activities have to be enforced. Of equal importance, however, is the maintenance of order outside the market. Licensed traders in a market will not be willing to cooperate in raising standards if they face competition from unlicensed operators outside who do not pay any of the costs involved in providing a proper service.

MARKET INFORMATION

Efficient market information can be shown to have positive benefits for farmers and traders. Up-to-date information on prices and other market factors enables farmers to negotiate with traders and also facilitates spatial distribution of products from rural areas to towns and between markets. Most governments in developing countries have tried to provide market information services to farmers, but these have tended to experience problems of sustainability.

Moreover, even when they function, the service provided is often insufficient to allow commercial decisions to be made because of time lags between data collection and dissemination. Modern communications technologies open up the possibility for market information services to improve information delivery through SMS on cell phones and the rapid growth of FM radio stations in many developing countries offers the possibility of more localised information services.

In the longer run, the internet may become an effective way of delivering information to farmers. However, problems associated with the cost and accuracy of data collection still remain to be addressed. Even when they have access to market information, farmers often require assistance in interpreting that information. For example, the market price quoted on the radio may refer to a wholesale selling price and farmers may have difficulty in translating this into a realistic price at their local assembly market. Various attempts have been made in developing countries to introduce commercial market information services but these have largely been targeted at traders, commercial farmers or exporters. It is not easy to see how small, poor farmers can generate sufficient income for a commercial service to be profitable although in India a new service introduced by Thomson Reuters was reportedly used by over 100,000 farmers in its first year of operation. Esoko in West Africa attempts to subsidize the cost of such services to farmers by charging access to a more advanced feature set of mobile-based tools to businesses.

MARKETING TRAINING

Farmers frequently consider marketing as being their major problem. However, while they are able to identify such problems as poor prices, lack of transport and high post-harvest losses, they are often poorly equipped to identify potential solutions. Successful marketing requires learning new skills, new techniques and new ways of obtaining information. Extension officers working with ministries of agriculture or NGOs are often well-trained in horticultural production techniques but usually lack knowledge of marketing or post-harvest handling. Ways of helping them develop their knowledge of these areas, in order to be better able to advise farmers about market-oriented horticulture, need to be explored. While there is a range of generic guides and other training materials available from FAO and others, these should ideally be tailored to national circumstances to have maximum effect.

ENABLING ENVIRONMENTS

Agricultural marketing needs to be conducted within a supportive policy, legal, institutional, macro-economic, infrastructural and bureaucratic environment. Traders and others cannot make investments in a climate of arbitrary government policy changes, such as those that restrict imports and exports or internal produce movement. Those in business cannot function if their trading activities are hampered by excessive bureaucracy. Inappropriate law can distort and reduce the efficiency of the market, increase the costs of doing business and retard the development of a competitive private sector. Poor support institutions, such as agricultural extension services, municipalities that operate markets inefficiently and export promotion bodies, can be particularly damaging. Poor roads increase the cost of doing business, reduce payments to farmers and increase prices to consumers. Finally, the ever-present problem of corruption can seriously impact on agricultural marketing efficiency in many countries by increasing the transaction costs faced by those in the marketing chain.

RECENT DEVELOPMENTS

New marketing linkages between agribusiness, large retailers and farmers are gradually being developed, *e.g.* through contract farming, group marketing and other forms ofcollective action. Donors and NGOs are paying increasing attention to ways of promoting direct linkages between farmers and buyers within a value chain context. More attention is now being paid to the development of regional markets (*e.g.* East Africa) and to structured trading systems that should facilitate such developments The growth ofsupermarkets, particularly in Latin America and East and South East Asia, is having a significant impact on marketing channels for horticultural, dairy and livestock products.Nevertheless, "spot" markets will continue to be important for many years, necessitating attention to infrastructure improvement such as for retail and wholesale markets.

5

Agricultural Economy

AGRICULTURAL ECONOMICS

Agricultural economics originally applied the principles of economics to the production of crops and livestock — a discipline known as agronomics. Agronomics was a branch of economics that specifically dealt with land usage. It focused on maximizing the crop yield while maintaining a good soil ecosystem. Throughout the 20th century the discipline expanded and the current scope of the discipline is much broader. Agricultural economics today includes a variety of applied areas, having considerable overlap with conventional economics.

ORIGINS

Economics is the study of resource allocation under scarcity. Agronomics, or the application of economic methods to optimizing the decisions made by agricultural producers, grew to prominence around the turn of the 20th century. The field of agricultural economics can be traced out to works on land economics. Henry Charles Taylor was the greatest contributor with the establishment of the Department of Agricultural Economics at Wisconsin in 1909 Another contributor, Theodore Schultz was among the first to examine development economics as a problem related directly to agriculture. Schultz was also instrumental in establishing econometrics as a tool for use in analyzing agricultural economics empirically; he noted in his landmark 1956 article that agricultural supply analysis is rooted in "shifting sand," implying that it was and is simply not being done correctly.

DEVELOPMENT

One scholar summarizes the development of agricultural economics as follows:

"Agricultural economics arose in the late 19th century, combined the theory of the firm with marketing and organization theory, and developed throughout the 20th century largely as an empirical branch of general economics. The discipline was closely linked to empirical applications of mathematical statistics and made early and significant contributions to econometric methods. In the 1960's and afterwards, as agricultural sectors in the OECD countries contracted, agricultural economists were drawn to the development problems of poor countries, to the trade and macroeconomic policy implications of agriculture in rich countries, and to a variety of production, consumption, and environmental and resource problems."

Agricultural economists have made many well-known contributions to the economics field with such models as the cobweb model, hedonic regression pricing models, new technology and diffusion models (Zvi Griliches), multifactor productivity and efficiency theory and measurement, and the random coefficients regression. The farm sector is frequently cited as a prime example of the perfect competition economic paradigm. Since the 1970s, agricultural economics has primarily focused on seven main topics, according to a scholar in the field: agricultural environment and resources; risk and uncertainty; consumption and food supply chains; prices and incomes; market structures; trade and development; and technical change and human capital;.

In terms of technical change, there have been increasingly rapid developments and innovations in the equipment designed for agricultural research. This equipment includes instruments for plant physiology research, and monitoring soil conditions and atmospheres.

AREAS OF CONCENTRATION

- Econometrics
- International development
- Community and rural development
- Food safety and nutrition
- International trade
- Natural resource and environmental economics
- Production economics
- Risk and uncertainty
- Consumer behaviour and household economics
- Health economics
- Labour economics
- Forestry economics
- Analysis of markets and competition
- Agribusiness
- Agricultural marketing
- Agricultural policy
- Industrial organization
- Marketing of agricultural products
- Rural economics
- Rural sociology

Agricultural economics tends to be more microeconomic oriented. Many undergraduate Agricultural Economics degrees given by US land-grant universities tend to be more like a traditional business degree rather than a traditional economics degree. At the graduate level, many agricultural economics programmes focus on a wide variety of applied microeconomic topics. Their demand is driven by their pragmatism, optimization and decision making skills, and their skills in statistical modelling. Graduates from Agricultural Economics departments across America find jobs in diversified sectors of the economy:

- Accounting
- Agriculture
- Breweries, distilleries, bottling plants
- Cigarette manufacturing
- Food processing - eg. flour mill
- Food manufacture - eg. cake factory
- Furniture manufacturing; production of linens, drapes, carpet
- Government and NGOs
- Information technology
- Leather tanning, footwear manufacturing, handbag production
- Logistics and supply chains
- Pulp and paper
- Sawmills, lumber mills, wood products
- Textiles processing and garment manufacturing.

INDIA'S AGRICULTURAL ECONOMY

In the early 1950s, half of India's GDP came from the agricultural sector. By 1995, that contribution was halved again to about 25 per cent. As would be expected of virtually all countries in the process of development, India's agricultural sector's share has declined consistently over time as seen in the table below.

Table. Share of agricultural output in India's GDP

Year	1950/51	1965	1976	1985	1991	1999
Percentage share	52.2	43.6	37.4	32.8	28.3	24.4

In the last five decades, the Government's objectives in agricultural policy and the instruments used to realize the objectives have changed from time to time, depending on both internal and external factors. Agricultural policies at the sectoral level can be further divided into supply side and demand side policies. The former include those relating to land reform and land use, development and diffusion of new technologies, public investment in irrigation and rural infrastructure and agricultural price supports. The demand side policies on the other hand, include state interventions in agricultural markets as well as operation of public distribution systems. Such policies also have macro effects in terms of their impact on government budgets. Macro level policies include policies to strengthen agricultural and non-agricultural sector linkages and industrial policies that affect input supplies to agriculture and the supply of agricultural materials.

During the pre-green revolution period, from independence to 1964-1965, the agricultural sector grew at annual average of 2.7 per cent. This period saw a major policy thrust towards land reform and the development of irrigation. With the green revolution period from the mid-1960s to 1991, the agricultural sector grew at 3.2 per cent during 1965-1966 to 1975-1976, and at 3.1 per cent during 1976-1977 to 1991-1992.

Acharya (1998) explains that the policy package for this period was substantial and consisted of:

(a) Introduction of high-yielding varieties of wheat and rice by strengthening agricultural research and extension services,
(b) Measures to increase the supply of agricultural inputs such as chemical fertilizers and pesticides,
(c) Expansion of major and minor irrigation facilities,
(d) Announcement of minimum support prices for major crops, government procurement of cereals for building buffer stocks and to meet public distribution needs, and
(e) The provision of agricultural credit on a priority basis. This period also witnessed a number of market intervention measures by the central and state Governments.

The promotional measures relate to the development and regulation of primary markets in the nature of physical and institutional infrastructure at the first contact point for farmers to sell their surplus products.

Acharya (1998) also notes that the rate of growth of productivity per hectare of all crops taken together increased from 2.07 per cent in the decade ending 1985-1986 to 2.51 per cent per annum during the decade ending 1994-1995. Similar evidence of an increase in yields, a partial measure of productivity gains given by output per unit of land area is seen below for various crops.

Although productivity gains were sustained in the 1990s after the liberalization process began, the yield rates for most of the agricultural products in India are far below comparable rates in a number of other countries. Except for sugarcane, tea, coffee and jute, India's yields are lower than the world average. It should be noted that India is ranked second both in area and output for sugarcane production and is the largest producer of tea and jute in the world. Although India is doing quite well in wheat production, the average yields in the Netherlands and Ireland are more than three times India's yield rates. In all other major crops, India's productivity performance seems to lag behind others.

Table. Yield for various crops (kg/ha)

	1950/51	1960/61	1970/71	1980/81	1990/91	1995/96	1998/99
Rice	668	1 013	1 123	1 336	1 740	1 855	1 905
Wheat	663	851	1 307	1 630	2 281	2 483	2 596
Coarse cereals	408	528	665	695	900	941	1 035
Pulses	441	539	524	473	578	552	661
Food grains	522	710	872	1 023	1 380	1 499	1 611
Oil seeds	481	507	579	532	771	851	948
Cotton	88	125	106	152	225	246	240
Sugarcane	33 422	45 549	48 322	57 844	65 395	68 369	69 288

WHY GLOBALIZE?

Globalization in the context of agriculture can be best discussed in the context of three components – improvement of productive efficiency by ensuring the convergence of potential and realized output, increase in agricultural exports and value added activities using agricultural produce, and finally, improved access to domestic and international markets that are either tightly regulated or are overly protected.

These components are linked in various ways. For example, productive efficiency would enhance value added activities in agriculture through agro-processing and exports of agricultural and agro-based products. These activities in turn would increase income and employment in the industrial processing sector. Thus globalizing agriculture has the potential to transform subsistence agriculture to commercialized agriculture and to improve the living conditions of the rural community.

Ahluwalia (1996) explains that this indirectly requires an improvement in agricultural growth from between 2 and 3 per cent in the past to about 4 per cent per year. Although initially, with respect to agriculture, there was no major policy reform package in the 1990s, it was however anticipated that the opening up of the agricultural sector to foreign trade, the move to a market determined exchange rate and reduction of protection for industry would, over time, benefit the agricultural sector.

Manmohan Singh (1995), the then Finance Minister, in his inaugural address at the 54th Annual Conference of the Indian Society of Agricultural Economics, brought to notice that a policy of heavy protection of the industrial sector operated to the disadvantage of the agricultural sector when industrial prices were raised relative to world prices and thus the profitability of investing in industry was raised relative to agriculture. This would lead to a shift of resources from agriculture to industry. A policy of heavy industrial protection also led to an appreciation of the exchange rate. Ahluwalia (1996) noted that over-valuation of the exchange rate (before the Indian rupee was devalued by 18 per cent in two phases starting in July 1991) discouraged agricultural exports more than industrial exports because Indian industrial policy had sought to offset the constraints faced by industries via a system of export incentives for market support. Agricultural exports on the other hand were denied any such incentives as they did not use imported inputs.

Ahluwalia (1996) argued that in the past, the agricultural sector was negatively protected because of the above two reasons and the fact that farmers were denied access to the world markets due to trade barriers. Exports of plantation crops and a few commercial crops were free from export restriction but exports of essential commodities, particularly food products, were subject to bans, quotas and other restrictions. Interestingly, Kruger and others (1991) showed that while many developed countries continue to protect agriculture, developing countries do not do so. However, no formal attempt or theoretical framework has yet been used to assess the extent of negative protection in Indian agriculture. The implementation of economic reform in the Indian agricultural sector has been a gradual process. These include an 87 per cent cut in tariff on agricultural products, sustenance of high-yield crop varieties, removal of minimum export price on selected agricultural products, a lift on quantity restrictions on the export of some crops and various land reforms related to tenancy rights and land ceilings.

PRODUCTIVITY GAINS FROM GLOBALIZATION AND ECONOMIC REFORMS

In the wake of India's efforts towards globalization and economic reforms, the expected benefits of total factor productivity (TFP) growth can be represented using the production frontier. The production frontier traces out the maximum output obtainable from the use of inputs.

Opportunities from globalization and economic reforms can lead to:

- Shift from A to B due to technical efficiency
- Shift from B to C on existing frontier due to input growth
- Upward shift from C to D due to technological progress

Which constitute various sources of TFP growth, can be linked with trade gains. The movement from A to B led by technical efficiency allows increases in output when inputs and technology are used to their fullest potential to obtain the greatest yield. Given that India has been involved in agricultural production for so long, there would be learning-by-doing gains that can help boost production given the expected increase in demand as India opens up. The increased production would enable a better utilization of inputs, especially that of advanced capital technology.

The reduction in the tariff rate for agricultural products from 113 per cent in 1990-1991 to 26 per cent in 1997-1998 is also expected to motivate local producers into rethinking their production techniques and efficiently utilizing the inputs and technology to keep costs of production down in order to remain competitive. The optimum or efficient use of land and water resources would then allow agriculture to respond to the demand for other products such as horticulture and livestock which is expected to increase following a rising trend in the per capita incomes of both rural and urban groups.

The move from an overvalued exchange rate to that of a market determined rate would also make agricultural exports cheaper and hence boost exports. The new trading opportunities would necessitate an increased use in the quantity of inputs to boost output and this allows for the movement from B to C along the existing production possibility frontier. Increased exports would bring about economies of scale and as Verdoon's law states, output growth would lead to productivity growth.

The scale of output under increased exports would justify the huge fixed costs underlying technologically advanced equipment and hence increase incentives to adopt high quality inputs. The use of such inputs would result in technological progress and this is represented by the shift from C to D. The reduction in tariff rates in industry from 1990-1991 to 1997-1998 range from 153 per cent to 25 per cent for consumer goods, 77 per cent to 18 per cent for intermediate goods and 97 per cent to 24 per cent for capital goods.

This means that farmers now have relatively cheaper access to imported new technology and better capital equipment as well as the option of adopting better farming techniques and this should lead to technological progress. In particular, the development of agro-processing as an instrument for agricultural and rural modernization will bring benefits, given its capital-intensive and technology-intensive nature.

Lower duty rates on plastics and metals also lower costs of packaging. These forms of cost efficiency should allow competitive pricing of products. In addition, external competition can be expected to motivate local producers into the production of improved quality intermediate inputs for agriculture.

The importance of technology in agricultural development was first demonstrated in the 1970s with impressive growth in yields following the introduction of new wheat and rice varieties. But this technology was limited to areas of assured irrigation as the new seeds also required heavy inputs of fertilizers and pesticides for optimal results. However, the potential for further extending this technology is not yet exhausted as there is scope for expanding irrigation further and improving the quality of irrigation in many areas. For further technological progress, genetic engineering and the biotechnology revolution provides a prospect of developing new varieties that can flourish with less dependence on water and chemical inputs. Such reduced dependence upon chemical fertilizers and pesticides is also desirable because of environmental considerations, which are an increasing concern.

It must however be acknowledged that the link between trade liberalization and productivity growth is two-way as they both feed on each other. The productivity gains can be obtained from openness but to benefit from openness via increased demand for exports, agricultural products need to be priced competitively. In other words, productivity growth is necessary to lower the costs of production.

INDIA'S ECONOMIC REFORMS IN AGRICULTURE

Although India's economic reforms were initiated in June 1991, the process of liberalization was implemented gradually and thus it is difficult to assess the full impact of the liberalization measures. Nevertheless, an attempt is made to discuss what is observable in terms of agricultural growth.

One observation is that the expected increase in exports due to liberalization simply did not occur. India's share in world exports was 0.6 per cent in 1997; India has to aim for at least 4 per cent by 2005 in order to meet the growing import demands for capital goods, raw materials and crude oil as well as to meet her external financial commitments. For the last decade or so, India's share in world exports of agriculture has been between 2 per cent and 3 per cent. India is not as competitive as the other countries and calculations show that India's crop yields have increased at a slower rate over the 1990s.

In addition, the agricultural sector's output growth decreased to 2.9 per cent during 1992-1993 to 1998-1999. Kalirajan and others (2001) explain that two important reasons for the slowdown are that there was no major breakthrough in developing new high-yielding varieties during the 1990s and there was a decline in the environmental quality of land which reduced the marginal productivity of the modern inputs. What could this mean in terms of the effectiveness of the policies of reduced protection to industry, a market determined exchange rate and the opening of the agricultural sector to foreign trade?

First, although the reduction to protection of industry is substantial, there is reason to believe that the reduction was not necessarily sufficient to benefit the agricultural sector whose tariffs were also drastically reduced. Hence, the expected shift in resources to agriculture did not occur. Second, is the apparent ineffectiveness of the market determined exchange rate in boosting exports. This is however not surprising as the exchange rate may not be a key factor determining agricultural export demand for India. In general, unlike manufacturing industries, agriculture did not benefit much from these two policies because the share of imported inputs in the value of agricultural production is small. It is likely that a change in the mindset and attitude of farmers has yet to take place and there are delays or hesitation in embracing India's openness.

Third, in opening up the agricultural sector to foreign trade, India has taken major steps towards trade liberalization since 1991, partly on its own initiative and partly from its commitments to WTO. Kalirajan and others (2001) provide a detailed review of these reform procedures. But why have the benefits from trade liberalization been slow to come? One reason is that prospects for growth in agricultural exports depend partly on domestic policies and partly on the removal of protectionist policies pursued by developed countries such as Japan and members of the European Union (EU).

An OECD report (1998) estimated that the producer equivalent subsidy in the OECD countries increased by US$ 9.3 billion from 1988 to 1993 and this subsidy as a percentage of the value of production in 1997 was 9 per cent in Australia, 20 per cent in Canada, 47 per cent in EU and 70 per cent in Japan. These protectionist practices do not seem likely to come to an early end. An UNCTAD report (1999) noted that 29 member countries of the OECD spent an average of US$ 350 billion a year in agricultural support between 1996-98. Schumacher (2000) further reports that the EU provides product-specific trade distorting domestic support to at least 50 different agricultural products. The implication of these reports is that food exports from India may not show a large increase given the international environment and the still-existing restrictions on exports in the major importing markets based on the self-sufficiency argument and food security. Other macroeconomic factors, such as the recession in developed countries in 1996-98 as well as the 1997 South-East Asian financial crisis, have clouded the possibilities of increasing Indian exports.

Another problem faced by Indian agricultural exporters is the protectionist measures in the form of non-trade barriers that developed countries use to restrict market access. This is by tightening requirements of quality, testing and labeling, and anti-dumping and countervailing measures. For example, in May 1997, the EU banned marine products from India citing unhygienic processing conditions. The extra costs of meeting the standards required in export markets as well as costs associated with changes in the production mix and transactions associated with exports may well be discouraging Indian exporters.

One existing problem of India's agricultural protection is the use of input subsidies. The general argument favouring this has been that it is necessary to encourage the use of particular inputs for production for various benefits. For India, Gulati and Sharma (1995) show that the input subsidy in per cent of GDP increased from 2.13 in the triennium ending 1982-1983 to 2.73 in the triennium ending 1992-1993.

But the benefits of these subsidies have accrued to only certain classes of farmers in some regions cultivating irrigated crops. Furthermore, highly subsidized prices of inputs such as irrigation water and electricity for pump sets have encouraged cultivation of water-intensive crops, over-use of water, ground water depletion/salinity and water logging in many areas. Subsidy for nitrogen fertilizer on the other hand has resulted in nitrogen phosphorous potassium imbalance and acted as a disincentive for use of the environmentally friendly organic manure. As a result, the linkage between food crops and non-food crops, which include fodder, has been reduced.

These adverse consequences are a drain on the fiscal burden of central and state Governments. Thus, if not properly monitored, input subsidies can be counterproductive and, in this context, protection to lower costs of production should be done selectively in the course of liberalization.

In fact, Agenda 21 of the United Nations Conference on Environment and Development in 1992 stressed that there is a need for integration of environmental considerations in the pricing of natural and other resources in such a way that prices reflect social costs.

Such a pricing policy will not only lead to a more efficient use of scarce resources but also result in subsidy reductions and improvements in environmental quality. The money saved from the reduction of subsidies can be spent in the development of rural infrastructures, agricultural research, farmers' education and other forms of support for agriculture.

AGRICULTURAL GROWTH AND PERFORMANCE

While the above analysis has provided a general view of the impact of economic reforms, this chaper examines agricultural growth and performance in the states of Bihar, Karnataka, Tamil Nadu and Punjab with their attendant policy implications.

The yields for various crops in these states differ greatly. While Tamil Nadu had the highest yield in rice, oil seeds and sugarcane, Punjab enjoyed the highest yields in wheat, coarse cereals, pulses and food grains. Karnataka on the other hand is seen to do well in cotton and Bihar performed quite well in pulses and coarse cereals. Further analysis and findings by Kalirajan and others (2001) show that Punjab had made remarkable achievements on the agricultural front while Bihar had remained stagnant in the last two decades, with Karnataka and Tamil Nadu showing moderate achievement. Clearly, differences in physical endowments, climatic conditions and institutional characteristics are some of the reasons for the varying productivity performance. Thus, having across the board economic reforms is likely to work less effectively than state-specific policy measures that enable each state's agricultural yields to reach their full potential. The comparative advantage of each state's agricultural production should be determined and with inter-state restrictions removed, total agricultural output would see a very significant increase.

For example, Karnataka with less favourable soil and water resources should be given incentives to concentrate on agro-processed products and corporate agriculture in horticulture, floriculture and animal husbandry, or to undertake watershed development to help with dry land agriculture. Many studies have indicated that with watershed areas, productivity growth has been mainly due to seed and fertilizer use. Thus, this state has to be given input subsidies for high yielding seed varieties but at the same time, the farmers need to be educated on the over use of chemical fertilizers.

With Bihar, agricultural performance is problematic on many fronts. First, although demographic pressure has increased and agricultural technology has improved, most of the uncultivated land is concentrated in southern Bihar, where irrigation facilities have not kept pace and the soil is of poor quality. Given the physiography of southern Bihar, wells are also unsuitable and thus the dominant mode of irrigation has been through tanks whose expansion and maintenance has been neglected. Second, the infrastructural facilities of Bihar have been lagging as seen by the infrastructure development. Due to infrastructural bottlenecks, availability of modern goods and services has not increased or their supply remains costly or unreliable.

Third, agriculture in Bihar is dominated by small and marginal farmers and the prevalence of mass poverty is largely related to the backwardness of agriculture. Fourth and importantly, the state agricultural policies in Bihar are in dire need of review. The semi-feudal production condition still exists in rural areas and the ineffective protection of tenancy rights has hindered agricultural growth. The slow pace of land consolidation reflects inadequate financial outlays and a shortage of manpower. Kalirajan and others (2001) note that marketing and extension services in Bihar are also rather weak compared to the other states.

Punjab on the other hand, was one of the few states which enjoyed the success of land reforms and the high priority of investment in rural infrastructure. Also, the irrigation base of the small and medium sized farms was comparable to that of large farms. In addition, the Punjab Agricultural University at Ludhiana contributed to the development of new seed varieties.

However, there are clear signs of a decline in crop yields since the 1990s and this has been associated with the increasing use of fertilizers and excessive water use which have increased the unit cost of production as a result of declining soil quality. Hence, care is needed when providing further input subsidies in fertilizer and water use.

Another related fact is the steep increase in wages in Punjab and in the absence of productivity increases, the cost increase has affected the profitability of farmers. With Tamil Nadu, the main crop has been rice as this state is blessed with two monsoons. But from 1992-1997, there has been a steady decline in the areas irrigated by canals and an increase in well-irrigated areas while the use of tanks remains an unreliable source of irrigation. However, major improvements in about 10 rice varieties released in the early 1990s can be expected to improve productivity growth in rice production although pests and diseases as well as imbalance in the use of fertilizers are major constraints. Thus Tamil Nadu could do with subsidies of pesticides and farmers should be educated on the more effective use of fertilizers to obtain high yields.

Interestingly, the cropping pattern of late has shown increasing substitution of food crops by commercial crops but there is concern that the benefits will reach farmers only with the development of adequate infrastructure such as roads and markets. However, that Tamil Nadu has a higher index than the all India average of infrastructure.

CHALLENGES OF GLOBALIZATION

It is important to realize that globalization poses many challenges to a developing country like India, which had relied on a state directed and regulated policy regime for more than four decades. In moving to a more open, market-based economy there are many transitional problems that the country has to manage. The Government must play a pro-active role in facilitating the globalization process so that the opportunity sets for the economic agents are widened and the adverse effects of globalization are minimized. The Indian Government must also prepare the necessary information base and develop its capacity to articulate India's concerns and policy trade-offs in the international forums for multilateral trade and environmental negotiations.

In addition, the Government should embark on an extensive programme to educate farmers on the need to meet the standards required in the export markets. In fact, India needs to seek technical assistance in creating the capacity for meeting such standards and to consider watershed developments for environmental considerations. Equally important is the need to disseminate information about possible export markets to farmers, so that market access is achieved at minimum cost. Given the requisite information about markets and profitability, the likelihood of farmers investing in post-harvest and processing technologies and storage and efficient transportation arrangements as well as developing supporting infrastructure is very high.

Although the brave and bold move by India to reduce the tariff rate for agricultural products from 113 per cent in 1990-1991 to 26 per cent in 1997-1998 deserves to be applauded, the question of whether India is ready to compete in world markets remains to be seen. The infant industry argument may still hold for India to shield itself from external competition but one can easily question the length of time that is required to that end. Also, a delay in opening up to foreign trade has the danger that local producers may become too complacent and never be ready for competition.

As India opens up externally, it is also expected to face vulnerability in the wider international price fluctuations and thus Acharya (1998) claims that a minimum price support scheme is important. These prices can also act as a signal to adopt modern inputs and invest in yield-raising infrastructure for increasing production. For instance, keeping basic staple food grains at reasonable prices would induce farmers to switch over to high value crops. However, during the 1990s, Kalirajan and others (2001) shows that procurement prices especially for rice and wheat have been increasing faster than the general price level. Such high prices along with guaranteed purchases by the Food Corporation of India have pushed up market prices. These higher prices are partly responsible for the large buffer stocks with the Food Corporation. If this trend continues, India's comparative advantage will be eroded.

With openness and high price instability, unstable export revenue can also be expected. One way of reducing such risk is for India to diversify her agricultural exports. For example, since 1990, even in commodities such as tea, coffee, cocoa and spices, where India is supposed to have a comparative advantage international prices have been unstable. Besides increasing the type of exports to obtain more export revenue, India should also seriously consider exporting more value added agricultural products through agro-processing such as processed vegetables, fruits, fish and meat products given that export or even local demand for basic agricultural products would decline as incomes rise.

The move to higher value added activities within the agricultural sector also spells greater opportunities for industrialization and vice versa as borne by Kalirajan and Shand's (1997) findings of a bi-directional relationship between agriculture and industry for most Indian states. On the other hand, Sivakumar and others (1999) establish empirical evidence of high forward linkages of agriculture due to the presence of agro-industries while Satyasai and Viswanathan (1999) show the significance of the spillover effects to the industrial sector via the intensive use of purchased inputs in the agricultural sector.

The lack or slow pace of internal or domestic liberalization is also seen to hinder the possible gains from external or trade liberalization. For example, although central zoning restrictions have been abolished, state government restrictions on inter-state and even inter-district restrictions on marketing and movement of goods still exist in many cases. This interferes with the benefits from crop specialization and economies of scale arising from comparative advantage.

The land market is another example of distortion whereby land ceilings exist preventing the operation of large-sized farms. This has led to the emergence of a large number of small economically unviable land holdings. The easy leasing of land should be permitted with assurance of resumption. Yet another problem lies with the insufficiency of credit to agriculture. From 1995-1996, the Rural Infrastructure Development Fund was set up to allocate funds for the completion of projects and the government has committed itself to strengthening the cooperative credit structure through substantial refinancing and restructuring of the Regional Rural Banks.

However, as mentioned earlier, due to varying institutional factors in the Indian states, these domestic reforms can be expected to yield quite different results.

Although India missed the opportunity to open up two decades ago, its attempts to do so now must be regarded as better late than never. Others such as Desai (1999) observe that, "the logic of the global economy as well as India's interests dictate that India become proactive in its liberalization policies. India must liberalize not because it has no choice but because it is the best choice".

His lament that India has adopted a 'victim mentality' when it really needs to adopt a 'winner mentality' has become less of a concern as over time, India has shown commitment to stay on the bandwagon of globalization. Having realized that globalization is a necessary but not a sufficient condition for high growth production, India has undertaken economic reforms, both internal and external. However, it must be ensured that these reforms are synchronized so that the pace of both reforms is set right in order to work hand in hand to promote agricultural productivity growth.

Thus, training the farmers and educating them appropriately to change their mindset and reorienting them to take up new activities or adopt foreign technology is of utmost importance. In this context, it is necessary to involve non-governmental organizations in training and mobilizing the rural poor to face the challenge of liberalization.

Also, with domestic economic reforms, more care needs to be exercised to draw up state-specific liberalization measures to maximize their benefits. Lastly, in the implementation of these reforms for successful globalization, one crucial element, not entirely within control is the need for good governance and stability in the political and economic environment. Political leaders who are the ultimate decision makers in these matters need to examine their own role dispassionately.

It is quite apparent that at this relatively early stage, there is little observable evidence of gains to India's agricultural performance after opening up. However, there could easily be benefits that have not yet surfaced, or are yet to be identified and perhaps too difficult or intangible to measure.

Whatever the case, it is highly likely that it is too soon to assess the full impact of globalization and economic reforms. Furthermore, the process of liberalization has been gradual and remains incomplete. For example, the complete removal of quantitative restrictions after March 2001 will have provided an opportunity for Indian farmers to tap world markets and, if they are successful, results should start to become evident soon. Export promotion via the development of export and trading houses as well as effective liberalizing export promotion zone schemes for agriculture are fairly recent measures and only time will tell as to how effective these measures are. Other possibilities such as agro-industry parks for promoting exports are also in the pipeline.

In conclusion, India has successfully set sail on the waters of globalization and economic reforms and even in the wake of economic and political instability, she has to carefully steer her course in order to reap the benefits of increased productivity growth in the agricultural sector.

AGRICULTURE AND ECONOMIC DEVELOPMENT

As a country develops economically, the relative importance of agriculture declines. The primary reason for this was shown by the 19th-century German statistician Ernst Engel, who discovered that as incomes increase the proportion of income spent on food declines.

For example, if a family's income were to increase by 100 per cent, the amount it would spend on food might increase by 60 per cent; if formerly its expenditures on food had been 50 per cent of its budget, after the increase they would amount to only 40 per cent of its budget. It follows from this that, as incomes increase, a smaller fraction of the total resources of society is required to produce the amount of food demanded by the population.

ADVANCEMENT IN FARMING

This fact would have surprised most economists of the early 19th century, who feared that the limited supply of land in the populated areas of Europe would determine that continent's ability to feed its growing population. Their fear was based on the so-called law of diminishing returns: that under given conditions an increase in the amount of labour and capital applied to a fixed amount of land results in a less than proportional increase in the output of food. This principle is a valid one, but what the classical economists could not foresee was the extent to which the state of the arts and the methods of production would change.

Some of the changes occurred in agriculture; others occurred in other sectors of the economy but had a major effect on the supply of food. In looking back upon the history of the more developed countries, one can see that agriculture has played an important part in the process of their enrichment. For one thing, if growth is to occur, agriculture must be able to produce a surplus of food to maintain the growing non-agricultural labour force. Since food is more essential for life than are the services provided by merchants or bankers or factories, an economy cannot shift to such activities unless food is available for barter or sale in sufficient quantities to support those engaged in them. Unless food can be obtained through international trade, a country does not normally develop industrially until its farm areas can supply its towns with food in exchange for the products of their factories.

Economic growth also requires a growing labour force. In an agricultural country most of the workers needed must come from the rural population. Thus agriculture must not only supply a surplus of food for the towns, but it must also be able to produce the increased amount of food with a relatively smaller labour force. It may do so by substituting animal power for human power or by gradually introducing labour-saving machinery. Agriculture may also be a source of the capital needed for industrial growth to the extent that it provides a surplus that may be converted into the funds needed to purchase industrial equipment or to build roads and provide public services. For these reasons a country seeking to develop its economy may be well advised to give a significant priority to agriculture. Experience in the developing countries has shown that agriculture can be made much more productive with the proper investment in irrigation systems, research, fertilizers, insecticides, and herbicides.

Fortunately, many advances in applied science do not require massive amounts of capital, although it may be necessary to expand marketing and transportation facilities so that farm output can be brought to the entire population. One difficulty in giving priority to agriculture is that most of the increase in farm output and most of the income gains are concentrated in certain regions rather than extending throughout the country. The remaining farmers are not able to produce more and actually suffer a disadvantage as farm prices decline. There is no easy answer to this problem, but developing countries need to be aware of it; economic progress is consistent with lingering backwardness, as can be seen in parts of southern Italy or in the Appalachian area of the United States.

PEASANT AGRICULTURE

One characteristic of undeveloped peasant agriculture is its self-sufficiency. Farm families in those circumstances consume a substantial part of what they produce. While some of their output may be sold in the market, their total production is generally not much larger than what is needed for the maintenance of the family. Not only is productivity per worker low under these conditions but yields per unit of land are also low. Even where the land was originally fertile, the fertility is likely to have been depleted by decades of continuous cropping. The available manures are not sufficient, and the farmers cannot afford to purchase them elsewhere.

Peasant agriculture is often said to be characterized by inertia. The peasant farmer is likely to be illiterate, suspicious of outsiders, and reluctant to try new methods; food patterns remain unchanged for decades or even centuries. Evidence, however, suggests that the apparent inertia may be simply the result of a lack of alternatives. If there is nothing better to change to, there is little point in changing. Moreover, the self-sufficient farmer is bound to want to minimize his risks; since a crop failure can mean starvation in many parts of the world, farmers have been reluctant to adopt new methods if doing so would expose them to greater risks of failure. The increased use worldwide of high-yielding varieties of rice and wheat since the 1960s has shown that farmers are willing and able to adopt new crops and farming methods when their superiority is demonstrated. These high-yielding varieties, however, require increased outlays for fertilizer, as well as expanded facilities for storage and distribution, and many developing countries are unable to afford such expenditures.

THE LABOUR FORCE

As economic growth proceeds, a large proportion of the farm labour force must shift from agriculture into other pursuits. This fundamental shift in the labour force is made possible, of course, by an enormous increase in output per worker as agriculture becomes modernized. This increase in output stems from various factors. Where land is plentiful the output per worker is likely to be higher because it is possible to employ more fertilizer and machinery per worker.

LAND, OUTPUT, AND YIELDS

Only a small fraction of the world's land area—about one-tenth—may be considered arable, if arable land is defined as land planted to crops. Less than one-fourth of the world's land area is in permanent meadows and pastures. The remainder is either in forests or is not being used for agricultural purposes.

GENERAL RELATIONSHIPS

There are great differences in the amount of arable land per person in the various regions of the world. The greatest amount of arable land per capita is in Oceania; the least is in China. No direct relationship exists between the amount of arable land per capita and the level of income; Europe has almost as little arable land per capita as Asia and less than Africa; Japan and the Netherlands have very limited amounts of arable land per capita.

The relationship between land, population, and farm production is a complex one. In traditional agriculture, where methods of production have changed little over a long period of time, production is largely determined by the quality and quantity of land available and the number of people working on the land. Until the early years of the 20th century, most of the world's increase in crop production came either from an increase in land under cultivation or from an increase in the amount of labour used per unit of land. This generally involved a shift to crops that would yield more per unit of land and required more labour for their cultivation. Wheat, rye, and millet require less labour per unit of land and per unit of food output than do rice, potatoes, or corn (maize), but generally the latter yield more food per unit of land. Thus, as population density increased, the latter groups of crops tended to be substituted for the former. This did not hold true in Europe, where wheat, rye, and millet expanded at the expense of pasture land; but these crops yielded more food per acre than did the livestock that they displaced.

As agriculture becomes modernized, its dependence upon land as well as upon human labour decreases. Animal power and machinery are substituted for human labour; mechanical power then replaces animal power. The substitution of mechanical power for animal power also reduces the need for land. The increased use of fertilizer as modernization occurs also acts as a substitute for both land and labour; the same is true of herbicides and insecticides. By making it possible to produce more per unit of land and per hour of work, less land and labour are required for a given amount of output.

RECENT TRENDS

Crop yields have increased dramatically since 1950, with a faster rate of growth in the developing than in the developed countries. Most of this increased output has been due to gains in yields rather than to the expansion of cultivated land. In Europe as well as in North and Central America, the total area under crops has declined; in South America it has increased by more than one-half and in Asia by more than one-third. The large increase in Oceania was due to immigration. The large decrease in Africa was due to a succession of droughts from the 1970s on. Grain yields in the developed regions of the world have increased consistently over the past several decades. In the rest of the world the pre-World War II yields were not achieved again until the mid-1950s. The increases in grain production were more than twice as high in the developing as in the developed countries.

Food production and total agricultural production exhibit nearly identical trends, and changes in food production can be taken therefore as indicative of changes in total agricultural production. Food supplies per capita in developing countries have increased at nearly the same rate as in developed countries, indicating a narrowing gap between food supplies and population growth in the developing countries.

THE ECONOMIC CONTRIBUTIONS OF AGRICULTURAL EXTENSION TO AGRICULTURAL AND RURAL DEVELOPMENT

Agricultural extension programmes are quite diverse from an international perspective. Most are managed as public sector agencies, usually located in the ministry of agriculture, but some are located in other ministries such as education or rural development. Many are managed by non-governmental organizations (NGOs). Many private firms and private organizations (for example, coffee-growers' associations) conduct extension programmes. Even within the most typical organizational structure, where extension is part of the government's ministry of agriculture, there is great variation in the degree of decentralization of management of extension services. In some countries, extension is decentralized, as in India, where it is a state subject. In most developing countries, however, governmental services are highly centralized, with varying forms of regional and subregional units designed to serve local areas.

Further, there is great variation in the skill level and agricultural competence of field staff. In some systems, field staff have little formal technical training in the agricultural sciences. In some cases, this is dictated by a village worker philosophy, in others by local language demands. But, in most cases, it simply is the result of the decisions to expand agricultural extension programmes rapidly during the 1950s and 1960s, when few highly trained agriculturalists were available.

Finally, this diversity of skills, management systems, and objectives has changed over time in many countries. Perhaps the major changes in the management and design of agricultural extension systems over the past four decades is associated with the training and visit (T&V) system introduced in the 1970s by Benor, Harrison, and Baxter (1984) and implemented in many countries with World Bank lending support.

Given this diversity, broad generalizations about the economic contribution of agricultural extension to agricultural development are not feasible. Many situation-specific factors impinge on the effectiveness of extension programmes. The fact that substantial reform and redesign of many extension programmes has taken place indicates that some of them were perceived by their supporters to have been less than fully effective. However, we now have a substantial body of economic studies of extension services in a number of countries; 75 studies of economic impacts of extension systems have been published to date. My task in this chapter is to review the findings of 57 of these studies and to draw out some of the lessons they have to offer.

INVESTMENT INDICATORS: AGRICULTURAL RESEARCH AND EXTENSION

Several relevant indicators of investment in "technological infrastructure". Agricultural extension, agricultural research, and human capital investments are included. The country groups are formed to reflect diversity in levels of "technology infrastructure." These country group categories will be used throughout this paper as a means of recognizing that extension programmes are conducted in different settings, and that their design, management, and effectiveness are conditioned by these settings. It is thus important to describe these country.

There are three Type 1 categories and three Type 2 categories for developing countries. The Type 1 categories cover countries which have not yet mastered production of a full range of goods and services using modem technology.

Type 1a includes approximately 20 countries (including Yemen, Laos, Surinam, Zaire) that lack basic infrastructure of all types. Governments have limited influence, and little manufacturing capacity exists in these countries. Type 1b includes approximately 30 countries (including Nepal, Papua New

Guinea, Haiti, Ethiopia, Burkina Faso) with rudimentary technological infrastructure. Some direct foreign investment has taken place, and this pro vides some access to foreign technology. Agricultural policy in these countries is often dominated by parastatal organizations.

Table. International Technology Investment Indicators: Investment Intensities (

Indicators	Technological Infrastructure Type							
	Type 1 Developing Countries			Type 2 Developing Countries			Industrialized Countries	
	1a	1b	1c	2a	2b	2c		
	Traditional Technology	First Emerge nce	Islands of Moderni-zation	Mastery of Conventional Technology	Transition to NIC-Hood	NIC-Hood	Recently	OECD
R&D/GDP								
Agriculture								
Public-NARs	.002	.004	.005	.006	.007	.010	.010	.015
Public-IARCs	.0005	.0005	.0005	.0005	.0005	0	0	0
Private	0	0	.0002	.002	.003	.005	.005	.015
Industry								
Public	0	.0001	.0001	.0002	.0005	.001	.003	.003
Private	0	0	.0002	.005	.007	.010	.015	.023
Extension/GDP								
Agriculture								
Public	.005	.005	.010	.010	.010	.010	.010	.010
Private	0	0	0	0	.001	.002	.005	.010
Expenditures/staff (1980 000 dollars)								
Research	47	47	47	40	45	50	70	95
Extension	2	2	4	4	10	15	35	35

Some higher education is provided. Most graduates are employed by government agencies, including agricultural extension services.

Type 1c includes 25 countries that have achieved partial modernization (including Sri Lanka, Tunisia, Kenya, Ivory Coast, Bangladesh). Modem agricultural practices have been introduced in most of these countries, and most have well-developed agricultural research and extension services. Most workers are literate. Universities have begun to train scientists and engineers. Graduates have begun to work outside the government. No significant private sector R&D capacity has yet been built in these countries.

The 20 Type 2 developing countries, on the other hand, have made sufficient investments in technology infrastructure to realize "followers" or "catch-up" growth. In other words, these countries are catching up to the developed countries.

The Type 2a countries (including India, Colombia, Mexico, Argentina, Turkey) have achieved modern engineering capabilities. Significant private sector R&D is undertaken. Universities are advanced. Most have not achieved the industrial, trade, technology, and macro-economic policy regimes to realize rapid growth.

The Type 2b countries (including Indonesia, Thailand, Malaysia, Chile, China) have achieved transition to newly industrialized country (NIC) status and the rapid growth associated with it. These countries have advanced technological capabilities and effective policy environments. Some of these countries (for example, Indonesia) have managed to move through the Type 2a stage in a relatively short period of time.

The Type 2c countries (including Hong Kong, South Korea, Singapore, Taiwan) are well-established NICs.

The two sets of indicators of investment intensities (investment/GDP) and one set of "price" indicators for these types of countries.

Agricultural research indicators show that public sector research capacity exists in most countries; perhaps 10 or so of the Type la countries would not have such capacity. In contrast, private sector R&D relevant to agriculture or to industry is not important in Type 1 countries. Such R&D becomes increasingly important as countries become more advanced (that is, for Type 2a to Type 2c). Research capacity for the industrial sector clearly lags behind research capacity for agriculture in all except the most advanced developing countries. Only the Type 2 countries have significant industrial R&D capacity.

Agricultural extension programmes, by contrast, serve almost all countries. For the Type la and 1b countries, these programmes represent the only modernizing investments of significance. Extension spending intensity exceeds research spending intensities until NIC-hood is reached.

The expenditures-staff data show that the real costs of supporting a research scientist are roughly constant across developing country groups and are roughly half the level prevailing in industrialized countries. For extension, however, the real cost of supporting extension field staff is very low in the poorer countries, in fact too low for efficiency in many cases. This ratio explains why many poor countries have very large extension staffs and large staff-farmer ratios. Extension staff are perceived to be low-cost producers of economic growth relative to researchers, and to some degree they are.

THE CONCEPTUAL FOUNDATION FOR EXTENSION IMPACT

Two conceptual themes are relevant to extension impact. The first is the awareness-knowledge-adoption-productivity (AKAP) sequence. The second is the "growth gap" interrelationship between extension, schooling, and research

The AKAP Sequence

It is convenient to visualize extension as achieving its ultimate economic impact by providing information and educational or training services to induce the following sequence:

A: Farmer awareness

K: Farmer knowledge, through testing and experimenting

A: Farmer adoption of technology or practices

P: Changes in farmers' productivity

Changes in farmer behaviour will be reflected in quantities of goods produced, the quantities of inputs used, and in their prices. These, in turn, can be measured as "economic surplus," which is the added value of goods produced from a given set of inputs made possible by the extension activities.

Studies of extension impacts have measured farmer awareness (and sources of awareness), knowledge (and testing of practices), adoption, and productivity. Not all studies have examined all parts of the sequence. Most have shown a statistical relationship between the quantity of extension services made available to farmers and increases in awareness, knowledge, adoption, and productivity.

While the AKAP sequence has a natural ordering, it is clear that real resources in the form of skills and activities by both extension staff and farmers are required to move along the sequence. Awareness is not knowledge. Knowledge requires awareness, experience, observation, and the critical ability to evaluate data and evidence. Knowledge leads to adoption, but adoption is not productivity. Productivity depends not only on the adoption of technically efficient practices, but of allocatively efficient practices as well. Productivity also depends on the infrastructure of the community and on market institutions.

Extension services affect each part of the sequence. They can be seen as both substitutes for and complements to the acquired skills of their clientele farmers. Empirical evidence indicates that they are, on balance, net substitutes for farmers' skills as reflected in farmers' schooling. For example, extension services are typically not the only sources of information (awareness). Skilled farmers can seek information on their own. Farmers with few skills may not do so. Extension information then may have a higher impact on farmers with less schooling. It appears, however, that the awareness-knowledge part of the sequence is where extension services are strong substitutes for farmer schooling. Through organized frequent contact, they "teach" farmers, and this is more than simply informing farmers.

The teaching versus informing distinction is also relevant to the "newness" of the information (that is, of the recommended practice or other technology) and of the nature of the practice or new technology. When technology is new (as for example with a recently released variety of rice) and is also "simple" to evaluate and adopt (where it is a matter of using new seed without altering other practices), information-awareness is relatively easily converted to knowledge and adoption. Farmers with few skills usually adopt such technology with a time lag. When the technological practice is more complex and requires substantial changes in activities and sometimes capital investment, teaching is required. Repeated messages clearly stated, followed up by field staff and often community organization, are required to proceed through the AKAP sequence in this case.

Productivity Gaps and Extension

The AKAP sequencing is, as noted above, related to the flow of new technical information and to the existing state of unadopted technology. We can see this interrelationship more clearly in the context of productivity "gaps."

These gaps provide a way to classify the contribution of extension activities and to show how research and extension are linked. A stylized sequence across technology types is depicted. This could also be visualized as a time sequence.

Extension programmes are designed to reduce both the practice gap, G(P), and the institutions gap, G(I). Extension programmes are not the only activities that reduce these gaps. Providing market information to farmers and developing organized farm groups reduce G(I). Information and teaching reduce G(P). Research programmes are generally required to reduce G(R), although extension programmes can facilitate the reduction of G(R) via facilitating the importing and local modification of improved technology developed elsewhere. Research programmes in most developing countries also modify and adopt imported technologies and germplasm.

Two of the gaps are closely linked. When G(R) is closed (that is, when the BPBI yields go up), G(P) is opened. (This may happen with G(I) also, but to a lesser extent.) Further, it should be noted that the size of the gap is an index of the potential impact of research or extension. As extension succeeds in closing G(P), diminishing returns set in. Successful research opens up new potential by increasing G(P). The relative mix of teaching versus informing is also related to these gaps. When the BPBI yield level has been constant for some time, the G(P) gap is closed mostly by teaching. When BPBI is increased, as by "green revolution" rice and wheat varieties, information and testing advice play a larger role. The pattern of gaps and yield levels across country groups is intended as a stylized pattern. It is roughly based on experience. For Type la countries, both G(I) and G(P) are depicted as large. These are traditional economies with not much new technology being produced that is relevant to them.

As economies move to Type 1b, improvements in institutions allow BP (and A) to rise even without new technology; BPBI remains unchanged. Extension can contribute to reducing both G(I) and G(P), and these contributions are qualitatively different from those required in more advanced country groups. There is little new technology (few new practices) in these countries, farmers have little schooling, and infrastructure is poor. The teaching and organizing activities of extension dominate here. As economies move to the Type 1c category, some new technology has been introduced (BPBI has risen), and the institutions gap has been further reduced. The practices gap, G(P), has been both opened (because BPBI increased) and closed because of continued teaching and because new practices now can be extended.

STATISTICAL METHODS AND ISSUES FOR ECONOMIC EVALUATION

The studies under review in this chapter sought to measure the impact of public agricultural extension programmes' activities in the following four areas: (1) farmer knowledge of technology and farm practices; (2) adoption or use of technology and practices; (3) farmer productivity and efficiency; and (4) farm output supply and factor demand.

Estimation of extension impact is subject to a number of problems which are also faced in the evaluation of other public sector investments. The approach commonly used is a statistical analysis relying on data measuring extension activities at the farm level. Alternatively, statistical analysis can be undertaken where observations refer to aggregate extension services supplied to a given region in a specific time period.

Studies assessing extension impact at the individual farm level that use a farm-level measure of extension may be affected by two basic estimation problems. The first is the problem of statistical "endogeneity" in extension-farmer interactions. 8 Early studies seeking to measure the impact of agricultural extension by identifying the extension variable as some form of extension contact often treated the extension contact as being unrelated to the farmers' actions and characteristics. However, it is likely that one of the characteristics of more productive farmers is the desire to acquire information about changing farm conditions or new technologies. Such farmers may be inclined to attend more demonstration days, read more literature, and seek extension contact. Analogously, extension agents themselves may also seek contacts with better farmers who would be good performers even in the absence of extension contacts.

In such cases, the extension contact variable is endogenous, and the estimates of extension impact on farmers' performance are likely to be biased upward, because some of the better performance credited to extension would in fact be the result of the superior attributes of the group which interacts with extension. The problem of endogeneity can, in principle, be handled econometrically by using two-stage procedures or simultaneous equations approaches, but this has been done in only a few of the studies undertaken so far.

The second source of potential bias is the problem of indirect or secondary information flows where knowledge which originates from extension contacts is passed on to other farmers who do not directly interact with extension personnel. The extension of interfarm communications is substantial, as shown in Birkhaeuser, Evenson, and Feder (1991), where data on farmers' sources of information were reviewed. This review showed that most farmers in areas receiving extension services report that other farmers are the main source of information. Except for the contact farmers in TandV extension areas who are singled out for extension contact by the nature of the programme, direct contact with extension personnel was typically not the major source of information to farmers. Information may, of course, be diffused (to other farmers) from farmers who were informed by extension agents. In such cases, there may be little difference in performance between farmers interacting directly with extension and other farmers, and an estimate of extension impact based on individual extension contacts would erroneously indicate zero extension effect. Generally the presence of interfarmer communications tends to cause an understatement of extension effects when the approach of defining extension impact by the number of direct contacts is used.

In the studies reviewed below, data from farm surveys and secondary sources on farmer awareness, adaptation, and productivity were related to the provision of extension services in different regions and time periods.Productivity is typically measured as production per unit of all inputs (including labour, land, and fertilizer), although in some studies an aggregate production function approach was used.

Estimated coefficients measure the *marginal product* of extension - the added production due to a one-unit addition to extension services supplied. The extension variables also typically have a time dimension. The adoption of improved practices will typically occur at some rate in the absence of extension services, depending on schooling and infrastructure. Extension both accelerates practice adoption and affects the long-run level of practice adoption. For Type la and 1b economies, extension may have a strong level effect if it is effective. For more advanced economies, the extension impact is primarily a speed-up effect. Most studies find speed-up periods of three to five years. Recent studies for Africa (Kenya and Burkina Faso) find significant level effects, implying that extension impacts in these economies are long-term impacts.

THE ROLE OF AGRICULTURE IN THE ECONOMY

The history of the Common Agricultural Policy (CAP) can be seen as the history of attempts to limit the cost of support. But the need for support has only been questioned recently. Not surprisingly, justifications change over time. But it would be wrong to see in agricultural support only the work of an influential political lobby. As the opinion poll of Eurobarometer in 1988 has shown the population at large in the European Community is in its majority prepared to reserve a particular treatment to agriculture. As long as the cost of this policy appears to be manageable, the public is prepared to see agriculture not as an economic activity like many others.

The CAP has even an added value for European integration. The CAP has played a crucial role in European integration from the very beginning. As the CAP is intervening in the markets, it had an integrating effect which has gone much beyond agriculture. The market organizations created under the CAP were not only the first regulations of the Community with legally binding effect on the citizens of the member states, they gave also the European Commission wide-ranging responsibilities in managing the policy. Thus, the CAP became the blue-print for the way European integration works. Its techniques have been applied in many other areas of Community policy.

THE CHANGING ROLE OF AGRICULTURE IN THE ECONOMY

To illustrate the changes agriculture has undergone two sets of figures may be useful. The Stresa Conference in 1958 worked on the basis that employment in agriculture on average in the six original member states was 25 per cent, whereas its contribution to GDP did not go beyond 14 per cent. The respective figures for the Community of 15 in 1994 are 5.7 per cent and 2.5 per cent. A fraction of the workforce in the 1950s is now producing a much higher quantity of food. It is well known that this is due to technical progress through mechanisation, better seed and breeding qualities, better resource management etc. It is, however, surprising to note that this tremendous technical and economic development has certainly led to a change in values considered important, but not to a reversal of attitude towards agriculture.

The early days of the CAP

Already the report submitted by the heads of delegations to the six Foreign Ministers in April 1956, the so-called Spaak-Report, suggested a special treatment for agriculture. The reasons were the particular social structure around the family farm, the volatility of production, low elasticity of demand and differences in yields, input-prices and revenues between the different regions.

The five objectives of the CAP as laid down in Article 39 of the Rome Treaty deal with the farmer and the consumer *i.e.*, the society at large. The farmer is expected to increase his productivity and thus his standard of living, whereas the consumer can rely on sufficient supplies at reasonable prices. Stability of the market is thought to benefit both, the producer and the consumer.

It is interesting to note that the 'specificity' of agriculture which played such a central role in the development of the CAP is not linked to climatic factors but to the place of agriculture in society and to regional disparities.

At the Stresa-Conference in 1958, which laid the basis for the development of the CAP, Ministers of the six original member states stressed the importance of the fanning population for social stability. The family farm was recognised unanimously as the way to provide this stability. Mr. Houdet, the then French Minister was the most outspoken when he said that the State has to ensure that all farmers enjoy the appropriate income and their proper place in the economy and in society. The then-Commission President Walter Hallstein mentioned independence and freedom based on the ownership of his farm as the particular virtues of the European farmer. Judging from the speeches made and the reports submitted, employment in agriculture was not a major concern.

It was widely recognised that increase of productivity would lead to a diminishing workforce in agriculture. Depopulation of rural areas in France was mentioned, but not by the French Minister. Nobody, except the Belgian representative, mentioned the most basic task of the farmer, *i.e.*, providing food for society. This, however, should not mislead us. It was not mentioned because it was so self-evident.

In post-war Europe the awareness that farmers produce an essential ingredient of life, namely food, was very strong. Despite industrialisation and urbanisation in Europe since the middle of the last century, the feeling that farmers had a special role in society as providers of food was wide-spread before the Second World War. The scarcity of supplies during the war could only enhance this feeling.

But there was even more to it. Rural life and thus the farmer had a symbolic value for all those who felt uncomfortable with or hostile to modernisation of society triggered by industrialisation and urbanisation. This is a recurrent feature of European civilization which can already be found during the Roman Empire.

Summing up there were mainly three reasons justifying a specific appreciation of agricultural activity:

- The farmer is producing the most basic goods for human livelihood;
- The farmer is providing social stability through his hard work and the particular structure of the rural society;
- The farmer is, in his production, subject to the volatility of weather conditions.

The Community dimension added a fourth reason which became very powerful over time:

- The regional disparities which the Community was supposed to overcome.

These four aspects and the ingredient of cultural nostalgia ensured a particular place for the farmer in society. As the farmer is ensuring the livelihood of society, society has to ensure the farmer's livelihood. In addition he was still representing the good old times, when European countries were basically rural societies.

In such a specific relationship between farmer and society economic efficiency is not the overwhelming consideration for agricultural policy although it is present in Article 39 of the Treaty, which sees increasing productivity as the main vehicle for growth of income.

THE IMPLEMENTATION OF THE OBJECTIVES OF THE CAP

Title II of the Treaty on the European Community does not rule out the option of meeting the objectives and other requirements of Article 39 by a mere co-ordination of agricultural policies pursued by the member states. This reflects the variety of support systems which existed in the 1950s in the different member states. But soon it became obvious that a common market for agricultural products could not function without a common policy. In order to enable free circulation of agricultural product, the support system had to be a uniform Community wide system. It was logical to finance such a system from Community and not member states resources and to establish a common import regime which in any case was called for by the requirements of a common commercial policy. In such a way the three fundamental principles of the CAP came to life: the common market, financial solidarity and Community preference.

The market organizations for all main production sectors which were created during the first 10 years of Community existence were, not surprisingly, biased towards stabilising markets and increasing the income of farmers which did not flow exclusively from increase of productivity. The consumer had to pay higher prices than world market prices, but he did not complain very much. In some member states prices even fell in comparison with the situation before the market organizations came into effect (*e.g.*, for cereals in Germany). But the prices of food-stuff became less and less relevant when the general income level increased and the share of food in household expenditure diminished. At the same time the offer of food in terms of variety of products and quality improved considerably.

Despite relatively high market prices the traditional structures changed. Employment in agriculture fell with increasing mechanisation and the average size of exploitations increased. But this did not affect the essence of the traditional family farm.

The family farm can be small or big, as long as it provides the livelihood for a family. But a farm will not provide the livelihood for the family if it is not a profitable undertaking. The term 'family farm' implies, however, that the farm is more than an opportunity to make money. The added value is the attachment to the traditional values: attachment to the land, keeping the farm in the family for generations, active involvement of the farmer and his family in the running of the farm. In the Community of six member states there were not many which could not be considered as family farms because the owners lived elsewhere. There were, of course, considerable differences in the size of the exploitations. But in all the six member states the relatively small family farm was predominant. This did not change fundamentally with the enlargements in 1973, 1981 and 1986. Membership of Britain increased the number of large holdings but this was somewhat counterbalanced by the rather small Greek and Portuguese farms, Denmark, Spain and Ireland being in the middle.

The family farm as the main type of Community farming and the flexibility this concept allowed, made it possible to treat farmers in the same way irrespective of differences in the size of the exploitation and income, in climatic conditions, etc. A notable exception was the additional support which farmers in mountainous and less favoured areas received. From hindsight the 1960s were a good time for the CAP when most of the market organizations were completed, had there not been the Mansholt Plan. In its Memorandum on the Reform of Agriculture in the European Economic Community of December 1968, the Commission sought to deal with two problems which had arisen: emerging surpluses and the falling back of income growth in agriculture compared with the rest of the economy.

The Memorandum talks about an explosive situation. It fears that the surpluses will undermine support for agriculture in society and that the growing disparity between farm and non-farm income will erode the contribution of agriculture to social stability. Unfortunately, the Memorandum with its proposals for reduction of arable areas and cow herds triggered the explosion: a very hostile reaction of the farming community. Not all was lost, however: The Mansholt Plan was the beginning of a structural policy on the Community level.

THE WATERSHED OF THE 1980S

Even without a Mansholt Plan Community agriculture was moving in the direction the plan had indicated, but much more slowly, of course. Going slow was possible because the British market opened up in 1973. Furthermore, the first oil-shock in 1972 increased world wide demand for agricultural products. But things were changing at the beginning of the 1980s and public perception of agriculture, too.

The second oil shock and the resulting world recession in 1979 made the bell ring for agriculture in the European Community. Ever increasing production which was met with shrinking demand led to huge stocks. These stocks could only be kept and disposed of with high costs. Around 70 per cent of the Community budget went into agricultural support. Newspapers started to write about butter and beef mountains and wine lakes. The impression that something was wrong with the CAP spread.

Agriculture also came under attack because of its negative impact on the environment due to the application of pesticides and fertilizers. Sympathy for farmers did not evaporate and the perception that the farmer was entitled to support did not wane. But support was more and more considered excessive and the budgetary implications unbearable.

This shift in public attitude led to a number of measures to redress the situation in the 1980s, including the establishment of milk quotas in 1984 and the stabilisation measures in 1988. The basic problems, however, were only addressed in the reform of 1992 which allowed the completion of the Uruguay Round Trade Negotiations within the General Agreement on Tariffs and Trade (GATT) at the end of 1993. During the GATT-negotiations agriculture came for the. first time under massive pressure from industry to reform the CAP.

As a satisfactory result for agriculture was a pre-condition for an overall deal, agriculture had to move. This does not mean that the 1992 reform was triggered only by the Uruguay Round, but it put additional pressure on agriculture to accept the reform of the CAP. At the same time, farmers realised that they had to respond to the increasing concern about the environment if they wanted to see continuing support by society. They also felt the need to respond to the increasing concern of consumers about the quality and safety of food.

This development is reflected in the initiatives the Commission took in order to reform the CAP. The Reflection Paper which the Commission published in December 1980 follows still traditional thinking. It cites among the positive aspects the consumer's security of supply at stable prices, the progress in agricultural techniques and the contribution of agricultural exports to the balance of trade. But it is worried about the budgetary consequences of production surpluses and regional disparities in benefits derived from the CAP.

The Green Book of 1985 on Perspectives for the CAP tries a new approach. It seeks to limit production increases by adjusting the mechanisms of the CAP without questioning the need for support. The Green Book still mentions food security as an important asset of the CAP but it puts less emphasis on the claim that agriculture is an essential element of social stability in general, it focuses instead on the social stability of rural areas. The attitude towards the protection of the environment is somewhat ambiguous reflecting changes of perception. The Green Book stresses the crucial role of agriculture for preserving the natural landscape and the environment but at the same time calls upon farmers to pay more attention to the environment. For the first time the perspective of agriculture providing raw materials for industry is set out in detail. On labour the Green Book wants to keep a substantial part of the workforce in agriculture, but it also recognises the need for increasing productivity further.

The systematic promotion of rural development on Community level has been one of the achievements of the reform launched in 1985. Since the reform of the so-called structural funds (regional fund, social fund, agricultural guidance fund) in 1988, the promotion of rural development is one of the objectives of structural policy (the so-called 5b programme). Further restructuring of Community agriculture will inevitably lead to a loss of population in rural areas. This may lead in certain regions to such a low level of population density that essential infrastructure and services cannot be maintained, resulting in the need to address this problem by strengthening rural development. In this context, it is recognised that agricultural activity is essential but not sufficient to keep rural areas economically viable.

The Basic Paper on the Future Development of the CAP, which the Commission submitted in February 1991 and which was the basis for the 1992 Reform, confirmed the role of agriculture in society as defined in 1985. But it pursued a new approach in order to tackle the surplus problem. It put an end to unlimited support for unlimited quantities of production in the cereal, oilseeds and beef sectors. Intervention prices were reduced and farmers were compensated for the income loss by direct payments linked to past production. In order to control supply, direct payment for cereal and oilseeds areas were made subject to set-aside requirements or limited to 90 animals per farm, and a stocking ratio of 2.5 livestock units per hectare in the beef sector.

THE EUROBAROMETER OF 1988

These developments were supported by public opinion. In 1987, the European Commission within its public opinion poll scheme conducted a survey on the attitude of the citizens of the Community towards agriculture which was published in 1988. One of the main results was that a majority of the population in each member state wanted the public authorities to support agriculture and felt that agricultural expenditure was not too high. The main reason given was the contribution of agriculture to the protection of the environment and to the preservation of the countryside. It was felt that the CAP prevents whole regions of Europe from becoming depopulated and deserted.

Strong support was expressed for healthier food. The majority in most member states supported production for non-food use. But it also became clear that the majority did not like surplus production and felt that support should go only to those who need it.

THE PERSPECTIVE OF THE 1990S

It is fair to assume that depending on overall growth the share of agriculture in GDP will further shrink. The same will happen with employment in agriculture. But these parameters are already fallacious when assessing the weight of agriculture in the economy, and even less so in society.

In economic terms, the impact of agriculture is much larger than the share in GDP suggests if one includes the up-stream and down-stream sectors. In society at large support for agriculture is still not challenged in principle as long as expenditure is kept under control and surplus production can be avoided. In his Agricultural Report 1996, published in April, the German Federal Minister of Agriculture cites secure supply of high-quality food-stuffs and renewable resources, environmental conservation, animal welfare, a well-tended landscape and pretty villages as the main reasons that society should support agriculture.

By the same token, he stresses the need for agriculture to be not only "environmentally friendly" but also "efficient, competitive, market oriented." His colleagues in the Council of Ministers would broadly agree with this statement, with the exception perhaps of the British and Swedish Minister. But even the British and Swedish Farmers' Union would be able to subscribe to the German Ministers statement. When comparing the statement with the situation in the 1950s and 1960s similarities and differences appear.

Food security has a different meaning. The European Community is a leading agricultural producer, exporter and importer. The negotiations of the Uruguay Round have brought home to a wider public that food security of the EC would not be at risk if it produced less. Therefore, the emphasis is shifting to the quality of food-stuff although the value added by high quality is in many cases not the work of the farmer. Under these circumstances the role of agriculture in maintaining the environment and providing the basis for rural development becomes more important.

Historically speaking, agriculture's role in the protection of the environment is ambiguous. Agriculture has been preserving and polluting the environment. Concern about the environment was already in people's minds before 1992, but the reform tried a new and more systematic approach in three ways:

- The reduction in price support was to be an incentive to more extensive farming by making massive use of inputs less profitable;
- The set-aside requirement for crops and the stocking ratio for bovine animals should enhance the favourable impact on the environment;
- A programme authorising payments for specific measures favourable to the environment.

But the new approach goes further than a mere prevention of pollution. European landscape is the result of centuries of human endeavour, not the least of the work by farmers. Thus, by cultivating the land, the farmer becomes the steward of the landscape in the European Union. In promoting and supporting this activity, the CAP responds to the expectations of the public at large.

But the farmer cannot preserve the landscape when rural areas are deserted. In the 1960s and 1970s the reduction of the number of people employed in agriculture was widely accepted as the inevitable consequence of higher productivity and higher income for those remaining in agriculture. It was only in 1988, when reform of the CAP was tried by means of so-called stabilisers, that the ensuing problem for the rural areas which may become deserted was clearly recognised. The 1992 reform gave a new push. Budgetary outlets for the promotion of rural development were increased. They are at the order of 1 billion Ecu/year now. A new programme 'Leader' was launched, which builds on local initiatives. It covers all areas, not only agricultural ones, but is particularly strong in agricultural areas.

Interestingly, the German Minister of Agriculture also mentions supply of renewable resources and animal welfare as contributions by agriculture which justify a special treatment. Both are in a wider sense related to the protection of the environment. Production of renewable resources helps to slow down the depletion of non-renewable resources. The reference to animal welfare tries to distinguish the German farmer from so-called 'industrial' methods as applied in feed-lots or 'egg-factories'. Animal welfare is, however, not a major concern in all the member states. Therefore, I doubt that it is an argument which can be used in all of the Community.

Further adjustments ahead?

Despite the 1992 CAP reform and its emphasis on a more market oriented approach to agricultural support, fanning is still not considered a normal commercial activity. This assessment is likely to change at least for those farms which are commercially viable and.shed the mantle of the family farm. The fact mat there are commercially viable farms in most member states which could compete under world market conditions is not common knowledge. But this is likely to change, mainly for two reasons:

- After the 1992 reform the budget transfers are much more visible and are substantial for farms with large crop areas. The sums involved will raise questions about their justification, in particular for big farms.
- Many East German exploitations which have been created after the down-fall of communism have a chance to become commercially profitable enterprises. These are large farms, specialising in crops with 1,000 to 3,000 hectares. It is significant that many of these exploitations are set up as incorporated companies. According to the Agricultural Report 1996 of the German government the average size is 1.721 ha and they cover 60 per cent of the agricultural area and 80 per cent of the herds in the new Bundesländer. It is interesting to note in this context that the Minister for Agriculture of Saxony, which is part of Eastern Germany, pleaded on January 9, 1997 for a reduction of market price support, for the elimination of supply control measures and for increased exports. This is not the view held by the Federal Government in Germany.

Since the public realises that there is a dichotomy between a commercially viable sector and the less competitive farms, support will not disappear, but it is likely to be more focused on the weaker part of the agricultural population as the Eurobarometer of 1988 has already indicated. With a commercially viable fanning sector becoming more visible, the attitude of the public to agriculture is bound to change at least towards this part of Community agriculture.

In its agricultural strategy paper on further enlargement of the European Union in December 1995, the Commission have come to the conclusion that a continuation of the 1992 reform is inevitable. This conclusion is not so much motivated by the perspective of future membership of Central and Eastern European (CEE) countries but by the dynamics of CAP and of Community agriculture. The results of the Uruguay Round will lead Community agriculture into an impasse if it does not become more competitive; *i.e.*, be able to export without export subsidies. It is by no means sure that the competitive farmers will no longer use the weaker ones as justification for support which benefits all, but the chances that the CAP will rather differentiate in the future have never been better.

THE SITUATION IN THE CENTRAL EUROPEAN FREE TRADE ASSOCIATION (CEFTA) COUNTRIES

Ten countries of Central and Eastern Europe have applied for membership to the European Union: the five CEFTA-countries (Poland, Czech Republic, Slovak Republic, Hungary, Slovenia), Romania, Bulgaria and the three Baltic states. For the purpose of this paper it appears preferable to focus on the CEFTA-countries, which show a certain similarity as far as the starting point of transition and the achievements of reform are concerned. Two of the Balkan States, Romania and Bulgaria, although keen to carry forward the transition process, are in a somewhat different situation.

In the CEFTA countries the share of agriculture in GDP is not particularly high, varying between 6.4 per cent for Hungary and 3.3 per cent for the Czech Republic, with an average of 5.5 per cent. The share of employment in agriculture is relatively high with the exception of the Czech and Slovak Republics. For Poland the figure is above 25 per cent, for Hungary and Slovenia above 10 per cent. The share of food in household expenditure is relatively high in all the five countries. It is still too early for an assessment of the structure of agriculture which will emerge from privatisation. In Poland and Slovenia a significant part of small private holdings have survived communism. Privatisation of land has made considerable progress in all of these countries. The emergence of a dichotomy between small farms and large farms is already obvious in Poland, and similar developments can be seen in other CEFTA countries, too. This suggests that in all these countries, after overcoming the difficulties of transition, at least part of the agricultural sector could be commercially viable.

Of course, this does not depend on the size of the farm only. Access to advanced technology, rural credit, management capabilities and a market oriented environment will also be crucial. For example, in the down-stream sector despite privatisation oligopolistic structures are still a problem.

There are two obstacles to grant or to increase support for agriculture: the relatively high share of food in household expenditure and budgetary constraints. Despite these powerful brakes support is increasing. The latest report of the OECD shows that in particular Hungary and Poland have increased support in PSE terms and that the Czech Republic is already at that level. Comparison of agricultural price levels in these countries with the price levels in the EC show that these prices are much closer to the EC level than the general income level. Whereas prices for main commodities in 1994 in these countries were roughly between 70 per cent and 50 per cent of the EC level, the wage level was only 20 per cent of the EC, and GDP per capita on a purchasing power parity basis around one third of the EC level. Slovenia has even higher prices than the EC.

The attractiveness of the CAP is not sufficient to explain these phenomena. It is not only the political pressure by parties under the influence of fanning interests. It is, as well, the acceptance by society that farming deserves particular treatment.

What is the 'added value' of agriculture which justifies this treatment? The importance of agriculture for food security is a traditional feature of agriculture's appreciation in society. COMECON was certainly not well suited to put this image of agriculture into question. Food self-sufficiency was an official objective of government policies during that period. The positive contribution of agriculture to the trade balance is for the countries concerned certainly a more rational way to approach the matter of food security. Given the relatively high level of unemployment in most of the CEFTA-countries, there is concern that restructuring of agriculture may aggravate the employment situation. Thus, the situation in CEFTA-countries is similar to the situation in the European Community in the 1950s and 1960s but for the employment situation. Rural development appears to be seen mainly under the employment aspect whereas the positive role of agriculture for the protection of the environment is hardly perceived. The farming population is seen as a stable part of society, in particular, where the part of employment in agriculture is still high.

Conclusions

What beliefs and values does the evolution of the CAP reflect ?

Over the last forty years five main beliefs, each attaching a specific value to agricultural activity, have shaped the CAP:

- The basic conviction that agriculture is essential to provide food security;
- The belief that agriculture has a specific contribution to make to the stability of society;
- The awareness that landscape in Europe is to a large extent the result of farmers' work over centuries;
- The recognition that agriculture is crucial for the preservation of the environment;
- The belief that the farming community is a homogeneous group, mainly composed of family farmers.

From the outset the CAP was, if not conceived, certainly implemented in such a way that it stimulated production and increased the income of farmers by means of higher production and prices. Thus it reflected a widespread belief in society that agriculture had to provide food security and social stability. When the CAP started, food scarcity during the war and the post-war period was still a vivid memory, and the ordeal of dictatorship and war had sharpened the desire for social stability. Of course, since the 1950s there was never any threat to food security in Western Europe. But there was a deep seated concern about food security and the oilseeds embargo applied by the US in 1972 did nothing to alleviate it. Only the surplus production of the 1980s dealt a severe blow to the idea that the CAP was essential to guarantee food security. Increasing and massive exports had no obvious link to food security in Europe either. The feeling that food security was perhaps less pressing when the self sufficiency ratio for all the basic products was above 100 grew. It also became obvious that this way of ensuring food security was very costly, indeed. But this does not mean that the concern for food security has become obsolete. European farmers will have to feed the European Union citizens in the future, and there is concern about global food security. But, food security is no longer the shining Grail to which people pay tribute.

The early CAP reflected the concern for social stability, too. Farmers had been traditionally conservative and were considered to be a pillar of societal stability. The CAP responded by providing a framework for agricultural activity and a price level above world market prices, which ought to ensure a fair standard of living in accordance with the objectives of the CAP as laid down in Article 39 of the Treaty. Over time, market price support was supplemented by support for structural improvements and for mountainous and less-favoured areas. In the 1950s the share of persons employed in agriculture in the original six member states of the European Community was rather high, on average higher than today in most of the countries of Central and Eastern Europe wanting to join the EU. Therefore, a stable farming community was an essential factor of social stability. But the contribution of the farming community to social stability diminished over time with a decreasing number of farmers. Farmers, at least in continental Europe, are still an important factor for rural areas and rural societies, but their role in society as a whole recedes.

The CAP-reforms of 1988 and 1992 reflect a shift of beliefs away from food security and social stability as agriculture's main service to society towards the value of agriculture in preserving the landscape and the environment. The reformed CAP makes support dependent on supply control in various ways, hardly a sign for concern about food security. Support price reductions are still compensated by direct payments. But as time goes by, these direct payments may be considered as an income support which will come under scrutiny, in particular in time of sluggish growth. Farmers will have to compete with other sectors of society for scarce budget resources. In such a debate their contribution to social stability will carry less weight than in the past. Will the new beliefs that agriculture is essential to preserve the countryside and the environment protect the direct payments?

Europeans were always attached to their countryside. Despite the attractions of urban life the negative side of urbanisation, like noise, air pollution, traffic congestion, has even reinforced this attachment. There was also never any doubt that Europe was largely shaped by the work of European farmers. There is a new growing awareness that this countryside may be in danger when agriculture recedes further. This puts agriculture in the position of providing a 'public good'.

It may be questioned whether agriculture is indispensable for preserving the European landscape. Keeping the countryside could be the work of specialised companies which are not interested in agricultural production. But that is not accepted, neither by farmers nor by the public at large. To the contrary, as farming alone under modem conditions can no longer keep rural societies alive, support for non-agricultural activities is increasingly being granted under programmes for the development of rural areas. However, the role of the farmer as the steward of the countryside is likely to weaken further the case for production and price support.

It may appear somewhat surprising that European agriculture has managed to jump on the band-wagon called 'protection of the environment'. For a long time agriculture has been perceived as a threat to environment, particularly under the CAP. The high support level was considered as an incentive to intensive agriculture.

Today the CAP is promoting extensive agriculture by methods reducing or eliminating the negative impact of agricultural activity. Thus the support provided enables agriculture in the European Union to qualify as protector of the environment. An even stronger contribution to the preservation of the environment is the production of renewable resources which would replace unrenewable resources as is the case when biomass replaces petrol. The shift in beliefs, however, will have its impact on the kind of support if not on its size. The more an environmental value is attached to agriculture, the more extensive forms of agriculture will be favoured and supported. This again is likely to weaken further the case for production support.

An issue which is stirring up a lot of controversy is differentiation among farmers. The debate started because of the question of how additional direct payments to compensate for further price cuts can be financed under existing budget guidelines. But the debate has a wider perspective: it is linked to the question whether farmers of future member states from Central and Eastern Europe should benefit from direct payments which were the result of prior reductions these farmers have never seen.

The CAP does not differentiate much between farmers or regions. It is obvious that in a single market price support has to be uniform across the Union. As direct payments introduced by the 1992 reform compensate for price reductions, the amounts could be seen as income support. Income levels vary between farmers and regions. The focus on preservation of the landscape and the environment as the main justification for support will reinforce the trend towards differentiation among regions, which is already present in the application of the programme to promote agricultural activity protecting the environment. The 1988 CAP reform included an income aid scheme, which was optional for member states. But member states were reluctant to use the programme because it was complicated and blurred the line between agricultural support and welfare. Therefore, differentiation which is likely to increase, will avoid the appearance of welfare payments. An indirect way to deal with differences in income would be to use criteria related to the size of the exploitation. Such an approach may alleviate but will not solve the problem of what to pay to the farmers of future member states. Differentiation based on income levels may have to be accepted eventually.

THE ROLE OF AGRICULTURE IN THE ECONOMY AND SOCIETY

Agriculture is that kind of activity which joins labour, land or soil, live animals, plants, solar energy and so on; and the Minister of Agriculture is the Minister of the Beginning of Life. So people who are involved with that kind of activity are involved in something special.

Attitudes towards agriculture change in relation to the level of development. In less developed countries there is a tendency to treat agriculture as normal life, as a regular or common type of activity. In highly developed countries, where only a small fraction of the people are employed in agriculture, there is a tendency to treat agriculture in a special way. Poland is between these two extremes so there is not the tendency to treat farmers as people chosen by God. This attitude is changing towards that held in Western Europe.

ASSESSING THE ROLE OF AGRICULTURE

It is interesting that the importance of agriculture in the economy in the six original member states in the late 1950s appears statistically very similar to Poland today in terms of agricultural employment as a share of total employment and the contribution of agriculture as a share of the total economy. That would mean that we are about 40 years back in development. I do not think so. It is a special situation, but times are different now from what they were in 1958, or slightly later, and changes are coming much more quickly. I think that closing the gap will take less time than it would have 40 years ago. But still it will take a lot. We observe in my country a kind of acceleration of development. It goes beyond my expectations. It is much faster than I expected five or six years ago, including in agriculture. Changes in the structure, area, size of the holdings, and so on, are occurring much faster than any predictions. So I hope it will not take 40 years to catch the train, but less.

If somebody would come to me and ask for special preferential credits for agriculture, or for any special treatment of agriculture, I would resign immediately. There is no reason to treat agriculture as a special part of the economy. Mining and ore extracting industries are in a terrible situation, construction industries, also. Why not treat construction industries or extracting industries in the same way, as a special case also?

His statements are those of a free market economist. He believes each sector of the economy should be treated the same, no preferences. Market forces should determine the situation of each sector. In the beginning, it was a widely-announced and popular policy. During the first one and one-half years it was in force, from 1990 to the middle of 1991, we did not have an agricultural policy. For one and one-half years there was no agricultural policy because there were only macro-economic policies. People like Mr. Leszek Balcerowicz accept some forms of macro-economic policy, but not sectoral policies. They simply hate those kinds of policies. The preference for macroeconomic policy - or free market policy - reflects the belief that a free market has advantages over a more controlled approach to managing the economy.

Specificity is the starting point for extra treatment of agriculture. If you believe agriculture is a special part of the economy that deserves special treatment, many believe you can start resolving Poland's problems. If this is true, others argue, this special treatment for agriculture may not be available for other parts of the economy. The problem of Specificity of agriculture or any one sector becomes a problem of political/economic resolution. If you say that there is Specificity in agriculture, it means that there are some values that are not attributed to other sectors, and you have to define those values.

BELIEFS AND VALUES: THE ROLE OF AGRICULTURE IN SOCIETY

The Specificity of agriculture absorbs the attention of many people in Poland. Why is it important? The reasons or beliefs attached to its importance vary.

In Western Europe there is a focus on rural life and its symbolic values for all who felt uncomfortable with the hostile modernisation of society triggered by industrialisation and urbanisation. I think that rural life is becoming attractive for highly developed countries.

In countries less developed, rural life means backwardness. It is something to be ashamed of. For example, Marx was very hostile in his attitude towards peasants. You may remember that he wrote about peasants or the rural life "as stupidity which should be absolutely destroyed. It should be transformed into industrial or socialist type production. The peasant economy and rural life are full of backwardness, stupidity" and so on. Since Marx was very popular in socialist countries for political reasons, rural life, traditional family values, individual holdings and all these things were viewed by Marxists as something old fashioned that should be modernised. So there is a legacy of the communist ideology - at least in the minds of people. It is quite idiotic.

BELIEFS JUSTIFYING A SPECIAL ROLE FOR AGRICULTURE

In the European Community several reasons are given to justify specific appreciation of agricultural activity:

- The farmer is producing the most basic goods for human livelihood.
- The farmer is providing social stability through his hard work and the particular structure of the rural society.
- The farmer's livelihood is subject to the volatility of weather conditions.

Also, the Catholic church is deeply rooted in the rural areas. Our peasants are 99 per cent Catholic. One of the famous figures in the history of the Polish church. Cardinal S. Wyszynski wrote several papers about the role of peasants strengthening Poland. In one of several papers about the roles of peasants stressing that the territory of Poland shifted several times, he wrote that where peasants speak Polish the nation is rooted in the soil. The nation is where the peasant is on the soil. This is often presented as a special value of farming and peasants.

A fourth reason used to justify the special treatment of agriculture is regional disparities. Regional disparities link with something that is called social justice. Income differences and also differences in the level of infrastructure development mean that the people do not have the same access to the basic services, to the basic achievements of the civilization. It is unjust. To change this disparity means to bring social justice to life. So it is important and worth stressing that rural policy and agricultural policy may add a special factor to the idea of social justice. It is becoming important in constructing rural policy in Poland and in other post-communist countries because during the last six years this disparity is growing very, very quickly. Looking at the differences in local income in gminas (gmina - the smallest administrative area or unit -similar to a county in the US), according to the latest surveys, the 10 per cent (or 295) gminas having the highest incomes compared with the 10 per cent (or 295) with the lowest, have per capita income ten times as high. This is a problem. Regional disparities do not exist only in the European Community, they have become a problem in Central Europe too.

THE ROLE OF ECONOMIC EFFICIENCY - THEN AND NOW

Interestingly, the values that are basic values for Common Agricultural Policy (CAP) and the European Community, a common market, financial solidarity and community preference, are the same values used to justify socialist integration like the COMECON (Council for Mutual Economic

Assistance). But in socialist countries it was not implemented, as we know. In the case of the European Community, the progress is quite good, but the goals not fully achieved.

The idea of raising farmers' incomes is a very important part of the CAP and the general agricultural policy in the community. There is a statement in Moehler's paper that in the fast years of European Community (EC) existence, "policy was aimed at increasing the incomes of farmers that did not flow exclusively from the increase of productivity."

As we know, the productivity has grown faster than the incomes of farmers in the Community during the last 25 years. Of course, there was income support in the community, but the money that was pumped into agriculture helped farmers to increase productivity tremendously. European agriculture is a great success in terms of productivity. This is an important issue in Poland, the Czech Republic, Hungary and so on, and will be for the next few years because they cannot pump money into agriculture to increase productivity. They start from a very low point in productivity. The difference between the productivity of Western European agriculture and Central and Eastern European agriculture will be there for many years. Addressing the productivity problem is becoming a number one goal for agricultural policy in Poland and in other countries. To increase productivity in agriculture is combined usually with competitiveness in agriculture. So the emphasis is on productivity plus competitiveness.

THE RELATIONSHIP BETWEEN THE FARMER AND SOCIETY

The role of prices of food-stuffs in agricultural products becomes less relevant when the general income level increases and the share of food in household expenditures decreases. It is only at 15 per cent of household spending in Western Europe, but in practice in Central Europe it is closer to 40 per cent. So it is still a very sensitive element. The price of food is politically and economically a very important and very sensitive issue. The mass media is criticising the behaviour and demands of farmers' organizations by pointing out that agriculture and food products constitute about 40 per cent of total spending in Polish households. When you increase prices, there will be inflation - a problem for so many families. There is a different perspective on this issue among farmers and consumers.

THE FAMILY FARM

The Role of Agriculture in the Economy and Society makes the point that despite all the development in Western European agriculture, it has not affected the essence of the traditional family farm. The family farms have changed a lot, but still there are mostly family farms, and there is something stable in the farms. This is confirmed by the experience of the post-communist countries. Even during the communist period, that the essence of the farm survived even in some countries where collectivisation was predominant. There were small household plots that represented the essence of the farm. This is difficult for general historians, economists or rural sociologists to understand.

In market economies added value is so important because added value is the measure of productivity. In the family farms added value is an attachment to the traditional values. It is true for peasant farms, of course, Russian economist A.V. Chayanov presented one of the best analyses of the peasant economy in literature, when he stressed the point that in the peasant farm you cannot create activity as a business firm creates routine activity. Profit is not the main goal; there is no maximisation of profit. There are some other very important goals from day to day. I think that added value is something additional to traditional values in family farms, especially peasant farms. It is not the same in commercial farms, especially those with hired labour.

Poland, it must be recognised, during the socialist period, was an exception, in the sense that they kept a significant peasant agriculture. Poland with its agriculture structure, was really unique in the centre of the east European environment. The development in all other places was quite different. In Bulgaria it was extreme. While it was not so extreme in Hungary where the private agriculture survived in the form of household plots. But the tradition of family farming was virtually destroyed everywhere, except Poland. It is a question whether it is possible to have a revival of these family farms or not. That is one of the crucial issues, what is happening with the farming structure, apart from Poland. Some contend Poland was a very interesting case because Poland kept the fragmented family farm, and did not let this agriculture develop.

Hungarian agriculture, for example, became much more efficient with the large-scale cooperative structure than the Polish one because, at least in the Hungarian structure, there was a scope for managerial independence. There was a quasi-market working. They could invest. They were free. While in Poland that structure was frozen, more or less. That is why Poland inherited an undeveloped agricultural structure.

What is happening right now? Many experts believe family fanning in the Western European way, will be the major way of fanning in Central and Eastern Europe. The family farm is coming back in a few countries, surprisingly, in the Baltics, where there are few people and a lot of land. However, not in the Czech Republic, nor in Hungary, nor in Slovakia do you see the family farm. Albania is China. What happened in Albania is the Chinese way and cannot be compared with the rest. This is very important to Western Europe's underlying values.

THE POLARISATION OF AGRARIAN STRUCTURE

Since the fall of communism some big farms are emerging in East Germany, the former Democratic German Republic (DGR). There is also an area in Western Poland that was within German boundaries before the war. Now, after the collapse of state farms, there is a growing number of really big farms. I remember from the report of the agency responsible for state sector restructuring and transformation that during the first five years of its operation more than 3,000 farms have emerged with an average size of over 500 hectares. Some of them have 1,000 or 1,500 hectares. These are very big farms. Why is this important? Because it causes a change in the attitude of the people towards agriculture. They ask why should we support agriculture and agricultural producers, especially those who have 1,000 hectares and are becoming like owners of *latifundia*. This is a contradiction of their concept of fanning as an area occupied by family farms, small and medium sized farms.

The picture of agriculture is changing because of the emergence of such big farms. It causes political discussion even in parliamentary circles. It is a proposal, and I hope it will be overcome, to accept legal regulation about the upper size limits for the family farms. First, there is a need to define what a family farm is.

How should government policy treat such farms? Should such farms have preferential treatment? Because the peasant party is a very strong party, the second largest in the parliament, and also the ruling party, together with the post-communist party, they can approve this proposal in the parliament. They have a majority, but there is some controversy within the party. For example, one of the leading figures in the peasant party, the Minister of Agriculture, is an owner of a very big, bigger than average, farm in Poland. His sons also have quite big farms. So there is a conflict of interests - also among the communists. There are some former directors of the state farms who now operate private farms with 2,000 - 3,000 hectares.

They are very good managers and can operate such farms. From my point of view as a scholar, I can say that any form of codification or stating in a legal document that such and such entity is a real family farm - or is not a family farm - is conceptually very difficult. This is one result from the emergence of a new kind of farm in Central Europe, the same as those in East Germany. It is a part of the problem that is called polarisation of the agrarian structure. All countries in Central and Eastern Europe have experienced this polarisation. On one side we had a large number of small farms with two hectares, for example, and the number of the farms is relatively stable. We also have a growing number of very big farms. But the number of farms in between those with six, seven or eight hectares is declining. They are all small or big. I remember in one of the American publications the saying, "Get big, or get out." It was very popular. It is becoming the same in Poland. You should get bigger, especially to face the competition in European agriculture. It contributes to the problem of social justice, social and economic policy, etc. Maybe we have the beginning of the strengthening of a bimodal agricultural system in Central Europe.

NEW CONCERNS AND SHIFTING VALUES

In the European Community the farming community has a hostile reaction to programmes that reduce arable areas. There is also a hostile reaction in Central Europe, but we have idle land, not because of government programmes, but because of the collapse of the state or co-operative farms.

It is a shame for the economic policy in Poland that 39 per cent of the land that was in the state farms is not cultivated at the moment because of problems with land transfer. According to the last record that I read, there were 4.2 million hectares, or 20 per cent of the total arable land, in state farms. One-third of those 4.2 million hectares, or nine per cent of the total land, is still not cultivated because the state farms collapsed, and the new operators have not entered yet because of many problems, legal, financial and so on.

The next point in Moehler's paper deals with the new values for agricultural policies. Such values as food safety, quality of food and environment are well known. Some people add ethics to that - ethics associated with agricultural production.

During the VIII European Congress of Agricultural Economics in Edinburgh, L. Mahe presented an interesting paper on the new protectionism driven by problems of ethics, environment and so on. New values are behind the agricultural, food and protectionism policies. Environment is becoming a very important aspect. There is progress in Central Europe in that one of the achievements of the last five or six years is the reduction in pollution. Sometime the reduction is 40 per cent or 50 per cent because of several reasons, but mostly because of the closing of the most polluting factories. Public awareness of the environmental problem is growing.

FOOD SAFETY AND QUALITY

The food safety is a new problem and it is becoming popularised. For example, some of our producers started to use food safety as an argument against the importation of goods. Farmers argue that imported goods have some chemical ingredients, and so on; therefore, "Buy Polish products - they are safer because they are produced in a more natural way."

ENVIRONMENT AND PRODUCTIVITY

When you compare the old major values and goals, for CAP and for programmes in Central Europe, I think that the list will be the same, but the emphasis will differ. In first place would be productivity and competitiveness. For example, food security is regarded as one of the important aspects, but because of the liberalisation of trade, it is not so important as it was some years ago. Preserving the natural landscape and environment is included, but is not considered as important as productivity. I think we have a unique chance to avoid some of the mistakes in development that were present in Western

Europe and to combine productivity with preserving the landscape and the environment. It is one of the lessons we should learn from the Western countries.

PROMOTING RURAL DEVELOPMENT

There is a shift from agricultural policy to rural policy in my country. In 1994, some government documents were presented about rural policy and the need to shift from agricultural policy to rural policy. This shift may portend new functions for agriculture. Rural development policy is discussed as multi-functional in rural areas, but agriculture is also multi-functional - essentially it is combining the functions for rural areas and agriculture. In the document presented by the Council of Europe, called the European Charter for Rural Areas, significantly distressed rural areas will be treated in completely new ways. We should appreciate rural areas and present agriculture as not a producer of agricultural good, but as a source of multi-functional activity. It is changing both in Western Europe and in Central Europe.

ECONOMIC SHOCK THERAPY FOR AGRICULTURE

The Organization for Economic Co-operation and Development (OECD) reports increased support in terms of Producer Subsidy Equivalents (PSEs) in Poland, Hungary and the Czech Republic. There is an important story about the PSEs. In the last three or four years it was presented in some reports that at the end of the socialist period, PSE in Poland was about 28 per cent. It was lower than in the European Community, but it was high. In 1990 and 1991, it dramatically dropped to minus 18 per cent. And in the next two years, increased to between 13 and 15 per cent. Relatively speaking, it is not high. It is higher in Hungary and probably higher in the Czech Republic.

The issue is not how high PSE is; it is the sharp change from 28 to minus 18, which caused farmers to demand increased protection, to do something. It was a part of the shock therapy - and shock therapy is not a good therapy. It produced negative consequences because the attitude of farmers towards liberalisation, a market economy, had shifted during the first years from one position to its opposite.

At the end of the socialist period the peasants or the leaders of the peasants supported the idea of a market economy. They thought that they could survive and develop with the new market environment. For the peasant the difficulties of the first years were a shock for them - especially because of the shock therapy, shock liberalisation, including the opening of the borders for any kind of activity, almost no tariffs, no border protection and also no support from the government. Government support declined, in real terms, five times in two years. So that is the problem of the transitional economy. What is the added value agriculture provides that justifies this special treatment? For me, the problem is how to measure that added value. It is so difficult to measure some reserves of the agricultural policy. There are some values agriculture provides - serving as a shock absorber. It plays such a role in my country. There were several cases in which agriculture played such a stabilising role, as well as the role of shock absorber. The role of agriculture is probably less important for highly developed countries, but for countries where 20 to 25 per cent of the people live from agriculture, it could serve as a shock absorber.

It is difficult to measure other aspects of the added value regarding the protection of landscape or environment. In Edinburgh a paper was presented on how to arrange contracts with farmers. When you have contracts, you must monitor the contracts. What is the contract? The farmer will protect the natural landscape, and we will pay him some money. Agricultural policy is becoming more complicated, in my opinion, because of the new values and new goals for agricultural policy.

ECONOMIC STRUCTURE AND PERFORMANCE IN THE FOOD AND AGRICULTURE SECTOR

Agriculture in the Central and Eastern European (CEE) countries is undergoing a fundamental transformation. The far reaching changes surpassing the reforms of earlier years characterise the agricultural economies of Central and Eastern Europe. Under present economic and political conditions the CEE countries have no alternative to the establishment of market-driven agricultural sectors based on private ownership. The process of reform has yet to be completed in any of the countries concerned, and it is now clear that the transformation is essentially more complicated and complex process than predicted by domestic and foreign experts and politicians. This paper intends to cover the process of change regarding the overall economic structure of agriculture and the change in underlying values and beliefs in order to get a better understanding of current developments as well as the forecasted course of developments in agricultural policies.

THE HERITAGE OF THE CENTRALLY PLANNED SYSTEM

In the 1950s and 1960s, in all CEE countries, the so-called socialist reorganization of agriculture was carried out. In practical terms this meant a collectivisation of the predominant small-holders agricultural system according to Soviet models. By the middle of the 1960s, the state farms and agricultural cooperatives became predominant in all the countries, except Poland. The organization of the socialist 'large-scale farms,' as they were called, was accompanied by the introduction of central planning, and by the establishment of the agricultural production system directed according to centrally determined and planned figures. The agricultural policies in this period in the respective countries were based on the premise that *agriculture forms an integral part of the centrally planned national economy*. The basic targets for agriculture were formulated in the national economic plans. The most important objective of agricultural policy was to produce the quantity of food needed for the planned level of personal consumption and to satisfy the industrial demand for agriculture products.

These general targets of course depended on the specific conditions and on the actual economic situation in each country. In spite of the similarities in the basic structure of the agricultural sector, no uniform agricultural policy spread throughout the countries regarding the implementation of the overall planned targets. In the classic system of centrally planned agriculture the overall national targets for agriculture were implemented by the integrated system of smaller-scale plans designed for specific sub-sectors of agriculture, for regions, and for individual farms. Obviously, in this system the freedom of decision making at the lower levels was rather limited. In some countries, for certain periods, even the production technology was assigned for individual farms. During the more recent years, in some countries, the goals of the overall targets were implemented by using indirect methods, and various instruments of market economy such as prices, profit motivation, etc. It has to be underlined that even in these cases (*e.g.*, Hungary and Poland in the late 1970s and 1980s) *the plan always had a priority over the market*, and the market instruments were used to implement the planned targets rather than as major instruments of controlling the agriculture.

The agricultural investment policy was also developed according to central plans and was determined by central planners. The scale of investment in agriculture or its share of the total at any time reflected the overall strategy of planners. In general, the development of industry was central to the economic policy, and part of the income generated by agriculture was often redistributed for the development of industry. Sometimes, however, when the planning authorities gave priority to the increase in personal food consumption, the agricultural sector would receive considerable investment resources, often much higher than its eventual contribution to the national income.

An important universal general characteristic of the agricultural policy was the *vigorous effort for self-sufficiency*. In each country, planners put an emphasis on meeting domestic demands for all commodities that could be produced in the given country as far as possible from domestic production. In those countries where natural conditions are favourable to agriculture, the maximisation of foreign currency receipts from agricultural foreign trade was also a rather important target for planners. Under this philosophy, quantity had a priority over quality, producing more was the crucial objective put before efficiency and competitiveness.

As a result of the priorities in plans, the domestic prices, and the income situation did not reflect actual economic performance. Domestic prices were mainly tools to aggregate various types of products rather than to reflect values and actual supply-demand relations. The whole system was based upon the *philosophy of low food price and low wages*. Food prices were kept artificially low and unchanged in order to be able to keep wages low and unchanged. As a result, the financial performance of a single farm had no real relation to the efficiency of production at that given unit. Production expenses were often higher than the procurement prices imposed by the state distribution system. Deficits or losses were always covered and farms were never liquidated.

The real markets for agricultural products and inputs did not exist, the only exception were the products of the household farms and those of small gardens. Both agricultural outputs and inputs were distributed rather than marketed. The domestic prices, therefore, had no market clearing role and had no direct relationship to the border price of the products. *Foreign trade obviously was also part of the central planning system* with the state foreign trade agencies holding a full monopoly. This system was in reality equivalent to a high level of protectionism, although this was implemented by the planners' decisions and not by the market economies' methods. Foreign trade decisions reflected planned targets and not domestic market situations.

The agricultural policy considered the so-called socialist large-scale enterprises, cooperative and state farms, to be the fundamental pillars of agriculture. With the exception of Poland, these farms were dominant in both land use and output. Private production was considered to be a temporary phenomenon which was allowed but not supported. Only the household plots belonging to the co-operative farm members were regarded as an integral part of the large-scale farming system. In spite of this hostile environment, private agriculture provided a significant portion of agricultural output due to the better utilisation of resources and benefits gained informally through their links with the large-scale farms.

After a relatively short time it became evident that the administrative central direction regulating the agricultural production limited the growth of the sector and the signs of the unsatisfactory performance of the system appeared relatively early in the 1960s. The first reform attempts date to this period.

The history of socialist agriculture is a history of reforms that proved to be unsuccessful or produced only partial results. In the first phase of the reform, in the 1960s, attempts were made to improve the planning methods and to develop the means to implement the planned targets. In the successive reforms, the decentralisation, the increase in farm independence, the need for the application of economic incentives became central. All of these reforms, however, could not step beyond the limits of the socialist economy as a whole, therefore they did not result in a visible improvement in agricultural performance. Even Hungarian agriculture, which was the most successful among all the centrally planned economies due to early implementation of indirect and market type control methods, showed signs of serious efficiency problems by the mid 1980s.

Table. Socialised farm enterprises and land held in the private sector before reforms (1989)

Country	State farms		Collective farms		Land in private sector (% of total land)**
	Number	Average area (ha)	Number	Average area (ha)	
Albania	70	2,400	420	1,270	-
Bulgaria*	536	9,692	NA	NA	10
Czechoslovakia	226	6,204	1,677	2,605	9
East Germany	465	945	3,904	1,370	11
Hungary	128	7,598	1,270	4,195	15
Poland	1,258	2,665	2,342	297	80
Romania	419	4,895	4,363	2,093	15
USSR	22,690	16,051	26,660	6,370	2

Note:

*Total number of agricultural units in 150 agro-industrial complexes.

**Land used in individual production, not necessarily privately owned land.

The aftermath of decades of socialism is largely similar throughout the region. The food and agriculture of CEE countries on the eve of the transition was characterised by:

- Large inefficient farms and food processing enterprises with excessive scale of production and eroded incentives for efficient and profitable production.
- High levels of food consumption relative to market economies of comparable prosperity.
- Subsidised food prices and excess demand for food and subsidised prices.
- Artificially controlled trade and pricing policy which distorted the sector priorities and incentives.
- Excessive transfer of financial resources which neither addressed the underlying structural problems nor resolved efficiency issues.
- A widespread monopoly in food processing and distribution.
- The absence of institutions required by private, market-based agriculture.

The decades of planned economy obviously had an impact on values and beliefs related to agricultural policy and to agriculture in general. While the almost half century of centrally planned agriculture was not long enough for a fundamental change in attitudes and aspirations, it was long enough to result in changes which need to be taken into account when the current development and future perspective in agriculture policies are being assessed. The major impacts are:

- Quality lost importance to quantity as a measure of success.
- Performance was given a new interpretation: subjective evaluation of planners became dominant.
- Limited understanding of market functioning.
- Declined entrepreneurship.
- Decreased confidence in Government and institutions, including the legal frame-work in general.
- Bias against private business and foreign ownership.
- Risk avoidance and a high degree of suspicion in both political and economic environments.
- Lack of trust in collective efforts and search for individual survival techniques.

PROGRESS IN CREATING A NEW ECONOMIC STRUCTURE IN AGRICULTURE

The political changes that affected the entire region in the late 1980s and early 1990s created new conditions for reforms in agriculture. The former goal of reforming the socialist agriculture system was replaced throughout the region by the *new goal of changing the regime and creating a market type agrarian economy based on private ownership*. The current ongoing agrarian reforms in CEE countries are aimed at realising the latter objective. The creation of a market economy and privatisation in agriculture involves the following actions:

- The creation of a new macro-economic framework for agriculture: price and market liberalisation.
- Land reform and creation of a new farming structure based on private ownership of land and productive assets.
- The creation of a competitive environment for agriculture: privatisation in agro-processing, input supply, services, and trade.
- The creation of a rural financial system, serving the needs of privatised agriculture and agricultural services.
- Institutional reform creating institutions and public services required by a market economy.

In theory, Central and Eastern Europe has the possibility of shaping a legislative and regulatory system which, combined with adequate capital input, can bring about the creation of the farming structure most suitable for the market circumstances. At present, the region has only partly taken advantage of this possibility. What is the reason for this? One of the most important features of the past few years has been the *predominance of politics* over economic rationality. In all the CEE countries, agricultural transformation is taking place fundamentally as a process guided by political, often short-term, motivations in which the economic considerations and the effort to create an efficient new structure, as rapidly as possible, play a secondary role. The beliefs inherited from the communist period, distorted and often illogical views of the market economy have played a significant role in Central and Eastern Europe. As a result of the *ideologically-charged* prejudices against the inherited structure and the overemphasis on seeking justice and reprivatisation, the output has fallen by a greater extent than necessary, the loss of fixed assets has been greater than justifiable, land ownership has become fragmented, and people not engaged in fanning have acquired a substantial share of land ownership. On the other hand, the political conservatism and the mistrust of private production and market relations often demand the continuation of subsidies in general and the preservation of the remaining larger-scale farms.

Beyond the common overall objectives the picture is far from being uniform in the region when it comes to the concrete forms and the implementation of the various tasks of transition.

There are substantial differences among countries in practically all the areas of reforms listed in the previous paragraph. However after five years of transition, it can be stated that the CEE countries have made significant progress in transforming their food and agricultural sector. It should be noted, however, that some countries, in particular the Czech Republic, Hungary and Poland, have made faster progress in their global transition, than in their agricultural sector reforms.

Table. Overview of the status of agricultural reforms in CEE countries

Country/area of reform	Price & market liberalisation	Land reform	Privatisation in agro-processing	Rural financing	Institutional framework	Overall score
Bulgaria	3	3	2	3	3	2.8
Czech Rep.	4	4	4	3	4	3.8
Estonia	5	3	4	3	4	3.8
Hungary	4	4	5	3	3	3.8
Latvia	3	4	3	3	3	3.2
Lithuania	3	3	3	3	3	3.0
Poland	4	4	3	3	4	3.6
Romania	3	3	2	3	2	2.6
Slovak Rep.	4	3	3	3	3	3.2
Slovenia	3	5	4	3	4	3.8
Average	3.4	3.5	3.1	2.8	3.1	3.36

Land reform

1. Sovkhoz/Kolkhoz type system dominates.
2. Legal framework for land privatisation and farm restructuring, yet only recent implementation.
3. Advanced stage of land privatisation, but large-scale farm restructuring is not fully complete.
4. Most land privatised, but titling is not finished and land market is not fully functioning.
5. Fanning structure based on private ownership and active land markets.

Privatisation of agro-processing and input supply

1. Monopolistic state owned industries
2. Spontaneous privatisation and mass privatisation in design or early implementation stage.
3. Implementation of privatisation programmes in progress.
4. Majority of industries privatised.
5. Privatised agro-industries and input supply.

Rural financial systems

1. Soviet type system. Agrobank sole outlet.
2. New banking regulations are introduced.
3. Restructuring of existing banking system, emergence of commercial banks.
4. Emergence of financial institutions serving agriculture.
5. Efficient financial system for agriculture and agro-industries.

Institutional framework

1. Institutions of command economy.
2. Modest restructuring of government and public institutions.
3. Partly restructured governmental and local institutions.
4. Government structure has been refocused while research, extensions, and education is being reorganised.
5. Efficient public institutions focused on the needs of private based market agriculture.

Looking at the analysis, we can conclude that:

- The reform process in agriculture is more difficult than projected;
- The results do not exactly meet original expectations;
- The Central European countries, namely Slovenia, the Czech Republic, and Hungary, have made the most significant progress, while the transitions in Lithuania, Bulgaria and Romania are still far from completion;
- The most significant progress can be seen in price and market liberalisation while rural financing, institutional reform, and privatisation of agro-processing seem to be lagging behind in the process.

It seems that the most critical issues which will determine the further progress in agricultural reforms are: a) the success in restructuring the agro-processing industries and trade; b) the speed in establishing an efficient support service environment, particularly a well functioning rural financial system to support privatised agriculture, farmers and processors; and c) the progress in creating public institutions consistent with the market economy, and particularly with EU standards. Efforts to strengthen liberal trade and market policies, and to consolidate private agriculture, need to be continued. A competitive and efficient agriculture cannot be developed in CEE countries without recovery of the global rural economy. The productivity increase in agriculture requires a massive reduction in the agricultural labour force. The experience of the Czech Republic and Hungary indicates that this can be achieved only if the non-farm employment opportunities are in place, and if the rural services are improved significantly. Unfortunately, in most countries, the non-agricultural rural business activities are developing slowly, making the agricultural transition process more difficult. More attention, therefore, needs to be given to complex rural development, including the improvement of rural infrastructure.

The transformation of agriculture in Central and Eastern Europe is still far from being finished. The results achieved, but also the difficulty of transition. The transformation was further hampered by the *decline in output* which accompanied the first phase of the transition process and can be attributed only partly to the economic crisis and decline in demand. The rebound effect of the transformation is also considerable. The shift from large-scale farming to privatisation is generally a *process of dismantling*, and not a process of constructing. This is especially true for the first stage of the process. The loss of a part of the stock of fixed assets and other disorders are an unavoidable side effect of such a process. Unfortunately, the behaviour of the governments concerned has tended to strengthen rather than counterbalance these unavoidable negative effects.

There is a lack of general management of the transformation and a lack of comprehensive, consistent and carefully contemplated *agrarian strategy*. Instead, the process of transformation is being handled in an *ad hoc* manner with frequent changes and amendments, some having retroactive effects and some made during the process of its implementation. The *legislation* that underlies the agrarian transformation is not based on a clearly pondered and formulated programme, realised through a broad social consensus. The implementation of some of the new laws is beyond the administrative capacity of a country's state administration.

A further serious problem affecting the transformation is the shortage of capital and *the conditions of the immediate economic environment* surrounding agricultural production. One of the most restrictive factors for restructuring agriculture is the insufficiently developed and the centralised and monopolistic nature of the trade and processing structures, accompanied by the primitive financial system. It would appear that the debate concerning land ownership and the transformation of large-scale farms has obscured the recognition of the real impact of transformation in these fields. In the absence of a functioning market, private agriculture is defenceless. But without a stimulating macroeconomic environment and availability of investment sources, the development of agriculture is inconceivable. For this reason, the strengthening of the market framework with the creation of fair market competition, a functioning rural banking system, an active land market, and competitive trade and processing industry are the conditions of vital importance for further advancement in the agricultural sector.

Table. Evolution of agricultural production (1994 indices; 1989 = 100)

Country	Gross agro product (value)	Cereals	Oilseeds	Sugar	Milk	Beef	Pork	Chicken
Poland	78.6	81	48	77	73	71	87	93
Hungary	65.6	75	83	84	70	70	59	78
Czech Rep.	72.2	93	151	66	64	68	84	83
Slovak Rep.	74.6	87	105	65	40	50	63	73
Slovenia	118.2	107	100	110	94	70	77	63
Romania	101.0	99	78	45	90	121	92	79
Bulgaria	70.2	73	126	18	53	79	52	39
Lithuania	47.7	80	42	66	51	54	33	44
Latvia	-	57	33	48	47	53	35	26
Estonia	56.4	43	200	-	60	37	30	28
All the CEE	-	84	80	68	67	70	73	74

IMPACT OF THE TRANSITION UPON INDIVIDUAL COMPONENTS OF ECONOMIC STRUCTURE IN AGRICULTURE

Role of government

The transition to a market agriculture based on private ownership requires a fundamental change in the role of government in the agricultural sector and in the economy as a whole. Direct government intervention in agriculture, such as establishing mandatory targets for production and/or delivery of goods and central allocation of investments and inputs, must cease. The government's role should extend beyond the management of the difficult transition process: the government should establish general rules and facilitate the conditions for smooth operation of markets and independent business entities. This role is no less important than the previous one; however, it requires a different philosophy, as well as different means and institutions.

Agriculture in the Central East European countries has made the first fundamental steps towards a market driven system based on private ownership. Under these circumstances, the most important task for the governments and for the agriculture policy specifically, is to liquidate the obstacles preventing the emergence of a more competitive and efficient production system. The governments need to create the *enabling environment for private sector growth*, rather than remain directly involved in production management, according to the practice of the former regime. Unfortunately, for most of the governments, the transition to this new role is not a smooth and continuous process. The emergence of the new role of the government is constrained by the inherited beliefs that the government should take care of everything. The mix of social and agricultural policies is an additional typical problem. The governments try to resolve transition related social problems in rural areas by agricultural policies, rather than by appropriate social measures. The influence of various interest groups, such as farm managers, newly-emerged private salesmen, and trade unions also slows down the emergence of the new role for the government.

As the consequence of the above-mentioned problems, during recent years, the agrarian sector in the region has been controlled by rules and regulations which are loosely connected and sometimes inconsistent. There are a great number of uncertainties affecting the farms: significant measures are often adopted with delay, and important agricultural policies are the outcome of bargaining over concrete individual situations. *There is a need for a clear agricultural policy framework*, for a comprehensive set of agricultural policy decisions oriented, not only towards managing the transition, but which will also create the foreseeable and reliable long-term conditions for the restructured and privatised agricultural sector. The major goals of the agricultural policy and the related system of instruments, with special reference to the budgetary consequences, have to be formulated for at least three to five years, similar to the practice of developed market economies, *e.g.*, the US Farm Bill.

In most countries the agricultural policy is obviously closely connected to the aspirations of various political and interest groups. Presently, the interests of producers and the trading sector are very clearly expressed. In most countries, however, representation of the consumers' interests is still very weak. The budgetary implications of agricultural policies are not well known. There is a strong need for a clear, transparent agrarian budget, which clearly illustrates the costs of various policy measures and shows who pays the bills. Subsidies are probably the most debated elements of the agricultural policies in the region. The general belief is that agriculture cannot function without subsidies and that the transition and the forthcoming EU accession, make the subsidisation of agriculture indispensable. There is less than a full understanding of benefits and shortcomings of the protective policies.

The dismantling of the bureaucratic structure inherited from the period of central planning is a very important task. Various ministries should not merely change their names, but need to undergo radical modification and/or merger. Units related to central command and direct intervention should be eliminated, while those remaining should be organised and managed to meet the needs of a free market system. As the transition proceeds, the whole structure can be simplified further, with fewer institutional units and fewer employees. Structural changes are needed in the regional units. In many countries, the large regional bureaucracies implementing central control, are still very strong. There is no need for then-existence in their present forms. A relatively small local administration would be adequate to enforce agricultural regulations, promote development, and provide extension and market information services.

Given the uniqueness of the region, its history and politics, it is not possible to define an exact structure of a modernised government administration of food and agriculture for the countries concerned. However, within a market-driven economic system the government should continue to play three roles: *regulation, provision of public goods and support services, and information and analysis*. Regulation should include food inspection, seed inspection, establishment and enforcement of grades and standards, establishment and control of phytosanitary standards, epidemiology and livestock disease control. The essential public services to be provided by the government should include domestic and foreign market information, agricultural research, farm advisory services, and higher education in agriculture. Finally, the government should monitor, review, and diagnose the implication of changing circumstances and develop options for public policy.

LAND REFORM AND FARM RESTRUCTURING

Inherited beliefs and values probably have been playing a more significant role in the course of changing the land ownership and the farm restructuring than in any other area of transition. On the one hand, among many people there is a deeply rooted belief in the superiority of private ownership and family farming over state and collective structures. While, on the other hand, the socially positive experience of collective fanning had an impact on the view of many other people. As a result, the land reform as it has been carried out in the region, reflects these conflicting views and beliefs.

Creating private land ownership

The creation of private land ownership is the starting point for land reform. It is around this issue that an intense political debate is raging in virtually all the countries in the region. Two closely linked tasks are involved here. One is the creation of a legal framework for private possession of land, and the other is me transfer of land into private ownership, the actual privatisation of land. In other words, it had to be decided who would receive land, by what procedures, and on the basis of what criteria. In Central East Europe, the legal recognition of private land ownership did not meet with strong opposition. Land in these countries has never been fully nationalised, and a tradition of private land ownership continued after World War II. The return to traditional values enjoys the full support of public opinion.

Eligibility, meaning the determination of who is entitled to receive land, is one of the major issues related to land ownership: one approach is to return land to the former owners (so-called restitution or reprivatisation of land), and the other is to distribute land to new beneficiaries regardless of the former ownership rights. The two approaches are not necessarily mutually exclusive: land remaining after the claims of the former owners have been satisfied can be distributed to other beneficiaries. Moreover, restitution is not always implemented by actually returning the original plot of land to the former owner: due to technical considerations a substitute plot may be allocated or, instead of land, money compensation may be offered.

When the policy of land distribution, regardless of the former ownership rights, or the distribution of unclaimed land, is adopted, the legislators must decide the exact configuration of the beneficiaries. A number of different definitions of beneficiary groups have been considered in the CEE countries. The possible combinations of beneficiaries are listed below in the order of progressively more encompassing eligibility for land and non-land assets:

- Only those who work the land ("land to the tiller"), usually including tiller-pensioners;
- All members of collective farms, including those not engaged in farming;
- All members and hired workers of collectives;
- All members and hired workers of the collective, as well as people employed in rural agricultural service enterprises and social services;
- The whole rural population;
- The whole rural population, as well as former rural residents satisfying certain criteria;
- Sell the land on an open market.

In Central and Eastern Europe, *reprivatisation* or *restitution* dominates, supplemented by various schemes designed to provide land to agricultural producers who were not former owners. The Central and Eastern European countries in general recognised the rights of the former land owners. Compensation was provided in various concrete forms, varying from actual land restitution to financial compensation to allowing purchase of alternative land if desired. Albania is an example of full and rapid privatisation of the land. There, the vast majority of land previously part of the co-operatives and state farms was divided among the rural population, according to the size of the family, regardless of the former ownership.

In Romania, the land of the co-operatives was transferred to former owners and co-operative members relatively quickly, in the space of approximately six months. The process is advancing more slowly in Bulgaria and the declared goals, which include the creation of private farms of a rational size, understandably complicate the process. In both, the Czech and the Slovak Republics, the former owners can recover their land if they decide to cultivate it. In Hungary, the compensation of former owners has been based on a voucher system. As a result of this process, the great majority of agricultural land in Central and Eastern Europe will be privately owned in the near future, although the legal settlement of ownership rights will still take some time. The Baltic states, like the CEE countries, adopted a strategy of restitution to the former owners. Individual land users who are not former owners have to purchase land from the state reserve or from the new owners who do not wish to farm. Otherwise, they keep their rights to use the land, but they do not obtain the ownership rights.

Farm restructuring

In the process of developing private ownership and property markets in the former socialist agricultures, it is strategically important to separate the privatisation of land and assets from the restructuring of agricultural production. Though the possibility exists for a fast and simultaneous privatisation of both, land and non-land assets (as the dismantling of the large-scale farms in Albania demonstrates), this method would be acceptable only under very specific political and economic circumstances. Under the typical circumstances, land and non-land assets are allocated to private owners without a complete fragmentation or dismantling of the existing farm structure. With land and assets transferred to individual ownership, the process of restructuring agricultural production is guided by the free will of the individual owners, who choose to assign their land and assets to the organizational form which best satisfies their needs and preferences. This has to be an open-ended dynamic process in which the new owners learn through the choices made regarding the management of their land, and adjust these choices over time in response to the experience gained.

The role of the government is to provide the framework for a smooth and flexible restructuring process. This process should be based on the individual decisions of the beneficiaries, who select freely among the various forms of farm management and land use. The government should, therefore, provide a legislative and policy framework which clearly recognises these various options and the freedom of choice of the beneficiaries, and reinforce it with an appropriate public information system.

Expeditious implementation of transfer and registration of land ownership depends entirely on allocation of sufficient resources for the registration and demarcation system. The physical registration process is hindered by the lack of suitable modem equipment and qualified personnel, but both, the equipment and training are a function of money and expertise, which can be provided by international donors. However, the frequent political opposition of special interest groups, that object to both private ownership of land and to the allocation process, represent a serious obstacle. The political opposition can be overcome only by educating the public as to their rights, options, and the exact nature of the process related to the change in the land ownership. This is a long-term process that requires political willingness and resolution to challenge the special interest groups. Farm restructuring involves the creation of a market-oriented, profit-motivated farming structure, based on clear individual ownership of land and assets, and an incentive system encouraging individual responsibility and rewarding individual effort.

The new farm structures may take a variety of forms. Some people will keep their land and shares of other assets and establish private (family) farms. Others will pool their land and assets and create small partnerships or farming cooperatives. Yet others may choose to lease their land to active agricultural producers and assume the role of 'inactive investors,' or, alternatively, focus on development of private farm support services. The former collectives will, thus, gradually break up into individual or smaller private farms or other fanning organizations in which the production will be based on privately owned land and assets. These new producers need to be supported by market services, some of which are provided by new private entrepreneurs (individually or in groups) while others may be based on former collective management structures that will redefine their role and become service firms or co-operatives.

The results of restructuring in the CEE countries indicate that large-scale farming still plays an important role at this stage of the reform. Yet the restructuring goes far beyond a mere 'changing of the sign on the door.' Most large-scale farms actually reorganised into several smaller functionally specialised units, built around the land and asset shares of their member-owners. The shareholders underwent fairly radical voluntary regrouping in the process of downsizing of the original farms. A significant degree of separation between ownership and management has been achieved in these new structures, which no longer guarantee employment to their shareholders. The new large farms in the CEE countries are thus definitely moving away from the traditional syndrome of the 'labour-managed firm' that in the past plagued the socialist economies.

The current features of the economic environment in most countries regarding restructuring are transitory. As the economic environment supportive of private activity outside the traditional farms improves, more holders of the land and asset shares will seek to withdraw shares and form new ventures, either individually or in combination with other shareholders. In view of the fact that the economic environment is still changing, and it is still severely behind in the development of market infrastructure in most of the Central and Eastern European countries, it is desirable that the options for restructuring remain open. Thus the recipients of the land rights who choose to remain in the collective-type enterprises should retain the right to exit with their land and assets at any point in the future. Whether the exit rights are protected, or property rights devolve to the enterprise when a shareholding company is formed, the withdrawal issues depend on the existing relevant laws, and on the enterprises' by-laws' treatment of such issues.

Table. Farm structure in East Central Europe and the Baltics

I. Share in total agricultural land(%)						
Country	Collective farms*		State farms**		Private farms***	
	Pre-1990	Current	Pre-1990	Current	Pre-1990	Current
Albania	75+	0++	23+	12++	3+	88++
Bulgaria	-	41	90	40	10	19
Czech Rep.	61	48	38	3	1	49
Slovak Rep.	68	63	26	16	6	13
Hungary	80	55	14	7	6	38
Poland	4	4	19	18	77	78
Romania	61	35	14	14	25	51
Estonia	45+	33	50+	NA	5+	67#
Latvia	57+	17	38+	2	5+	81#
Lithuania	62+	33	30+	NA	8+	67#
II. Average size (hectares)						
Albania	300+	0++	2.000+	400++	NA	1.4++
Bulgaria	-	750	13,000	1,100	0.4	0.6
Czech Rep.	2,561	1,430	6,261	498	4.0	16.0
Slovak Rep.	2,654	1,665	5,162	2,455	0.3	1.0
Hungary	4,179	1,702	7,138	1,976	0.3	1.9
Poland	335	400	3,140	2,000	6.6	6.7
Romania	2,374	170	5,001	2,002	1.5	1.8
Estonia	3,689+	567	3,816+	NA	0.5	2.1#
Latvia	3,900+	706	4,200+	547	0.5	5.8#
Lithuania	3,000+	567	3.300+	NA	0.5	3.1#

Note:

* Collective farms and co-operatives pre-1990, private producer co-operatives and associations currently.

** State managed or controlled farms pre-1990, remaining state farms and state held farm enterprises currently.

*** Household plots and small individual farms pre-1990, individual (including part-time) farms and other business entities (limited-liability partnerships, joint-stock companies, etc.) currently.

+ Based on pre-1990 statistical abstracts of the respective countries.

++ World Bank data.

Includes subsidiary household plots.

The rate of the increase in the number of individual private farms across the region is far behind the original expectations (both local and international) and behind the possibilities provided by the legal framework. Different explanations, based on theoretical considerations and speculations, are given by the politicians and experts. The World Bank surveys conducted among the rural population indicate that the reluctance to becoming a private farmer is due to the lack of supportive conditions for private fanning. The cautious behaviour is thus an expression of the sober and realistic judgement of the situation of the peasantry. The present generation of CEE peasants has only limited or no skills necessary for successful private family farm management. However, in today's CEE, the most restrictive factor inhibiting the expansion of private farming remains the relatively modest income compared to the high risks involved. Progress down the path towards a market economy will gradually establish the conditions that encourage further transformation of the fanning structure.

LAND MARKET

The allocation and distribution of land to individuals in the process of land reform is only the fast step towards the creation of a new and efficient farming structure. The system must facilitate the future exchange of land parcels between the individuals who wish to establish a farm of the size suitable for efficient farming, given the skills and resources in each particular case. The establishment of such farm sizes requires a functioning land market, where land parcels can be bought and sold, leased in and out, or exchanged. Without such markets, land will remain locked into an inherently inefficient distribution pattern despite private ownership.

This was the case of Poland throughout the Communist period, where, in the absence of a land market, privately owned land could not be transferred to more efficient and more enterprising users. If the land markets are allowed to develop, land owners who are the least risk averse and most motivated to farm individually will be able to increase their initial holding over time by renting or purchasing land from other, less active owners. Some beneficiaries may not want to farm individually at the first stage (or ever), and to avoid the underutilisation of valuable land resources, a legally clear process must be defined for inactive land owners to lease or sell land to those who want to farm. Procedures must also be developed to facilitate the consolidation of privately owned land into larger parcels as their new owners make the medium term decision about land utilisation.

The development of a land market is still in the first stages, even in the most advanced CEE countries, which have already reached an advanced stage in the formulation and implementation of a legal framework for privatisation. However, the prevailing ideas about the rules that will govern land market functioning and private land ownership are extremely varied and not entirely reassuring. There are strong reservations about the possibility of the emergence of large, privately owned farms. Many people fear the repetition of the old system of big estates and absentee owners, or at least rapid concentration of land ownership in a few hands. As a result, practically all countries impose various limitations to private land ownership.

Current and pending legislation in all countries of the region circumscribes owners' rights in the important areas of land transfer, land utilisation, and size of holdings. Typical restrictions on land transfer in various countries vary from an outright ban on buying and selling of land (indefinitely or during a long moratorium) to restrictive provisions which allow alienation only to residents in the same village, to local farm enterprise, or through intermediation of the state. Even the right to lease land to other farmers is still not universally accepted.

Restrictions on land use typically include the obligation to farm the land, and to observe sound conservation practices. Failure to observe this rule can result in expropriation. Restrictions on size of holdings are universal. The maximum size of private farms is limited to 20-30 hectares in Bulgaria, 50 hectares in Latvia, and 100 hectares in Romania. Restrictions on the rights of owners are intended to prevent absentee ownership and underutilisation of land resources. They are also intended to prevent new owners from becoming victims or beneficiaries of speculative land transactions. The restrictions on property rights reflect fundamental ambiguity regarding private land property, even where private ownership is legally recognised. The restrictions on land ownership rights are among the most contested issues of agricultural reform.

Various restrictions related to ownership rights, while often understandable in the historical and political context, are inconsistent with the principles of a market economy. These restrictions may become a serious obstacle to the shaping of an efficient economic structure based on private ownership and, in the final analysis, harm the competitiveness of agriculture in the region. Restrictions regarding land buying and selling and the size of holdings are not consistent with the need to stimulate the emergence of efficient farms. The requirement that land in agricultural use at the time of transfer of ownership remains indefinitely in agricultural use can inhibit much needed adjustments in land use patterns.

Because of legal restrictions on transactions and because land rights, distribution has not been completed in a number of countries. Land markets are still not functioning and the correct market values of land are ignored. Valuation of land is often performed on the basis of administratively decided formulas and models.

What are the prospects for the future?

It is now clear that, contrary to the original expectations, the transformation of large-scale agriculture will be a lengthy process. It is extremely difficult to foresee what type of units will emerge as dominant. The future agriculture in the region probably will be characterised by a variety of forms of farming operations. Private farms, restructured cooperatives, commercial farms of various sizes, and small part-time subsistence farms producing mainly for their own consumption will coexist side by side. Applying the major lessons of the first phase of the transition to assess the prospects for the future (without attempting to make more concrete predictions), the following issues deserve attention:

- The region's agriculture must adjust to the basically new internal and external circumstances. Rapid expansion of internal demand cannot be expected and capturing a permanent share of external markets requires a basic improvement in efficiency, quality, and competitiveness. In this situation, the level of output will remain for some time lower than that reached in previous years, which means that agricultural production cannot become the engine for general economic recovery. What is even more important, private production is emerging under conditions of strict efficiency requirements, hard budget constraints, and intense market competition, and this process will have losers as well as winners.
- A relatively large number of private farms may not be able to operate profitably, despite the initial support that many of them receive. It would be a serious mistake to keep these farms alive with further inflow of resources from the state budget. State support, if it is given at all, should serve to improve the general income conditions of agriculture and to enhance international competitiveness. A stratum of private farmers dependent on central government support could easily become an economic and political burden.
- The transition to a profitable and efficient agricultural structure should be completed as quickly as possible. The establishment of unequivocal and clear land-ownership relations is essential. In the future, the development of the farming structure should be shaped basically by the market relations and market tools, principally through an efficiently functioning land market. This means minimising the restrictions on land ownership and land use, as well as on ownership and lease rights. The transfer of ownership and lease rights should be simple and cheap. The greater the government intervention in the operation of the land market, the slower and more costly the shaping of the new farming structure will be. The leasing of land will have a very important role in shaping an efficient farming structure, and the conditions for leasing must be determined fundamentally by the supply and demand relation.
- It is unlikely that family fanning of the Western European type and scale will become the generally accepted model, not even over a longer period. The countries in the region cannot afford to provide state subsidies over a long period in order to keep alive farms that are not viable under free market conditions. For this reason, the

desirable farm size and form of farming must be capable of generating self-sustainable profits without subsidies or with relatively small support. These farm sizes almost certainly will be many times larger than the average size of independent, family based private farms in CEE today (2-20 hectares), and possibly larger than farm sizes in some of the Western European countries.

- Given the traditional dual structure of former socialist agriculture and the emphasis on the household plots in the process of reform, it seems likely that the distinction between the subsistence fanning and the commercial farming will persist for some time in the future throughout the region. The success of the ongoing process of land reform and farm restructuring will be judged by the gradual increase in the number of commercially viable small and medium-size farms and continued downsizing of the traditional large-scale structures. Persistence of a dual structure of small subsistence plots and large subsidy-dependent capital-intensive farms will surely signal failure of the process.
- In international practice, agricultural production co-operatives have proven less efficient than individually owned farms, especially in environments with adequately functioning markets. The development of market forces will lead to further disintegration and transformation of many agricultural production co-operatives that still exist in these countries. Because of the social and political implications, the changes should be achieved not through administrative intervention but through the market and efficiency instruments, just as in the case of privately owned farms.
- The strengthening and the expansion of processing, marketing, purchasing, and other service co-operatives is expected to gain momentum in the future. These service cooperatives will provide a viable support framework for the growth of private farms which will be detached from the collective sector still undergoing the process of transition. The development of service cooperatives calls for government attention and support.
- Rural finance is one of the components of the farm support system which at present is still not developed in any of the CEE countries. The traditional financial system in the socialist countries is not oriented to meet the needs of the new private agriculture. Debt writing off and interest rate subsidies will be eliminated as agriculture moves towards the market. Both banks and borrowers will have to pay attention to the concept of creditworthiness and will have to take loan repayment seriously. A mortgage finance system will have to be developed as soon as private land ownership and land titles are

instituted and the legal restrictions on the use of land as collateral are removed. Rapidly some new, creative financial solutions may emerge as alternatives to the scarcely developed banking system. This could include suppliers' and marketers' credits associated with farmers' product delivery. Development of village-level credit societies and mutual credit arrangements is also a possibility for meeting the financial needs of a small private farmer.

- Finally, special attention should be paid to the social aspects of the transformation in agriculture. The decline in production and the change of ownership structure are obvious sources of social problems and tensions. Agriculture will be able to contribute to the solution of these problems only within the framework of efficient and profitable production. The fundamental objective of agrarian policy thus should be the achievement of an efficient and profitable farming structure based on private ownership. The governments should avoid the pitfalls of using agrarian policy to resolve social issues, and leave the solution of rural social problems to social policy instruments. Any deviation from this direction will have tangible negative economic consequences.

There are many problems to be solved, but the basic final message is not the one of despair. Most of the countries concerned are not poor and most of them can be self-sustaining, creditworthy and have access to markets. But *we must recognise the complexity of the changes taking place; the social costs; the political strains; the investments needed in nation-building*. It is necessary to face the reality, the future tasks and problems, and take into account the time required to solve them. In particular, the following points should be emphasised:

- Dedicated people are at work trying to change the system and make the new approaches work. But there is fear of the future.
- This is not a question merely of revealing a pool of latent entrepreneurial talents. Generations of experience have taught risk avoidance, suspicion, and survival techniques aimed at circumventing the system.
- The scale of the problem is unprecedented.
- All previous discussions about sequencing have become academic.
- Speed is essential but also difficult. Efforts need to be focused on:
  - Actions which will show results quickly (retail privatisation, SME development in opportunities);
  - Actions which will offer opportunity to more people and involve them in the process.

No one today is in a position to project where the agricultural sector of various Central and Eastern European countries will be at the end of the decade. We can only state with some certainty that:

- They will not all have evolved in the same way, nor at the same pace.

- Not all transformations which have been initiated will be fully completed.
- Progress will not be linear. There will be setbacks.
- Institutional weaknesses, social structures, attitudinal changes will slow down structural change and growth in agriculture.

To avoid disappointments, it is better to be realistic from the beginning. The early euphoria envisioned a quick transformation, followed by an equally quick supply response. But developments thus far, and the experience of both the Central and Eastern European countries and East Germany, suggest otherwise. Gradually, the perception is growing that the process will be slower than anticipated and that, consequently, the social and political strains on very fragile systems, greater. However, one also has to be convinced that the whole task is not impossible. The region has all natural, economic and human resources to become a fully integrated and prosperous part of the developed world and the European Union, specifically, in the foreseeable future.

BIOBASED ECONOMY – SUSTAINABLE USE OF AGRICULTURAL RESOURCES

The biobased economy can be to the 21st century what the fossil-based economy was to the 20th century. Agriculture has the potential to be central to this economy, providing source materials for commodity items such as liquid fuels and value-added products (chemicals and materials). At the same time, agriculture will continue to provide food and feed that are healthful and safe, which may give rise to some situations of trade-offs.

The use of agricultural raw material in a biobased economy is not new. However, now agriculture has to compete with alternative land uses in order to claim the status of socially responsible entrepreneurship. Conservation of valuable landscapes, habitats, biodiversity have come to the forefront of some policy makers' agenda. The public-good benefits that could accrue from the biobased economy are compelling. They include increased security in some countries (such as USA), economic advantages to farmers, industry, rural communities, and society, environmental benefits at the global, regional, and local levels, and other benefits to society in terms of human health and safety.

How should this economy develop so that whatever is done is done well? This question requires examining some of the issues related to sustainability of this economy. Such an investigation has not taken place and thus, there is a need to explore this aspect of the biobased economy. In this chapter, opportunities and challenges facing the bioeconomy are introduced, primarily through a review of the literature. Major concentration of this study is on the agricultural feedstocks for use in the production of liquid transportation fuels, and related products. Some attention is also paid to production of biogas for electricity and heating purposes.

DEFINITION OF BIOBASED ECONOMY

As an alternative, researchers working in the agriculture, forestry, and fisheries sectors recognize the use of biobased products for competing with the fossil-based industry, commonly referred to as the 'biobased economy'. This economy uses renewable bio-resources, biological tools, eco-efficient processes that contribute to GHG emission reductions to produce sustainable bioproducts for medical treatments, diagnostics, and more-nutritional foods, energy, chemicals and materials while improving the quality of the environment and standard of living. Biobased resources are materials derived from a range of plant systems, and may include starch, sugar, wood, cellulose, lignin, proteins etc. These resources are produced from different sources such as, biomass, crop residue, dedicated crops and crop processing by-product.

The major commodity produced in the biobased economy is energy, in the form of liquid fuels (ethanol and biodiesel) and biogas. The types of energy generated from these products include uses in transportation, heating, electric appliances etc. Agricultural and forest products are generally used in the production of the above biofuels.

Generally, agricultural activity generates a variety of feedstocks for the production of bio-products, particularly bioenergy. Main feedstocks of agricultural activity are from crop biomass including crop residues and livestock waste. Canada, possessing about 67.5 M ha of agricultural farmland, has the potential to offer feedstocks for bioenergy (including biofuels). Of this area, 31.87 M ha are planted each year to grow starch (wheat, barley, corn and oat), oil (rapeseed, soybean and flaxseed) and forage crops (Rye, fodder corn and tame hay), with a total carbon content of about 33.5 Mt C/yr, and an energy content of about 2 exajoules (EJ) yr^{-1} or 2 times 10^{18} J yr^{-1}. Additionally, agricultural crop residues were estimated to contain about 56 Mt C/year. Although some of this residue may be incorporated into the soil to maintain soil fertility and carbon content, the recoverable portion contains 14.6 Mt C/yr and has an energy potential of 0.52 EJ/yr. To this estimate, one can add livestock wastes in Canada, which could produce over 3 billion m^3 of biogas which is equivalent to energy of 0.065 EJ/yr.

DEFINITION OF SUSTAINABILITY

What is Sustainability?

Sustainability is inherently about durability and endurance. The World Commission on Environment and Development defines it as "the capacity to meet the needs of the present without compromising the ability of future generations to meet their own needs". It emphasizes strategies that promote economic and social development to meet human needs in ways that avoid environmental degradation, overexploitation or pollution. At the 2005 World

Summit it was noted that this requires the reconciliation of economic, environmental and social demands - the "three pillars" of sustainability.

SUSTAINABILITY IN THE CONTEXT OF BIOBASED ECONOMY

The biobased economy can contribute to a more sustainable society, not only because it leads to an economy no longer primarily dependent on fossil fuels for energy and industrial raw materials, but also by generating less waste, by a lower energy consumption and by using less water. In addition, the biobased economy provides also for the established industries the opportunity for further growth in a sustainable way. However, does it mean that the production and use of bioenergy is intrinsically sustainable? The Environmental Audit Committee (EAC) found that although biofuels can reduce GHG emissions from road transport, most first generation biofuels have a detrimental impact on the environment overall. In addition, most biofuels are often not an effective use of bioenergy resources, in terms either of cutting GHG emissions or value-for-money.Stoeglehner and Narodoslawsky (2009) answered this question from an ecological footprint perspective. They found, by comparing different technologies, that biofuels are considerably more sustainable than fossil options presently in use. Yet, to what extent biofuel use is sustainable remains open as this can only be answered in a regional context taking other land use demands, visions and values into account.

Major utilitarian frameworks define and identify sustainable choices as those that maximize per capita utility subject to an ethical constraint that per capita utility will not decline over time. The utilitarian framework can be applied to derive sustainable outcomes in the context of biofuels, and in particular to identify which biofuels to produce and to what extent, by assuming that utility is derived from the consumption of food, fuel (fossil fuel and biofuel) and other private goods and is maximized subject to budget constraints, land availability and various sustainability constraints. Biofuels would be considered a sustainable substitute if they can compete with fossil fuels in a free market setting at prices that internalize all environmental costs of production, minimize damages to the environment and allow food and other goods and services to be available such that overall utility is non-decreasing over time. The production of any type of biofuel is likely to involve trade-offs among these multi-dimensional aspects of sustainability. The degree to which biofuels can accommodate the three pillars of sustainability, taking account of potential tradeoffs among these pillars, needs to be evaluated

Economic Sustainabililty

The economic sustainability of biofuels depends on the costs of production and market price of supply. The sustainability of the corn ethanol industry depends on its ability to deal with volatility in both gasoline and corn prices.

Variability in the price of corn could lead to cycles of boom and bust for the biofuel industry with the impact of supply shocks being exacerbated when inventories are low. The oil price, commercially viable technology to produce cellulosic biofuels, and trade barriers also affect economic viability of the biofuel industry. The rising oil price has contributed to higher corn prices because of increased cost of production of corn, in addition to its demand. Besides the supply-side considerations, the demand for ethanol and the availability of infrastructure to deliver the ethanol produced to the blenders are the driving forces behind the biofuel industry sustain expansion.

Environmental Sustainability

Biofuels are occasionally claimed as being carbon neutral and fossil-fuel free, but serious concerns about the carbon benefits of current biofuels have been raised. Actually, biofuels consume a significant amount of energy that is derived from fossil fuels. Equally important is the fact that production of biofuels has other environmental impacts, such as soil erosion due to tilling, eutrophication due to fertilizer run-offs, impacts of exposure to pesticides, habitat, and biodiversity loss due to land-use change, etc., which have not received the same attention as GHG emissions. Conversely, the grain used for ethanol feedstock production is often the poor quality, impure grains which are mostly unsuitable for either human or livestock, and which also do not require as much pesticide. In contrast to grain-based ethanol, cellulosic biofuels from perennial grasses (such as switchgrass) have the potential to produce more biofuel per hectare of land and thus have smaller indirect land use effects. While, the environmental benefits of cellulosic biofuels depend on the mix of feedstocks use, the location and management practices used to grow them are equally important. There might also be some trade-offs between environmental benefits and most profitable methods of producing cellulosic feedstocks.

Social Sustainability

sKhanna et al. (2009) consider that the social sustainability of biofuel depends on the distribution of biofuel costs and benefits across countries, income groups, and rural and urban areas. One should keep in mind that human rights, health and equity are also important issues that are related to social sustainability.

Higher crop prices in response to increased demand of biofuel will improve farm incomes. However, the higher commodity price may be capitalized into land rent and prices of inputs, which will reduce the future benefit to farmers. Cost of food to consumers may also increase, which may create a heavy burden on the urban poors. The development of biofuel production may also bring to the forefront equity and gender-related issues, such as labour conditions on plantations, constraints faced by small holders and the disadvantaged position of female farmers. All of these could affect the welfare of the society and sustainability.

THE CRITERIA AND INDICATORS FOR ASSESSING THE SUSTAINABILITY OF BIOENERGY DEVELOPMENT

An indicator can be used to quantify a specific impact of bioenergy production (*e.g.* the rate of soil erosion). Ideally, to evaluate the sustainability of bioenergy use, the impacts of bioenergy production, conversion and trade must be analysed using an integrated approach, taking account of the three dimensions of sustainable development: people (social well-being; the social impacts), planet (maintaining environmental quality; the environmental impact), and profit (economic viability of bioenergy production and its welfare impacts; and other economic impacts). The production and use of bioenergy can only be deemed sustainable if the net impact is positive. Practically applicable criteria and/or indicators are required to monitor and assess the sustainability of bioenergy production and use.

Various ongoing initiatives aim to ensure the sustainability of bioenergy production and use through certification, a form of communication that assures the buyer of bioenergy that the supplier complies with specific sustainability criteria. The European Union and several individual countries, most notably the UK and The Netherlands, are currently developing certification systems. Other countries, for example Brazil, are linking biofuel certification with tax reductions and other incentives to stimulate sustainable bioenergy use. Also, various non-governmental organisations are formulating sustainability criteria.

Areas of concern and sustainability criteria in Smeets's study, criteria in parentheses are not translated into cost

Smeets (2008) analysed to what extent implementing a sustainability certification system affects the management system (costs) of bioenergy production and availability (quantity) of land for energy plantations. The certification system takes account of twelve sustainability criteria and accompanying indicators. However, this certification system lacks the important criterion of "GHG emissions". A project group "Sustainable Production of Biomass" was established in 2006 by the Interdepartmental Programme Management Energy Transition to develop a system for biomass sustainability criteria for the Netherlands for the production and conversion of biomass for energy, fuels and chemistry. A set of generic sustainability criteria and corresponding sustainability indicators was formulated.

The need to secure the sustainability of biomass production and trade in a fast growing market is widely acknowledged by many stakeholder groups and setting standards and establishing certification schemes are recognized as possible strategies that help ensure sustainable biomass production and trade. McBridge et al. (2011) have developed a selection criteria framework for bioenergy sustainability. There seems to be a general agreement that it is important to include economic, social and environmental criteria in the development of a biomass certification system. However, mutual differences are also visible in the strictness, extent and level of detail of these criteria, due to various interests and priorities and geographic constraints. The development of biomass certification systems is still in its infancy and largely in development. Therefore, it is worthwhile to consider in this preliminary phase which ways can be followed if the strategy to be taken in the development of a reliable and efficient biomass certification system.

ENVIRONMENTAL IMPACTS OF BIOBASED ECONOMY

Agriculture involves a large human manipulation of the biosphere that impacts the environment. For all the impacts considered, Engstrom et al., (2007) noted that agriculture affects the environment through: eutrophication of water resources, GHG emissions, and loss of biodiversity. On a life cycle analysis basis the impacts are even larger but much of that environmental harm is associated with fossil fuel use. In addition to direct fossil fuel use for agriculture, agriculture production involves further fossil fuel use for energy-intensive inputs like N fertilizers and for transportation of inputs to the farm and products from farm to market.

Bioenergy production is an important existing bioeconomy initiative whose current and potential environmental impacts have been studied extensively. Bioenergy production may cause eutrophication of water, increases ecosystem and human exposure to toxins, causes loss of biodiversity, degrades air quality, and increases acidification of the ecosystem.

Informed decisions by society require comparative studies of environmental impact of alternatives. For agriculture, the most useful information for decision–makers is not the damage from agriculture to the environment but the comparative measures of environmental harm between food types, production practices, and/or geographical situations. This information facilitates making choices that best balance food need with acceptable environment damage. A similar situation exists for bioenergy. The comparative values of environmental impact between energy sources are required to make sound choices in bionergy. Thus, the problem becomes a multi-objective, albeit limited, optimization across the considered alternate energy sources or across considered alternative ways to provide energy-related functions, such as km of passenger travel.

GREENHOUSE GAS EMISSIONS

Reducing GHG emissions compared to fossil-fuel alternative is often considered the environmental value of biofuels. Several standards require that biofuels provides GHG emission reductions at least 60 per cent lower than those for competing fossil fuel.

The estimated GHG benefits of bioenergy are complex, variable, and controversial. Most biofuel production systems provide GHG benefits, typically at least 30 per cent less than fossil fuels. Some favourable systems such as biodiesel from palm oil and ethanol from sugarcane in Brazil can achieve life-cycle reduction of 50 per cent to 90 per cent. Second generation biofuels using biomass crops and crop residues have been estimated to achieve GHG reductions greater than 50 per cent. However, some studies argue that the GHG emissions associated with bioenergy production are underestimated and that there is no net GHG savings for many biofuels.

Considering changes in soil carbon associated with crop production can reduce GHG emissions. Where there is an increase in land carbon stocks this reduces net GHG emissions and, if the carbon stock change is sufficient, GHG emission can become negative, *i.e.* a net removal.

Searchinger et al. (2008) included indirect land-use change (ILUC) from major increases in ethanol production from US corn. There are large GHG emissions from the land use change, particularly from clearing of forests. They calculated that it would take 150 years of biofuel production before the aggregate GHG emission reductions from ethanol compared to fossil-fuel gasoline are larger than the GHG emission from biofuel-induced ILUC. Fargione et al. (2008) estimated that the GHG effects of ILUC increases the GHG emission for ethanol from US corn by 17 to 420 times. However, the analysis of Searchinger et al. (2008) has attracted criticism that it oversimplifies trade effects, neglects the effect of increases in yield over time, and the use of alternatives pathways to ethanol from feedstock other than corn.

Kløverpris et al. (2010) used a global trade model to show that land use impact is complex and depends on where feedstock production is taking place. Gains in productivity are more feasible in some regions than others. For example, Denmark has high yield and restrictions on use of fertilizer and pesticides so opportunity for increased production is lower than countries with lower initial yield and fewer restrictions on farming activities. Feasible increases in yield of crops can overcome the ILUC associated with bioenergy. Schmidt et al. (2009) determined that selection of location for sourcing food to replace that lost from bioenergy is important to ILUC effects. For example, exports of Canadian rapeseed oil to Europe would displace palm oil from tropical countries where palm plantations threaten the rain forests in those countries. Similarly, by strengthening the market demand for field crops in the Canadian Prairies, the demand for biofuel feedstock will increase the area seeded to crops, rather than left fallow, a practice that is known to increase wind erosion.

LAND USE AND BIODIVERSITY

Gomiero et al. (2010) have argued that agreed limits to human appropriation of ecosystem services and global net primary productivity are needed. The world will not be able to support biofuels and food production when loss of agricultural land for transportation, industry, and settlements are considered. Appropriation of net primary productivity beyond the current 50 per cent is unsustainable. They point out that the area impact of biofuel is already much larger than that of fossil fuels considering their relative impacts on energy supply. Fibre and bioenergy needs will exacerbate the pressure on global biodiversity from conventional food production. Bioenergy is a tradeoff between GHG reductions and biodiversity.

Land use impact is not only how much land but also what land and how land is used. Dale et al. (2010) present a potential scenario of increases in biofuel production with increases in biodiversity, mostly through increase production of perennial biomass crops included vegetation mixtures more similar to natural prairies. Solid biofuels for commercial and industrial applications could be an effective and sustainable way to grow the bioeconomy. The use of biomass pellets – which can be produced from wood, switchgrass or straw, would not only create new market oppourtunities for the forest and agricultural industries, it would reduce dependence on coal as well as the GHG emissions associated with coal use. Sophisticated geographical analysis involving land use, habitats, and sensitive ecosystems allows for design of bioenergy production that minimizes potential biodiversity impact. However, Gomiero et al. (2010) note that efficient biofuel production requires monoculture and mechanization for land near the biofuel plants to achieve maximum efficiency. Such production practices could be detrimental to biodiversity.

Bioenergy feedstock production will affect land use which can impact biodiversity to varying degrees, depending on the crop type and the region. Growing grain crops probably has the greatest detrimental impact on biodiversity if these crops are managed more intensively, with increased inputs and fewer rotations. Growing perennial herbaceous crops on marginal land can often reduce biodiversity loss compared to using the land for row crops such as corn. However, Dyer et al. (2011) found that if the marginal land is natural grassland, such as much of the rangeland in Western Canada, rather than the result of land degradation, even a perennial feedstock crop (such as switchgrass) could result in the loss of extensive areas of natural habitat. When cattle are displaced by feedstock crops (ILUC), they may be grazed at unsustainable stocking rates or in rangeland not previously used for grazing. Good geographic planning of bioenergy development can protect high-carbon high-biodiversity compared to letting market forces determine land use.

SUSTAINING LAND PRODUCTIVITY

Crop residues are an attractive feedstock for bioenergy since they do not reduce food production, are available in large quantities, and are relatively low cost. However, crop residue protects the soil from erosion and maintains soil organic matter. The removal of 20-30 per cent of crop residue is probably sustainable although residue removal will eventually require additional fertilizer to replace nutrients removed. The balance between the residue removal rate and long-term soil health is a challenge.

Soil erosion is affected by crop type and its production practices. Generally, increased bioenergy production increases erosion risk. The choice of crops is important, especially if maize replaces grass and forages. Production practices, such as winter cover crops where appropriate, can mitigate erosion risk.

EUTROPHICATION

Nutrient loss through run-off leads to eutrophication of water bodies. This is largely a consequence of fertilizing crops for bioenergy feedstock. Consequently, bioenergy can increase eutrophication compared to fossil fuels even in highly optimized production systems. The use of perennial biomass crops for bioenergy feedstocks can decrease contamination of water with nutrients compared to annual crops Similarly, removal of crop residue can increase nutrient contamination from surface run-off.

ECONOMIC IMPACTS OF BIOBASED ECONOMY

The economics of biofuels critically depend on the price of fossil fuels, price of feedstocks, the cost of conversion (including investment needs) and the revenues generated by the by-products. Storage, transport and logistic costs also need to be included. Two major sources of revenue from biofuel production are sale of the fuel, and sale of by-products, which may include dry distiller's grain and sollubles (DDGS), glycerine and carbon dioxide, as well as rapeseed or soybean meal.

Investigations by $(S+T)^2$and Edna Lam Consulting (2005) for ethanol and biodiesel production suggest that these products cannot compete with fossil-based products without a subsidy. The impact of biofuel production on various sectors of the society is also very different. Benefits are realized by the ethanol industry, but at the cost of state revenues, and consumer expenditures. But with new markets that respond differently than conventional food markets, the rural economy is enhanced. Society as a whole benefits from the country's reduced reliance on crude oil imports and reduced economic costs for mitigating GHG emissions.

JOB CREATION AND RURAL DEVELOPMENT

Brazil is one of the examples of successful job creation from bioenergy industry. The bioenergy industry offers direct or indirect employment opportunities -. Employment generation from a biofuel plant differs between the two stages: construction stage and operations stage. During the construction phase, employment impacts are large but temporary in nature. Plant operation generates fewer but permanent jobs. For example, Haig (2006) estimated that the impact of producing 2 billion litres of ethanol on the rural economy would generate 6,645 jobs in rural Canada.

Urbanchuk (2006) has found that local ownership of biofuel plants maximizes the rural development potential. He estimates that the full contribution to the local economy of a farmer-owned co-operative ethanol plant is likely to be as much as 56 per cent higher than the impact of an absentee-owned corporate plant. This is attributed to two main factors unique to farmer owned plants:

1. A larger share of operational expenditures is made in the local community; and
2. The distribution of dividend payments to farmer-owners of a co-operative ethanol plant represents additional income to farmers and their families.

Meanwhile, if a market for selling carbon credits could be established, this would provide another source of revenue to farmers.

IMPROVED TRADE BALANCE

The activities associated with the biobased economy such as the expansion of biofuel would cause, in some cases, substantial increase in exports of agriculture commodities due to a diversified set of agricultural products. In addition, a biobased economy is economically viable in a longer term perspective. In a study of Thailand, although the costs of biofuel production may exceed the cost of importing equivalent petroleum, domestic production of biofuels allows virtually all of the money to stay within the country's economy, and thus, adds to the balance of payment for the country.

ESTABLISHMENT OF NEW INDUSTRIES

An increase in feedstock production for biobased industry results in an increased production of by-product and residues that are in turn utilized as raw materials for several other sectors, such as livestock production, cosmetics and pharmaceutical industries, among others. Input providing industries, such as agricultural equipment manufacturing firms and fertilizer industries, will expand to supply additional goods and services to support the increased biomass production activity. Byproducts and inputs can be important criteria for feedstock crop choices. For example, soybean-based biodiesel was shown to have a lower carbon footprint than rapeseed-based biodiesel due to both providing more livestock feed byproduct than rapeseed oil and being a legume that does not require N-fertilizer input.

The oil price plays an important role in determining the economics of biofuels. If the world oil price remain high, biofuels will be more financially viable even without government support. The remote areas (or countries) usually have the comparative advantage of labour, but due to poor facility and transportation system, prices of oil may be markedly higher than the international prices. In these cases, if biofuel production and processing are located near consumption centers or can be transported to them at relatively low costs, they can be competitive against imported fossil fuels.

FISCAL EFFECTS OF BIOFUEL DEVELOPMENT

Biofuel development can affect several levels of governments through one or a combination of three pathways:

1. Provision of public subsidies;
2. Generation of new and different sources of government revenues; and
3. Change in government expenditures.

Under current fossil based fuel prices, biofuels are not competitive. Many jurisdictions have accepted the need for public subsidies to enhance the public cause. However, biofuel support programmes can act as a substitute for other agricultural programme subsidies. For example, the U.S. ethanol tax credit, according to Gardner (2003), has served to displace some of the government deficiency payments related to corn. The financial impact on government is likely to include both positive and negative components. There is a cost to government for any incentives provided to the biofuel industry, but there will also be tax revenues that flow to government from the income generated by these operations. Intuitively, if subsidies are retired at some point in time, the benefits from the programme would exceed costs to government.

In the case of an energy importing country, impact on the government would be through replacement of petroleum imports. However, this cost should be weighed against government spending to develop the biofuel industry. In some countries such as Brazil, development of the biofuel industry has resulted in a net benefit even after all government support expenditures are included.

SOCIAL IMPLICATIONS

There are mainly two major social benefits of biobased industry: increased standard of living and increased social cohesion and stability. While the biobased industries help create income generation and other positive impacts, their effectiveness depends on a number of other factors. These may include: whether the industry can provide full-time jobs or part-time and night shift jobs; total employment created per energy unit or per amount of land; number of households or people employed in a region; whether skilled or unskilled labour are required, etc.

SOCIAL BENEFITS

Improved Quality of Life in Rural Areas

The increased income in a household or community would further help increase a community's or individual's accessibility to good education, health care, resources (*e.g.* water, land), food products and employment opportunities etc. Biobased industry, being located in rural areas, may provide many of these benefits by establishing livelihood opportunities for the local people. In addition, increased income may help strengthen the cohesion or stability of a community.

Improved Human Health

The biobased economy may also play an important role in improving human health and safety. For example, sugarcane bagasse used for making paper and fibreboard would otherwise be burnt in the field releasing harmful air pollutants. In addition, improved air quality will reduce diseases such as asthma, and biodegradability characteristics of biobased products, compared to petroleum-based alternatives, are an added advantage. Finally, the local energy security created by bioenergy sector especially biogas will help replace the use of firewood which otherwise would cause air pollution creating negative impact on health of people. In poor countries, increased family incomes would make health care more affordable.

Poverty Alleviation

Although liquid fuels are currently being developed for transportation, modern technologies to convert biomass into energy promises to be a more directed way to alleviate poverty, especially in remote oil-dependent regions. Some of this would happen through providing employment opportunities in regions where alternatives are scarce or non-existent.

Economic and Social Impacts on Indigenous People

Well-planned biofuel projects could allow indigenous communities to generate capital and maintain or rebuild livelihoods based on the sustainable use of natural resources. In Canada, there is evidence that aboriginal communities and organizations have seldom been incorporated into rural/regional economic development planning, and biobased economy could offer them this opportunity.

SOCIAL COSTS ASSOCIATED WITH BIOBASED ECONOMY

Some of the social challenges that may arise from biobased industry include changes in land-use rights, food insecurity, and destruction of traditions, among others.

Land-Use Change and Impcats on Land Access

Changes in land use due to increased expansion of agricultural lands for the cultivation of biofuel crops may affect land access and rights of local people. In addition, increased economic value created for agricultural biomass may attract agricultural producers to shift from food or cash crops to feedstock. This change would indirectly affect many others whose livelihoods are partially or completely dependent on food crops. Further, land values tend to rise when policies and market incentives are provided to convert lands for biofuels production. This increased land value may displace poor people from their land.

Food Security and Cultural Impacts

Several studies have argued that increasing demand for biofuel feedstock will pose some serious threats to the food security of people. In general, development of biobased industry could affect food security in two ways. One, higher food prices caused from the demand of feedstock for biofuel production will limit the purchasing power of the poor or marginalized people. Two, higher land-use change, such as diverting crop lands to biofuel feedstock production, can have major negative effects on local food security and on the social and cultural dimensions of land use. Increased livelihood opportunities from biobased industry would lead to destruction of traditional economic or cultural activities, such as hunting, fishing and trapping. Additionally, using food and feed crops for ethanol production would increase the prices of other food items which are derived directly (*e.g.* breads, cereals) or indirectly (*e.g.* chicken, eggs, milk) from these biofuel crops. Although higher food prices represent higher income for farmers, they will affect those whose livelihoods are not linked to agriculture (*e.g.* urban poor).

Social Impacts of Rapid Growth

Biofuel development could occur over a very short period of time and could change the social fabric of communities. New industrial developments always bring about some costs to communities. According to Finsterbusch (1980) some of these costs include:

1. new residents are frustrated by crowded housing (mainly trailers) and lack of amenities – especially recreational opportunities;
2. These conditions aggravate family relations and lead to family tension, child abuse and neglect, and delinquency; and
3. Reported cases of depression, alcoholism, and attempted suicide greatly increase, as do mental health cases.

Researchers have provided documentation of a general increase in crime, drug abuse, mental illness, child abuse, and related problems in communities among both new and long-time resident resulting from a rapid growth over a very short period of time.

PROVIDING THE BALANCE TO SUSTAINABILITY – TRADE-OFFS TO BE MADE

Biobased economy cannot provide all of society's material and energy needs. One therefore, needs to look at the value of displaced food production in social-economic context to know if trade-offs are worthwhile. Other possible trade-offs that may exist are:

1. Between economic and environmental goals of the society;
2. Between environmental and social objectives of the society; and
3. Between economic and social objectives.

ENVIRONMENT AND ECONOMY

Traditionally, there has been a view that investments for mitigation of environmental damage (environmental protection) is a cost that takes resources away from investments that would increase production efficiency. Consequently, there are trade-offs between environment and the economy. Many countries have developed (or proposed) policies for reducing GHG emissions, such as subsidies, carbon tax, import tariffs for biofuels, and mandates for quantities to be produced or blended.

These policies may promote investments in environmental protection and related technology development, while they can also distort markets and are subject to political decisions that may make them unsustainable. At the same time, some policies strive at maximizing the economic benefit, but will cause environment degradation.

An example of this is the U.S. volumetric tax credit for cellulosic biofuels, that does not differentiate across feedstocks and rewards monocultures of high-yielding biofuels per unit of land and are therefore unlikely to create incentives for maintaining biodiversity.

Climate Change Mitigation vs. Energy Security

Biofuels are attractive to governments which can diversify energy budget and reduce their exposure to international oil market to maintain economic sustainability. Corn-based ethanol in the United States and sugarcane-based ethanol in the Brazil have been built successfully with this objective in mind. While the well–to-wheel environmental benefits are different, such as sugarcane-based ethanol and cellulosic biofuels may achieve significant reduction of GHG, the corn-based ethanol performs poorly due to intensive fossil fuel input.

GHG vs. other Environmental Goods

Besides GHG emission reduction, there are many other environmental benefits associated with a biobased economy, such as decreasing soil erosion, water eutrophication, loss of biodiversity, that should be considered. Treating GHG emissions as the only environmental cost, with no concern for other environment threats, can probably result in the other environmental goods and services, such as soil, water and biodiversity, becoming the unintended casualties. Decision makers need to include the full range of desired environmental outcomes in the design of appropriate and robust biofuel policies.

ENVIRONMENT AND SOCIETY

Emphasis on biofuels as renewable energy sources has developed globally. The use of food crops for biofuel production raises major nutritional and ethical concerns. As a result some trade-offs may exist. One such trade-offs is use of agricultural commodities for food vs. for fuel production.

The food versus fuel debate arises because increased use of land and water for bioenergy production reduces the availability of these resources to produce food for human consumption. The competition is direct in terms of first generation biofuel production that uses feedstocks of cereal grains (*e.g.* corn, wheat, etc.), oilseeds (*e.g.* rapeseed, soybean, palm oil), or other crops (*e.g.* sugar cane) that are conventionally used for food. However, even if the bioenergy feedstock crop is not suitable for food directly, it uses land that could be used for food production.

Secure and affordable food is basic to social sustainability. However, bioenergy may be at the origin of social benefits in providing better quality of life for rural population. It also has great potentials to mitigate environmental impacts. Therefore, if bioenergy is seen as a net environmental benefit, then the extent to which bioenergy production threatens the supply of secure and affordable food becomes an environment and society trade-off. However, if bioenergy is seen as environmental benefit, then the trade-off becomes between society and environment.

ECONOMY AND SOCIETY

Usually, it is hard to clearly distinguish between economic and social issues. While economic sustainability emphasizes the economic feasibility and viability, society sustainability focuses more on distribution, human health, human rights and equity. Some social conflicts hide behind the economic benefit maximization. For example, the smaller scale operations generally have higher cost. However, the social sustainability policy goals for biofuels include promotion rural development and inclusion of small farmers. This trade off is important as many commodity dependent developing countries are characterised by a high proportion of small producers.

If an industrialized form of bioenergy crop cultivation is practiced, then the land required will most probably be controlled by large land owners or national companies. From maximization of the economic profits, crop cultivation tends to be industrialized which in turn will affect small landowners and poor people's right and welfare. Land ownership should be equitable, and land-tenure conflicts should be avoided. This requires clearly defined, documented and legally established tenure rights. To avoid leakage effects, poor people should not be excluded from the land. Customary land-use rights and disputes should be identified. A conflict register might be useful in this context.

SWOT ANALYSIS OF BIOBASED ECONOMY DEVELOPMENT

A Strength-Weakness-Opportunities-Threats (SWOT) analysis of the biobased economy is developed which would help decision makers understand strengths and need for developing appropriate policies to overcome limitations for such developments in the future. One can see whether taking an action or building a project based on biobased economy depends on consideration of many positive and negative factors.

Table. Relevant factors identified in each SWOT category

	Internal	**External**
Positive	Strengths Energy security Job creation and rural development Improved trade activities Establishment of new industries Reduce GHG emissions	Opportunities Renewable energy requirement Policy encouragement and technology development
Negative	Weakness Food security Economic viability Environmental impact uncertainty Equity concerns	Threats Rise in fuel and food price Natural hazards and Crisis on financial market

How to get win-win outcomes from biobased economy development? A map and related policies are urgently needed for the global biofuels industry that supports sustainability. Preventing environmental degradation and social-economic disruption from activities associated with bioenergy supply is seen as a basic principle of sustainability. Vermeulen et al. (2008) mentioned that it may be better for the EU to miss its target of reaching 10 per cent biofuel content in road fuels by 2020 than to compromise the environment and human wellbeing. The “decision tree”, which is developed by Vermeulen et al. (2008), can guide the interdependent processes of deliberation and analysis needed for making tough choices in biofuels to balance the tradeoffs between environment, economy and society.

6

Farm Equipments and Farm Machinery

FARM IMPLEMENTS AND EQUIPMENT

Running a modern farm is not easy; it takes the mastery of dozens of skills and long hours of work in all kinds of weather. It also takes a wealth of specialised tools and equipment. Farm implements go back thousands of years, all the way to the invention of agriculture; equipment, as distinct from implements, goes back at least to the industrial revolution. For most farmers, these implements are as important as the land itself, because there is no way they can get the harvest in without them. They also represent a huge investment, especially the heavy equipment, and no business can afford to throw away a major investment, especially not an investment that's integral to their business. This is why it is so important to maintain farm implements and equipment. Proper maintenance extends the life of equipment, and reduces the expenses required to keep a farm going. It may seem expensive to someone who does not budget for regular maintenance, but the cost is always less than paying for catastrophic repairs.

HISTORY OF FARM IMPLEMENTS AND BASICS OF FARMING

Simple farm implements go back as far as the invention of agriculture, which was probably some 10 000 years ago around Catal Hüyük in what is now Turkey. Farming, in simplest terms, is nothing more or less than growing specific plants to order. The farmer has to break the ground for planting, plant the seeds, and then harvest and thresh the crop. The first implements were little more than sticks used to break up the soil, although they soon developed into the precursors of hoes and spades. The next big step was the plough, which quite literally took breaking earth out of the hands of the farmer. The first ploughs were little more than bigger sticks pulled by animals, but with the introduction of metal ploughshares and the horse collar, ploughs advanced steadily. Harvesting and threshing was also originally done by hand, with scythes, sickles, and flails. It took a lot of work, but the invention of agriculture enabled all of civilization that followed.

THE BASICS OF FARMING

While people may think that farming starts with just putting a seed in the ground and letting it grow, there are a lot more steps involved, many of them before anyone even thinks of touching a seed. To start with, the ground has to be prepared for the seeds, and that takes work. Only then can a farmer even think of planting. In fact, for centuries, the hardest work was that which needed to be done at the very beginning. Planting by hand is much easier than ploughing. The basic steps of farming and the order in which they are normally performed:

Step	Description
Ploughing	Turns over the soil to aerate and bury weeds
Harrowing	Breaks up the soil to prepare the seedbed for planting
Planting	Introduces the seed to the ground
Reaping	Harvests the crop
Threshing	Beats the grain to loosen the chaff for removal
Winnowing	Separates the wheat from the chaff

There are a very large number of steps involved in the art of farming, and many of them involve a great deal of physical labour. In fact it was that physical labour that was one of the driving forces of the agricultural revolution that started in the 18th century and led to the development of farm equipment as opposed to implements.

FARM EQUIPMENT

Farm equipment, as opposed to implements, can be considered those tools that are too big to be used by hand. In other words, the farmer carries implements, equipment carries the farmer. Most modern farm equipment is either motorised, or comes as an attachment to a piece of farm machinery such as a tractor. While the development of equipment took a huge step forward after about 1800, its history goes back to at least ancient Rome.

TRACTOR ATTACHMENTS

Tractor attachments are among the most common pieces of farm equipment available. They range from ploughsand harrows, to seed drills and spreaders.

It is the variety of these attachments, as well as others such as the backhoe, that make the tractor such a valuable piece of farm equipment. Depending on the farm, the tractor may be the single core piece of farm equipment around which everything else revolves.

Combine Harvester

The combine harvester is a single piece of equipment that fulfils three tasks in succession; reaping, threshing, and winnowing. They get the name because they combine all three aspects of harvesting into a single piece of equipment. First invented in the 1830s, they have since become a vital part of the modern farm. The earliest combine harvesters were horsedrawn, with some of the bigger ones requiring teams of over a score of horses. Later models were designed to be pulled by tractors, or in the case of the larger units, to be self-propelled. The one thing all versions have in common is that they are both complex and expensive, making proper maintenance that much more important.

FARM EQUIPMENT MAINTENANCE

Regular maintenance is vital to any piece of equipment. Not only does it cost less overall than having to repair catastrophic damage, but it is infinitely easier to budget for a small regular expense on an ongoing basis, than a potentially much larger expense that could hit at any time with no warning. Regular maintenance not only keeps the farmer in control of these expenses, but also decreases the chance of surprise if a major failure does occur. There are a number of components that show visible signs of wear long before they fail, which means the person doing regular maintenance can see the signs and is then able to take preventative action beforehand.

CLEANING AND INSPECTION

One of the most important aspects of maintenance is cleaning; it is also the first step. Regular cleaning not only prevents damage by ensuring nothing gets stuck in the machinery, which can be particularly important with equipment like combine harvesters which have a lot of moving parts, but is also a necessary precursor to inspecting the equipment or implements for wear and damage. Once the equipment is clean, it is easy to see what needs repair, what needs replacing, and what is in perfect condition and needs nothing more than the cleaning.

Maintenance and Repair

After cleaning and inspection, the next stage is taking the time for maintenance and any repairs if necessary. Every piece of machinery has some parts that wear out through normal use, and farm equipment is no exception. Proper maintenance involves inspecting and replacing those parts as they wear out, and making sure all fluids are topped up. It also means making all those little repairs, the ones that do not appear to make any immediate difference, but add up over the long run. A tractor can work perfectly well with one broken lug nut, but break a second one, and the tractor has an issue. Keeping everything else in good condition helps minimise the collateral damage from an otherwise minor breakdown.

BUYING FARM IMPLEMENTS AND EQUIPMENT ON EBAY

No matter how well you maintain your farm equipment, there is a point at which you have to spend money; whether on parts for further maintenance, or to replace implements or equipment that has reached the point where the maintenance cost is greater than the replacement cost. Whatever you need to buy, eBay is a superb place to buy it. You can find anything, all you need to do is put your terms in the search box, there is one on every page, and watch the results appear. Once you've got all your results on screen, you can use the filters in the sidebar to narrow them down to just what you need. You can filter by everything from price to brand, as well as seller location. Then, once you've eliminated everything that does not meet your needs, you can use the sort function to ensure the ones you want most are at the top of your list.

After you have found everything required for your maintenance needs, the next thing to do is to determine which of eBay's many sellers you want to do business with. The place to start is their profile page, where you can see everything from their feedback to their location. Some sellers also offer bundles, or allow you to pick up purchases in person. The latter can be a great saving when buying equipment.

FARM MACHINERY AND IMPLEMENTS

BALER

New Holland balers are recognised as the most reliable and dependable balers, and with good reasons too. They have a heavy duty draw bar, big drive lines, tough plungers, gear driven knotters and heavy duty feeding system, all designed to give you years of reliable service.

The New Holland baler is compatible with a minimum of 35 PTO HP. Tractor and it runs best with a New Holland Tractor.

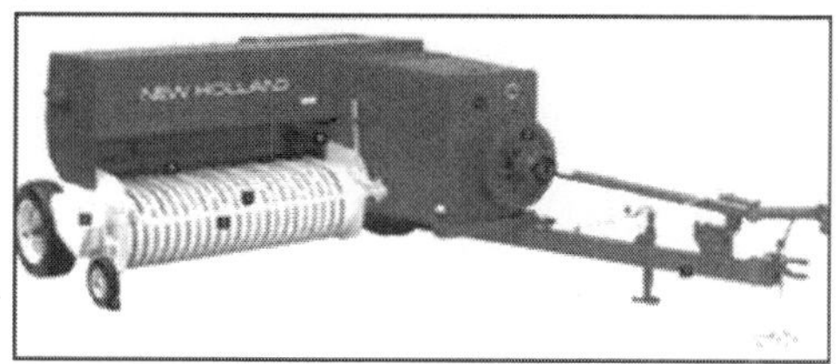

- Baler

- Mowers-Conditioners
- Crop Chopper®

Who are the users of baled straw/trash?

- Biomass power generation and co-generation plants of sugar industries require bulk quantity of crop residue to be burnt as fuel for electricity
- Paper industries use this as raw material for paper production
- Dairy/cattle owners use it for fodder application and urea treatment projects of straw
- Packaging industry, mushroom industry, straw/board manufactures, natural manure (wormi compost) and many more

The New Holland baler helps a farmer clear his crop residue in the shortest possible time so that the land can be prepared for the next crop, ensuring effective and timely sowing season after season.

Extraordinary Features of the Baler

1. The New Holland baler's heavy-duty drawbar, attached directly to the axle, keeps towing stresses away from the bale chamber. It has a transport positions and two field positions for use of various tractor widths.
2. The large 56 cms flywheel of the New Holland baler allows smooth power transfer.
3. The baler's heavy-duty clutch is protected by a shear bolt arrangement. It ensures gentle load engagement, long life and greater load capacity.
4. The wide pick up arrangement of the New Holland baler has multiple rows of curved teeth that can even handle a short file crop. The pick up can be raised or lowered according to different train and crop density, either manually or through hydraulics.
5. A floating wind guard of the New Holland baler gets the window under control and forms it into a smooth may for positive feeding.
6. The baler's chain-driven pick-up incorporates a slip clutch to prevent overload.
7. The pick-up gauge wheel of the New Holland baler compensates for uneven field conditions as it protects pick-up teeth.
8. The New Holland baler's high capacity feeding system is equipped with six tines. They provide a positive feed action into the baler chamber.
9. The baler's large feed opening handles the big capacity pick up and feeder.
10. The sturdy plunger of the New Holland baler with high speed and large stroke ensures optimum bale shape and density. A sharp blade mounted on the plunger cuts the crop evenly.

COMBINE HARVESTER

The combine harvester, or simply combine, is a machine that harvests grain crops. The name derives from its combining three separate operations comprising harvesting—reaping, threshing, and winnowing—into a single process. Among the crops harvested with a combine are wheat, oats, rye, barley, corn (maize), soybeans and flax (linseed). The waste straw left behind on the field is the remaining dried stems and leaves of the crop with limited nutrients which is either chopped and spread on the field or baled for feed and bedding for livestock.

Fig. A Lely open-cab combine.

Fig. Harvesting oats in a Claas case 570 combine with enclosed, air-conditioned cab with rotary thresher and laser-guided hydraulic steering

Combine harvesters are one of the most economically important labour saving inventions, enabling a small fraction of the population to be engaged in agriculture.

HISTORY

Scottish inventor Patrick Bell invented the Reaper in 1826. The combine was invented in the United States by Hiram Moore in 1834, and early versions were pulled by horse or mule teams. In 1835, Moore built a full-scale version and by 1839, over 50 acres of crops were harvested. By 1860, combine harvesters with a cutting width of several metres were used on American farms. In 1882, the AustralianHugh Victor McKay had a similar idea and developed the first commercial combine harvester in 1885, the Sunshine Harvester. Combines, some of them quite large, were drawn by mule or horse teams and used a bullwheel to provide power. Later, steam power was used, and George

Stockton Berry integrated the combine with a steam engine using straw to heat the boiler.Tractor-drawn, combines became common after World War II as many farms began to use tractors. These combines used a shaker to separate the grain from the chaff and straw-walkers (grates with small teeth on an eccentric shaft) to eject the straw while retaining the grain. Early tractor-drawn combines were usually powered by a separate gasoline engine, while later models were PTO-powered.

These machines either put the harvested crop into bags that were then loaded onto a wagon or truck, or had a small bin that stored the grain until it was transferred to a truck or wagon with an auger.

In the U.S., Allis-Chalmers, Massey-Harris, International Harvester, Gleaner Manufacturing Company, John Deere, and Minneapolis Molineare past or present major combine producers.

In 1911, the Holt Manufacturing Company of California produced a self-propelled harvester. In Australia in 1923, the patented Sunshine Auto Header was one of the first center-feeding self-propelled harvesters. In 1923 in Kansas, the Curtis brothers and their Gleaner Manufacturing Company patented a self-propelled harvester which included several other modern improvements in grain handling.

Both the Gleaner and the Sunshine used Fordson engines. In 1929 Alfredo Rotania of Argentina patented a self-propelled harvester. In 1937, the Australian-born Thomas Carroll, working for Massey-Harris in Canada, perfected a self-propelled model and in 1940 a lighter-weight model began to be marketed widely by the company. Lyle Yost invented an auger that would lift grain out of a combine in 1947, making unloading grain much easier.

In 1952 Claeys launched the first self- propelled combine harvester in Europe; in 1953, the European manufacturer CLAAS developed a self-propelled combine harvester named 'Herkules', it could harvest up to 5 tons of wheat a day. This newer kind of combine is still in use and is powered by diesel or gasoline engines. Until the self-cleaning rotary screen was invented in the mid-1960s combine engines suffered from overheating as the chaff spewed out when harvesting small grains would clog radiators, blocking the airflow needed for cooling.

Fig. A John Deere 9410 Combine with grain platform attached.

Fig. A John Deere Titan series combine unloading corn.

A significant advance in the design of combines was the rotary design. The grain is initially stripped from the stalk by passing along a helical rotor instead of passing between rasp bars on the outside of a cylinder and a concave. Rotary combines were first introduced bySperry-New Holland in 1975.

In about the 1980s on-board electronics were introduced to measure threshing efficiency. This new instrumentation allowed operators to get better grain yields by optimizing ground speed and other operating parameters.

COMBINE HEADS

Combines are equipped with removable heads that are designed for particular crops. The standard header, sometimes called a grain platform, is equipped with a *reciprocating knife cutter bar*, and features a revolving reel with metal or plastic teeth to cause the cut crop to fall into the auger once it is cut. A variation of the platform, a "flex" platform, is similar but has a cutter bar that can flex over contours and ridges to cut soybeans that have pods close to the ground. A flex head can cut soybeans as well as cereal crops, while a rigid platform is generally used only in cereal grains.

Some wheat headers, called "draper" headers, hi use a fabric or rubber apron instead of a cross auger. Draper headers allow faster feeding than cross augers, leading to higher throughputs due to lower power requirements. On many farms, platform headers are used to cut wheat, instead of separate wheat headers, so as to reduce overall costs.

Dummy heads or pick-up headers feature spring-tined pickups, usually attached to a heavy rubber belt. They are used for crops that have already been cut and placed in windrows or swaths. This is particularly useful in northern climates such as western Canada where swathing kills weeds resulting in a faster dry down.

While a grain platform can be used for corn, a specialized corn head is ordinarily used instead. The corn head is equipped with snap rolls that strip the stalk and leaf away from the ear, so that only the ear (and husk) enter the throat. This improves efficiency dramatically since so much less material must go through the cylinder. The corn head can be recognized by the presence of points between each row.

Occasionally rowcrop heads are seen that function like a grain platform, but have points between rows like a corn head. These are used to reduce the amount of weed seed picked up when harvesting small grains.

Self-propelled Gleaner combines could be fitted with special tracks instead of tires or tires with tread measuring almost 10in deep to assist in harvesting rice. Some combines, particularly pull type, have tires with a diamond tread which prevents sinking in mud. These tracks can fit other combines by having adapter plates made.

FARM EQUIPMENT MECHANIZATION AND TECHNOLOGY

Farm mechanization is an important element of modernization of agriculture. Farm Productivity is positively correlated with the availability of farm power coupled with efficient farm implements and their judicious utilization. Agricultural mechanization not only enables efficient utilisation of various inputs such as seeds, fertilizers, plant protection chemicals and water for irrigation but also it helps in poverty alleviation by making farming an attractive enterprise. The Department of Agriculture and Cooperation is following multi-pronged strategy for promoting Farm Mechanization.

The Dept. is implementing a scheme for Promoting Agricultural Mechanization through "Outsourcing of training and demonstrations of newly developed equipments". The objective of the scheme is to create awareness about agricultural equipment and machinery among the end users and other stakeholders. Through this scheme, State Governments organise demonstration of improved/newly developed agricultural/horticultural equipment as identified by them at farmers' fields so that the farmers get acquainted about their use and utility for production of different types of crops. In the year 2010-2011, an outlay of ₹ 10.91 Crores has been made. Out of total outlay, ₹ 50 lakhs is earmarked for North Eastern States.

Post Harvest Management being one of the thrust areas for the Dept., a scheme on "Post Harvest Technology and Management" is being implemented, with an outlay of ₹ 40.0 crore during XI Plan period. Under the scheme the technologies developed by ICAR, CSIR and those identified from within the country and abroad for primary processing, value addition, low cost scientific storage and transport of agricultural produce are promoted to minimize wastage during post harvesting processes. The main components of the PHT&M scheme are establishment of low cost Post Harvest Technology (PHT) units for transfer of primary processing technology, supply of PHT equipments to end users with

Government assistance, demonstration of PHT technologies and training of farmers, entrepreneurs and scientists. The outlay of 2010-2011 is ₹ 7 crores out of which ₹ 50 lakh is earmarked for North Eastern States.

Beside above interventions, the Department is promoting Farm Mechanization by making agricultural equipment available among farmers at cheaper rates. A level of 25-50 per cent subsidy on procurement cost is made available under revised "Macro Management of Agriculture (MMA)" scheme for different categories of equipment. The subsidy on tractors and power tillers is available on the models approved by the department under institutional financing. Besides tractors and power tillers, combine harvesters are also available to the farmers as per approved pattern of subsidy. As an individual farmer may not be in a position to purchase high cost equipment on his own, Self Help Group of farmers (SHGs), user groups, cooperative societies of farmers etc are also made eligible for assistance under the programme.

As a result of different programmes implemented by the Government of India over the years and equal participation from Private Sector, the level of mechanization has been increasing steadily over the years. This is evident from the sale of tractors and power tillers, taken as indicator of the adoption of the mechanized means of farming, during the last five years.

Table. Year wise sale of tractors and power tillers

Year	Tractors Sale (Nos.)	Power Tillers Sale (Nos.)
2004-05	2,47,531	17,481
2005-06	2,96,080	22,303
2006-07	3,52,835	24,791
2007-08	3,46,501	26,135
2008-09	3,42,836	35,294
2009-10(till Nov,2009)	2,39,789	18,375

This has resulted in increase in total farm power availability from 0.295 kW/ha in 1971-72 to 1.502 kW/ha in 2005-06 (estimated).

TRAINING OF FARMERS AND TECHNICIANS

The Farm Machinery Training and Testing Institutes (FMTTIs) located at Budni (Madhya Pradesh), Hissar(Haryana), Garladinne (Andhra Pradesh), and Biswanath Chariali (Assam), have been imparting training to farmers, technicians, retired/retiring defence personnel etc., in the selection, operation, maintenance, energy conservation and management of agricultural equipments. These Institutes have also been conducting testing and performance evaluation of various agricultural implements and machines. During the year 2009-10, 5278 persons were trained till 31st March, 2010 against the annual target of 5600 in different courses. The target of training during the Eleventh Plan has been increased to 28,000 from 25,000 during the Tenth Plan. To supplement the efforts of the FMTTIs in human resource development, outsourcing of the training through the SAUs, Agricultural Engineering colleges, polytechnics, etc., has been approved during the X plan. For training of farmers, the identified institutions are reimbursed ₹ 5200 per trainee per month, which also includes a stipend of ₹ 1200 and to and fro travel expenses by normal mode of transport. The target of training by outsourcing during the Eleventh Plan is 10,000 persons. The physical target for the 2009-10 is to train 2000* farmers.

TESTING AND EVALUATION OF FARM MACHINERY AND EQUIPMENT

The Institute at Budni has been authorized to conduct tests on tractors and other agricultural machines; while the institute at Hissar conducts tests on self-propelled combine harvesters, irrigation pumps, plant protection equipment, agricultural implements and other machines including issuing of CMVR certificate of Combine Harvester. The Institute atGarladinne has been authorised to test power-tillers and also conduct tests on various agricultural implements/equipment components. This Institute is being developed as a specialty institute for meeting the mechanization demand in rain-fed and dry land farming systems. The institute at Biswanath Chariali (Assam) tests bullock-drawn implements, manually operated equipment, tractor drawn implements, self propelled machines and small hand tools. For the Eleventh Plan, the target of testing has been kept as 550 machines/tools.

The four FMTTIs altogether have tested 211 machines of various categories, including tractors, power-tillers, combine harvesters, reapers, rotavators and other implements, till 31st March, 2010 against the target of 110 for the year 2009-10. During the remaining eleventh five year plan2009-10, 2010-11, 2011-12 an outlay of ₹ 17.33 crores for development of FMTTI, Budni for purchase of major equipments and machines for parallel test line has been made.

Demonstration of Newly Developed Agricultural/Horticultural Equipment

For enhancing production and productivity, as well as for reducing the cost of production, the induction of improved/new technology in the agricultural production system is inescapable. Therefore, with this aspect in view, the demonstration of newly developed agricultural equipment including horticultural equipment at farmers' fields has been included as a component of the restructured scheme. The scheme "Promotion and Strengthening of Agricultural Mechanization through Training, Testing and Demonstration is being implemented during the Eleventh Plan. This scheme envisages conduct of demonstration of improved/newly developed agricultural/ horticultural equipment, identified by the State Governments/Government Organizations at farmers' fields, to acquaint them about their use and utility for production of different types of crops.

During the year 2009-10 (up to 31st December 2009), the number of demonstrations conducted is 7407 covering 2634 hectares with field machines and 13638 hours with stationary machines in which over 1,51,134 farmers participated.

FARM MECHANIZATION PROGRAMMES UNDER MACRO MANAGEMENT

Assistance in the form of subsidy at the rate of 25-50 per cent of the cost with permissible ceiling limits is made available to the farmers for the purchase of agricultural equipment including hand tools, bullock-drawn/power-driven implements, planting, reaping, harvesting and threshing equipment, tractors, power-tillers and other specialised agricultural machines under the centrally sponsored scheme of Macro Management of Agriculture.

The feedback from the State Governments indicates that during the year 2009-10 (up to 31st March, 2010) 7008 tractors, 4560 power tillers, 103118 hand tools, 2501 bullock-drawn implements, 9401 tractor-driven implements, 4781 self-propelled/power-driven equipment, 5280 plant protection equipment, 5881 irrigation equipment, and 2630 gender-friendly equipment were supplied to the farmers.

STATE AGRO INDUSTRIES CORPORATIONS:

The Government of India had advised the State Governments in the year 1964, to set up State Agro Industries Corporations (SAICs) in the public sector to act as catalysts in providing access to industrial inputs to farmers, for their use in agriculture. Thus, 17 SAICs were set up in the joint sector with equity participation of the Government of India and the respective State Governments of Andhra Pradesh, Assam, Bihar, Gujarat, Haryana, Himachal Pradesh, Jammu and Kashmir, Karnataka, Kerala, Madhya Pradesh, Maharashtra, Orissa, Punjab, Rajasthan, Uttar Pradesh, Tamil Nadu and West Bengal during 1965 to 1970. Many of the State Governments have increased their equity participation as a result of which the Government of India, at present, is a minority shareholder. SAICs have since expanded their basic functions by commencing manufacture and marketing of agricultural inputs, implements, machines, after-sales-service, promotion and development of agro-based units/industries. The Government of India is implementing a policy of disinvestment of its shares in SAICs with a view to giving greater decision making power to the state governments by allowing transfer of its shares to state governments on following guidelines:

- Where the net worth of the SAIC is positive, the Government of India would be willing to consider offering its shares to the State Governments at a price 25 per cent less than the book value of the shares on the basis of the latest available audited balance sheet.
- In the case of SAICs whose net worth is negative, the Government of India would be willing to pass on its stake for a token consideration of ₹ 1000 for the value of the shares.

So far, the Government of India's shares in SAICs of Gujarat, Karnataka, Uttar Pradesh, Tamil Nadu, Rajasthan and West Bengal have been transferred to the State Governments concerned. The State Governments of Madhya Pradesh, Assam and Jammu and Kashmir have since agreed, in principle, for transfer of the Government of India's shares held in these SAICs.

Activities in the North-Eastern States:

A FMTTI has been established at Biswanath Chariali in the Sonitpur district of Assam, to cater to the needs of human resource development in the field of agricultural mechanization and also to assess the quality and performance characteristics of different agricultural implements and machines in the region. The Institute imparted training to 520 persons and tested 15 machines up to 31st March, 2010 during the year 2009-10.

The Scheme on 'Post Harvest Technology and Management'

The scheme on "Post Harvest Technology and Management" is being implemented with an outlay of ₹ 95.00crore during XI Plan period. Under the scheme the technologies developed by ICAR, CSIR and those identified from within the country and abroad for primary processing, value addition, low cost scientific storage and transport of cereals, pulses, oilseeds, sugarcane, vegetables and fruits and the crop by-product management shall be given a boost. The Scheme will basically focus on the lower end of the spectrum of post harvest management and processing.

The main components of the PHT&M scheme are as under:

a. Establishment of units for transfer of primary processing technology, value addition, low cost scientific storage, packaging units and technologies for by-product management in the production catchments under tripartite agreement.
b. Establishment of low cost Post Harvest Technology (PHT) units/ supply of PHT equipments with Government assistance.

c. Demonstration of technologies.
d. Training of farmers, entrepreneurs and scientists.

Gender Friendly Equipment for Women

Under the Central Sector Scheme – 'Promotion and Strengthening of Agricultural Mechanization through Training, Testing, and Demonstration, during 2009-10, about 2630 gender-friendly equipment have been distributed amongst farm women. Under the scheme for Outsourcing of Training and Demonstration of Newly Developed Agricultural Equipment, including Horticultural Equipment at Farmers' Fields, separate physical targets have been fixed and 10 per cent of the funds have been allocated for women farmers. During 2009-10 a total of 484 women have been imparted training at Farm Machinery, Training, and Testing Institutes (FMTTIs). A list of about 30 identified gender-friendly tools and equipment developed by the Research and Development Organisation for use in different farm operations has been sent to all states and UTs. for popularizing them. State governments have been directed to earmark 10 per cent of total funds allocated for the training for women farmers.

THE SIGNIFICANCE OF MACHINERY IN AGRICULTURE

Having established the fact of the extremely rapid development of the production of agricultural machinery and of the employment of machines in Russia's post-Reform agriculture, we must now examine the social and economic significance of this phenomenon. From what has been said above regarding the economics of peasant and landlord farming, the following conclusions may be drawn: on the one hand, capitalism is the factor giving rise to, and extending the use of, machines in agriculture; on the other, the application of machinery to agriculture is of a capitalist character, *i.e.*, it leads to the establishment of capitalist relations and their further development.

Let us dwell on the first of these conclusions. We have seen that the labour-service system of economy and the patriarchal peasant economy inseparably connected with it are by their very nature based on routine technique, on the preservation of antiquated methods of production. There is nothing in the internal structure of that economic regime to stimulate the transformation of technique; on the contrary, the secluded and isolated character of that system of economy, and the poverty and downtrodden condition of the dependent peasant preclude the possibility of improvements. In particular, we would point to the fact that the payment of labour under the labour-service system is much lower (as we have seen) than where hired labour is employed; and it is well known that low wages are one of the most important obstacles to the introduction of machines. And the facts do indeed show us that an extensive movement for the transformation of agricultural technique only commenced in the post-Reform period of the development of commodity economy and capitalism. The competition that is the product of capitalism, and the dependence of the cultivator on the world market made the transformation of technique a necessity, while the drop in grain prices made this necessity particularly urgent.

To explain the second conclusion, we must examine landlord and peasant farming separately. When a landlord introduces a machine or an improved implement, he replaces the implements of the peasant (who has worked for him) with his own; he goes over, consequently, from labour-service to the capitalist system of farming. The spread of agricultural machines means the elimination of labour-service by capitalism.

It is possible, of course, that a condition laid down, for example, for the leasing of land is the performance of labour-service in the shape of day-work at a reaping machine, thresher, etc., but this will be labour-service of the second type, labour-service which converts the peasant into a day labourer. Such "exceptions," consequently, merely go to prove the general rule that the introduction of improved implements on the farms of private landowners means converting the bonded ("independent" according to Narodnik terminology) peasant into a wage-worker – in exactly the same way as the acquisition of his own instruments of production by the buyer-up, who gives out work to be done in the home, means converting the bonded "handicraftsman" into a wage-worker. The acquisition by the landlord farm of its own implements leads inevitably to the undermining of the middle peasantry, who get means of subsistence by engaging in labour-service: We have already seen that labour-service is the specific "industry" of the middle peasant, whose implements, consequently, are a component part not only of peasant, but also of landlord, farming.

Hence, the spread of agricultural machinery and improved implements and the expropriation of the peasantry are inseparably connected. That the spread of improved implements among the peasantry is of the same significance hardly requires explanation after what has been said in the preceding chapter. The systematic employment of machinery in agriculture ousts the patriarchal "middle" peasant as inexorably as the steam-power loom ousts the handicraft weaver.

The results of the employment of machinery in agriculture confirm what has been said, and reveal all the typical features of capitalist progress with all its inherent contradictions. Machines enormously increase the productivity of labour in agriculture, which, before the present epoch, was almost entirely untouched by social development. That is why the mere fact of the growing employment of machines in Russian agriculture is sufficient to enable one to see how utterly unsound is Mr. N.–on's assertion that there is "absolute stagnation" in grain production in Russia, and that there is even a "decline in the productivity" of agricultural labour. We shall return to this assertion, which contradicts generally established facts and which Mr. N.–on needed for his idealisation of the pre-capitalist order.

Further, machines lead to the concentration of production and to the practice of capitalist co-operation in agriculture. The introduction of machinery, on the one hand, calls for capital on a big scale, and consequently is only within the capacity of the big farmers; on the other hand, machines pay only when there is a huge amount of products to be dealt with; the expansion of production becomes a necessity with the introduction of machines.

The wide use of reaping machines, steam-threshers, etc., is therefore indicative of the concentration of agricultural production – and we shall indeed see later that the Russian agricultural region where the employment of machines is particularly widespread (Novorossia) is also distinguished by the quite considerable size of its farms. Let us merely observe that it would be a mistake to conceive the concentration of agriculture in just the one form of extensive enlargement of the crop area; as a matter of fact, the concentration of agricultural production manifests itself in the most diverse forms, depending on the forms of commercial agriculture. The concentration of production is inseparably connected with the extensive co-operation of workers on the farm. Above we saw an example of a large estate on which the grain was harvested by setting *hundreds* of reaping machines into operation simultaneously.

"Threshers drawn by 4 to 8 horses require from 14 to 23 and even more workers, half of whom are women and boys, *i.e.*, semi-workers.... The 8 to 10 h. p. steam-threshers to be found on all large farms" (of Kherson Gubernia), "require simultaneously from 50 to 70 workers, of whom more than half are semi-workers, boys and girls of 12 to 17 years of age". "Large farms, on each of which from 500 to 1,000 workers are gathered together simultaneously, may safely be likened to industrial establishments," the same author justly observes. Thus, while our Narodniks were arguing that the "village community" "could easily" introduce co-operation in agriculture, life went on in its own way, and capitalism, splitting up the village community into economic groups with opposite interests, created large farms based on the extensive co-operation of wage-workers.

From the foregoing it is clear that machines create a home market for capitalism: first, a market for means of production (for the products of the machine-building industry, mining industry, etc., etc.), and second, a market for labour-power. The introduction of machines, as we have seen, leads to the replacement of labour-service by hired labour and to the creation of peasant farms employing labourers. The mass-scale employment of agricultural machinery presupposes the existence of a mass of agricultural wage-workers. In the localities where agricultural capitalism is most highly developed, this process of the introduction of wage-labour along with the introduction of machines is intersected by another process, namely, the ousting of wage-workers by the machine. On the one hand, the formation of a peasant bourgeoisie and the transition of the landowners from labour-service to capitalism create a demand for wage-workers; on the other hand, in places where farming has long been based on wage-labour, machines oust wage-workers. No precise and extensive statistics are available to show what is the general effect of both these processes for the whole of Russia, *i.e.*, whether the number of agricultural wage-workers is increasing or decreasing.

There can be no doubt that hitherto the number has been increasing. We imagine that now too it is continuing to increase: firstly, data on the ousting of wage-workers in agriculture by machines are available only for Novorossia, while in other areas of capitalist agriculture (the Baltic and western region, the outer regions in the East, some of the industrial gubernias) this process has not yet been noted on a large scale. There still remains an enormous area where labour-service predominates, and in that area the introduction of machinery is giving rise to a demand for wage-workers. Secondly, the growth of intensive farming (introduction of root crops, for example) enormously increases the demand for wage-labour. A decline in the absolute number of agricultural (as against industrial) wage-workers must, of course, take place at a certain stage in the development of capitalism, namely, when agriculture through out the country is fully organised on capitalist lines and when the employment of machinery for the most diverse agricultural operations is general.

As regards Novorossia, local investigators note here the usual consequences of highly developed capitalism. Machines are ousting wage-workers and creating a capitalist reserve army in agriculture. "The days of fabulous prices for hands have passed in Kherson Gubernia too. Thanks to... the increased spread of agricultural implements..." (and other causes) "*the prices of hands are steadily falling* " (author's italics).... "The distribution of agricultural implements, which makes the large farms independent of workers and at the same time reduces the demand for hands, places the workers in a difficult position".

The same thing is noted by another Zemstvo Medical Officer, Mr. Kudryavtsev, in his work *Migrant Agricultural Workers at the Nikolayev Fair in the Township of Kakhovka, Taurida Gubernia, and Their Sanitary Supervision in 1895*. "The prices of hands... continue to fall, and a considerable number of migrant workers find themselves without employment and are unable to earn anything; *i.e.*, there is created what in the language of economic science is called a reserve army of labour – artificial surplus-population". The drop in the prices of labour caused by this reserve army is sometimes so great that "many farmers possessing machines preferred" (in 1895) "to harvest with hand labour rather than with machines"! More strikingly and convincingly than any argument this fact reveals how profound are the contradictions inherent in the capitalist employment of machinery!

Another consequence of the use of machinery is the growing employment of female and child labour. The existing system of capitalist agriculture has, generally speaking, given rise to a certain hierarchy of workers, very much reminiscent of the hierarchy among factory workers. For example, on the estates in South Russia there are the following categories: a) *full workers*, adult males capable of doing all jobs b) *semi-workers*, women and males up to the age of 20; semi-workers are divided again into two categories: aa) 12, 13 to 15, 16 years of age – these are semi-workers in the stricter sense of the term – and bb) *semi-workers of great strength;* "in the language used on the estates, 'three-quarter' workers," from 16 to 20 years of age, capable of doing all the jobs done by the full worker, except mowing. Lastly, c) semi-workers rendering *little help*, children not under 8 and not over 14 years of age; these act as swine-herds, calf-herds, weeders and plough-boys. Often they work merely for their food and clothing.

The introduction of agricultural implements "lowers the price of the full worker's labour" and renders possible its replacement by the cheaper labour of women and juveniles. Statistics on migrant labour confirm the fact of the displacement of male by female labour: in 1890, of the total number of workers registered in the township of Kakhovka and in the city of Kherson, 12.7 per cent were women; in 1894, for the whole gubernia women constituted 18.2 per cent (10,239 out of 56,464); in 1895, 25.6 per cent (13,474 out of 48,753). Children in 1893 constituted 0.7 per cent (from 10 to 14 years of age), and in 1895, 1.69 per cent (from 7 to 14 years of age). Among local workers on estates in Elisavetgrad Uyezd, Kherson Gubernia, children constituted 10.6 per cent.

Machines increase the intensity of the workers' labour. For example, the most widespread type of reaping machine (with hand delivery) has acquired the characteristic name of "lobogreyka" or "chubogreyka," since working with it calls for extraordinary exertion on the part of the worker: he takes the place of the delivery apparatus. Similarly, intensity of labour increases with the use of the threshing machine.

The capitalist mode of employing machinery creates here (as everywhere) a powerful stimulus to the lengthening of the working day. Night work, something previously unknown, makes its appearance in agriculture too. "In good harvest years... work on some estates and on many peasant farms is carried on even at night", by artificial illumination – torchlight. Finally, the systematic employment of machines results in traumatism among agricultural workers; the employment of young women and children at machines naturally results in a particularly large toll of injuries. The Zemstvo hospitals and dispensaries in

Kherson Gubernia, for example, are filled, during the agricultural season, "almost exclusively with traumatic patients" and serve as "field hospitals, as it were, for the treatment of the enormous army of agricultural workers who are constantly being disabled as a result of the ruthless destructive work of agricultural machines and implements". A special medical literature is appearing that deals with injuries caused by agricultural machines. Proposals are being made to introduce compulsory regulations governing the use of agricultural machines. The large-scale manufacture of machinery imperatively calls for public control and regulation of production in agriculture, as in industry.

Let us note, in conclusion, the extremely inconsistent attitude of the Narodniks towards the employment of machinery in agriculture. To admit the benefit and progressive nature of the employment of machinery, to defend all measures that develop and facilitate it, and at the same time to ignore the fact that machinery in Russian agriculture is employed in the capitalist manner, means to sink to the view point of the small and big agrarians. Yet what our Narodniks do is precisely to ignore the capitalist character of the employment of agricultural machinery and improved implements, without even attempting to analyse what types of peasant and landlord farms introduce machinery. Mr. V. V. angrily calls Mr. V. Chernyayev "a representative of capitalist technique".

Presumably it is Mr. V. Chernyayev, or some other official in the Ministry of Agriculture, who is to blame for the fact that the employment of machinery in Russia is capitalist in character! Mr. N.–on, despite his grandiloquent promise "not to depart from the facts", has preferred to ignore the fact that it is capitalism that has developed the employment of machinery in our agriculture, and he has even invented the amusing theory that exchange reduces the productivity of labour in agriculture !

To criticise this theory, which is proclaimed without any analysis of the facts, is neither possible nor necessary. Let us confine ourselves to citing a small sample of Mr. N.–on's reasoning. "If," says he, "the productivity of labour in this country were to double, we should have to pay for a chetvert (about six bushels) of wheat not 12 rubles, but six, that is all". Not all, by far, most worthy economist. "In this country" (as indeed in any society where there is commodity economy), the improvement of technique is undertaken by individual farmers, the rest only gradually following suit.

"In this country," only the rural entrepreneurs are in a position to improve their technique. "In this country," this progress of the rural entrepreneurs, small and big, is inseparably connected with the ruin of the peasantry and the creation of a rural proletariat. Hence, if the improved technique used on the farms of rural entrepreneurs were to become socially necessary (only on that condition would the price be reduced by half), it would mean the passing of almost the whole of agriculture into the hands of capitalists, it would mean the complete proletarisation of millions of peasants, it would mean an enormous increase in the non-agricultural population and an increase in the number of factories (for the productivity of labour in our agriculture to double, there must be an enormous development of the machine-building industry, the mining industry, steam transport, the construction of a mass of new types of farm buildings, shops, warehouses, canals, etc., etc.). Mr. N.–on here repeats the little error of reasoning that is customary with him: he skips over the consecutive steps that are necessary with the development of capitalism, he skips over the intricate complex of social-economic changes which necessarily accompany the development of capitalism, and then weeps and wails over the danger of "destruction" by capitalism.

THE EMPLOYMENT OF MACHINERY IN AGRICULTURE

The post-Reform epoch is divided into four periods as regards the development of agricultural machinery production and the employment of machinery in agriculture. The first period covers the years immediately preceding the peasant Reform and the years immediately following it. The landlords at first rushed to purchase foreign machinery so as to get along without the "unpaid" labour of the serfs and to avoid the difficulties connected with the hiring of free workers. This attempt ended, of course, in failure; the fever soon died down, and beginning with 1863-1864 the demand for foreign machinery dropped.

The end of the 70s saw the beginning of the second period, which continued until 1885. It was marked by an extremely steady and extremely rapid increase in machinery imports from abroad; home production also grew steadily, but more slowly than imports. From 1881 to 1884 there was a particularly rapid increase in imports of agricultural machinery, due partly to the abolition, in 1881, of the duty-free import of pig-iron and cast-iron for the needs of factories producing agricultural machinery. The third period extended from 1885 to the beginning of the 90s. Agricultural machinery, hitherto imported duty-free, now had an import duty imposed (of 50 kopeks gold per pood). The high duty caused an enormous drop in machinery imports, while home production developed slowly owing to the agricultural crisis which set in at that time. Finally, the beginning of the 90s evidently saw the opening of a fourth period, marked by a fresh rise in the import of agricultural machinery, and by a particularly rapid increase of its home production.

There are, unfortunately, no such complete and precise data on the production of agricultural machinery and implements in Russia. The unsatisfactory state of our factory and-works statistics, the confusing of the production of machinery in general with the production of specifically agricultural machinery, and the absence of any firmly established rules for distinguishing between "factory" and "handicraft" production of agricultural machinery – all this prevents a complete picture of the development of agricultural machinery production in Russia being obtained. These data show the vigorousness of the process in which primitive agricultural implements are giving way to improved ones (and, consequently, primitive forms of farming to capitalism).

In 18 years the employment of agricultural machinery increased more than 3.5-fold, and this was mainly because of the expansion of home production, which more than quadrupled. Noteworthy, too, was the shifting of the main centre of such production from the Vistula and Baltic gubernias to the south-Russian steppe gubernias. Whereas in the 70s the main centre of agricultural capitalism in Russia was the western outer gubernias, in the 1890s still more outstanding areas of agricultural capitalism were created in the purely Russian gubernias.

It is necessary to add, regarding the data just cited, that although they are based on official (and, as far as we know, the only) information on the subject under examination, they are far from complete and are not fully comparable for the different years. For the years 1876-1879 returns are available that were specially compiled for the 1882 exhibition; they are the most comprehensive, covering not only "factory" but also "handicraft" production of agricultural implements; it was estimated that in 1876-1879 there were, on the average, 340 establishments in European Russia and the Kingdom of Poland, whereas according to "factory" statistical data there were in 1879 not more than 66 factories in European Russia producing agricultural machinery and implements.

The enormous difference in these figures is explained by the fact that of the 340 establishments less than one-third were counted as possessing steam power, and more than half as being operated by hand labour; 236 establishments of the 340 had no foundries of their own and had their castings made outside. The data for 1890 and 1894, on the other hand, are from *Collections of Data on Factory Industry in Russia*. These data do not fully cover even the "factory" production of agricultural machinery and implements; for example, in 1890, according to the *Collection*, there were in European Russia 149 works engaged in this industry, whereas Orlov's *Directory* mentions more than 163 works producing agricultural machinery and implements; in 1894, according to the first-mentioned returns, there were in European Russia 164 works of this kind, but according to the *List of Factories and Works* there were in 1894-95 over 173 factories producing agricultural machinery and implements.

As for the small scale, "handicraft" production of agricultural machinery and implements, this is not included in these data at all. That is why there can be no doubt that the data for 1890 and 1894 greatly understate the actual facts; this is confirmed by the opinion of experts, who considered that in the beginning of the 1890s agricultural machinery and implements were manufactured in Russia to a sum of about 10 million rubles, and in 1895 to a sum of nearly 20 million rubles.

Let us quote somewhat more detailed data on the types and quantity of agricultural machinery and implements manufactured in Russia. It is considered that in 1876 there were produced 25,835 implements; in 1877 – 29,590; in 1878 – 35,226; in 1879 – 47,892 agricultural machines and implements. How far these figures are exceeded at the present time may be seen from the following: in 1879 about 14,500 iron ploughs were manufactured, and in 1894 – 75,500. "Whereas five years ago the problem of the measures to be taken to bring about the wider use of iron ploughs on peasant farms was one awaiting solution, today it has solved itself. It is no longer a rarity for a peasant to buy an iron plough; it has become a common thing, and the number of iron ploughs now acquired by peasants every year runs into thousands."

The mass of primitive agricultural implements employed in Russia still leaves a wide field for the production and sale of iron ploughs. The progress made in the use of ploughs has even raised the issue of the employment of electricity. According to a report in the*Torgovo-Promyshlennaya Gazeta*, at the Second Congress of Electrical Engineers "considerable interest was aroused by a paper read by V. A. Rzhevsky on 'Electricity in Agriculture.'" The lecturer illustrated by means of some excellent drawings the tillage of fields in Germany with the aid of electric ploughs, and, from the plan and estimates he had drawn up at a landowner's request for his estate in one of 3 the southern gubernias, cited figures showing the economies to be effected by this method of tilling the land.

According to this plan, it was proposed to plough 540 dess. annually, and a part of this twice a year. The depth of furrow was to be from 4 1/2 to 5 vershoks. The soil was pure black earth. In addition to ploughs, the plan provided for machinery for other field-work, and also for a threshing machine and a mill, the latter of 25 h.p., calculated to operate 2,000 hours per annum. The cost of completely equipping the estate, including six versts of overhead cable of 50-mm. thickness, was estimated at 41,000 rubles. The cost of ploughing one dessiatine would be 7 rubles 40 kopeks if the mill were put up, and 8 rubles 70 kopeks with no mill. It was shown that at the local costs of labour, draught animals, etc., the use of electrical equipment would in the first case effect a saving of 1,013 rubles, while in the second case, less power being used without a mill, the saving would be 966 rubles.

No such sharp change is to be noted in the output of threshing and winnowing machines, because their production was relatively well established long ago. In fact, a special centre for the "handicraft" production of these machines was established in the town of Sapozhok, Ryazan Gubernia, and the surrounding villages, and the local members of the peasant bourgeoisie made plenty of money at this "industry". A particularly rapid expansion is observed in the production of reaping machines. In 1879, about 780 of these machines were produced; in 1893 it was estimated that 7,000 to 8,000 were sold a year, and in 1894-95 about 27,000. In 1895, for example, the works belonging to J. Greaves in the town of Berdyansk, Taurida Gubernia, "the largest works in Europe in this line of production" *i.e.*, in the production of reaping machines, turned out 4,464 reapers. Among the peasants in Taurida Gubernia reaping machines have become so widespread that a special occupation has arisen, namely, the mechanical reaping of other people's grain.

Similar data are available for other, less widespread, agricultural implements. Broadcast seeders, for example, are now being turned out at dozens of works, and the more perfect row drills, which were produced at only two works in 1893 (*Agriculture and Forestry*, 360), are now turned out at seven works, whose output has again a particularly wide sale in the south of Russia. Machinery is employed in all branches of agriculture and in all operations connected with the production of some kinds of produce: in special reviews reference is made to the extended use of winnowing machines, seed-sorters, seed-cleaners (trieurs), seed-driers, hay presses, flax-scutchers, etc.

In the *Addendum to the Report on Agriculture* for 1898, published by the Pskov Gubernia Zemstvo Administration, the in creasing use of machinery is noted, particularly of flax scutchers, in connection with the transition from flax production for home use to that for commercial purposes. There is an increase in the number of iron ploughs. Reference is made to the influence of migration in augmenting the number of agricultural machines and in raising wages. In Stavropol Gubernia, agricultural machinery is being employed on an increasing scale in connection with the growing immigration into this gubernia. In 1882, there were 908 machines: in 1891-1893, an average of 29,275; in 1894-1896, an average of 54,874; and in 1895, as many as 64,000 agricultural implements and machines.

The growing employment of machines naturally gives rise to a demand for engines: along with steam-engines, "oil engines have latterly begun to spread rapidly on our farms", and although the first engine of this type appeared abroad only seven years ago, there are already 7 factories in Russia manufacturing them. In Kherson Gubernia in the 70s only 134 steam-engines were registered in agriculture, and in 1881 about 500. In 1884-1886, in three uyezds of the gubernia (out of six), 435 steam threshing machines were registered. "At the present time (1895) there must be at least twice as many". The *Vestnik Finansov* states that in Kherson Gubernia, "there are about 1,150 steam-threshers, and in the Kuban Region the number is about the same, etc.... Latterly the acquisition of steam-threshers has assumed an industrial character.... There have been cases of a five thousand-ruble threshing machine with steam-engine fully covering its cost in two or three good harvest years, and of the owner immediately getting another on the same terms. Thus, 5 and even 10 such machines are often to be met with on small farms in the Kuban Region. There they have become an essential accessory of every farm that is at all well organised." "Generally speaking, in the south of Russia today, more than ten thousand steam-engines are in use for agricultural purposes".

If we remember that the number of steam-engines in use in agriculture throughout European Russia in 1875 1878 was only 1,351 and that in 1901, according to incomplete returns, the number was 12,091, in 1902 – 14,609, in 1903 – 16,021 and in 1904 – 17,287, the gigantic revolution brought about by capitalism in agriculture in this country during the last two or three decades will be clear to us. Great service in accelerating this process has been rendered by the Zemstvos. By the beginning of 1897, Zemstvo agricultural machinery and implement depots "existed under the auspices of 11 gubernia and 203 uyezd Zemstvo boards, with a total working capital of about a million rubles". In Poltava Gubernia, the turnover of the Zemstvo depots increased from 22,600 rubles in 1890 to 94,900 rubles in 1892 and 210,100 rubles in 1895. In the six years, 12,600 iron ploughs, 500 winnowing machines and seed-sorters, 300 reaping machines, and 200 horse-threshers were sold. "The principal buyers of implements at the Zemstvo depots are Cossacks and peasants; they account for 70 per cent of the total number of iron ploughs and horse-threshers sold. The purchasers of seeding and reaping machines were mainly landowners, and large ones at that, possessing over 100 dessiatines".

LAND DEVELOPMENT MACHINERY

TRACTOR DRAWN LEVELLER

The leveler consists of frame, 3-point linkage, cutting or scraping blade, and thick curved sheet closed from sides to form a bucket. The scraping blade is made from medium carbon steel or low alloy steel, hardened and tempered to about 42 HRC. The blade is joined to the curved sheet with fasteners and can be replaced after being worn out or becoming dull. The working depth of the implement is controlled by hydraulic system of the tractor.

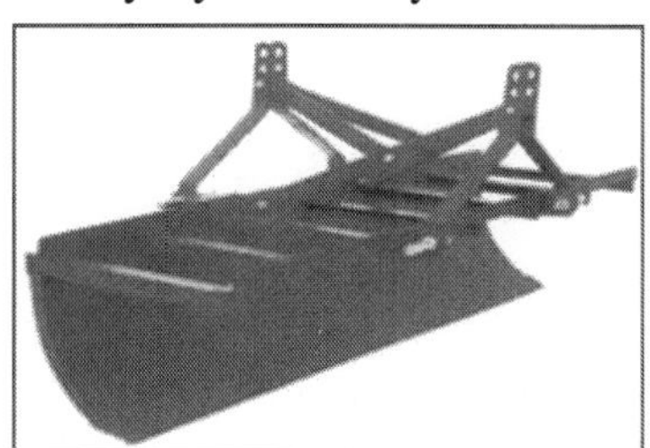

Features

- Specifications:
- Length(mm) : 1840
- Width (mm) : 700
- Height (mm) : 700

- Size of Blade (mm)
  - Length
  - Width
  - Thickness
- : 1830
- : 75
- ; 8
- Weight (kg)
- Power requirement (hp/kW): 90
- : 35/26.25, tractor

Uses: Leveling of fields and pulling or pushing loosened soil from one place to other.

LASER GUIDED LAND LEVELLER

Features

The laser land leveler consists of a laser transmitter, a laser receiver, an electrical control panel, a twin solenoid hydraulic control valve, two wheels and a leveling bucket. The laser transmitter transmits a laser beam, which is intercepted by the laser receiver mounted on the leveling bucket. The control panel mounted on the tractor interprets the signal from the receiver and opens or closes the hydraulic control valve, which raises or lowers the bucket. Some laser transmitters have the ability to operate over graded slopes ranging from 0.01 per cent to 15 per cent and apply dual controlled slope in the field. The leveling bucket can be either 3-point linkage mounted or pulled by the tractor's drawbar. Bucket dimensions, number of wheels and capacity will vary according to the available power source and field conditions.

Specifications:

Laser Source	: < 5mw 635nm
Operating diameter(m)	: (Above) 800
Grade range (per cent)	: –10 to +15 Dual Axes
Grade accuracy (per cent)	: 0.015, 3 mm@30 m
Remote control type	: Full 2-way communication
Power requirement (hp/KW): 60/45	

Uses:

It is used for precision leveling of land in one or both the directions.

Approx Cost: ₹ 3,50,000

SUB-SOILER

Features: It consists of beam made of high carbon steel, beam supports which are flanged at upper and lower edges for rigidity, hollow steel adaptor welded to the bottom end of the beam to accommodates share base having square section, share plate made from high carbon steel and shank drilled and counter bored for set board which secures the base in the adaptor. Share plate is made from high carbon steel, hardened and tempered to suitable hardness. Two symmetrically located bolt holes allow reversibility of share. The working depth of the sub-soiler is controlled by hydraulic system and linkage of tractor.

Specifications:

Length (mm)	: 600
Width (mm)	: 490
Height (mm)	: 1325
Maximum working depth (mm)	: 535
Weight (kg)	: 62
Power requirement(hp/kW)	: 55/41.25

Uses: It is used to break hardpan of the soil, loosening of the soil and helps the water to seep into the soil for improving drainage. A mole ball can be attached to create a small tunnel in the soil, which serves as drainage channel for water.

REVOLVING HAYRAKE EQUIPMENT

About the first contrivance for raking hay by horse power consisted of a stick eight or ten feet long with double-end teeth running through it, and pointing in two directions.

These rakes were improved from time to time, until they reached perfection for this kind of tool. They have since been superseded by spring-tooth horse rakes, except for certain purposes. For pulling field peas, and some kinds of beans, the old style revolving horse rake is still in use.

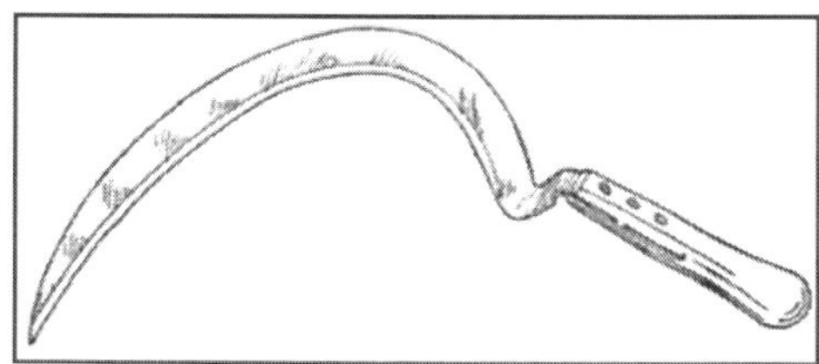

Fig. Grass Hook, for working around borders where the lawn-mower is too clumsy.

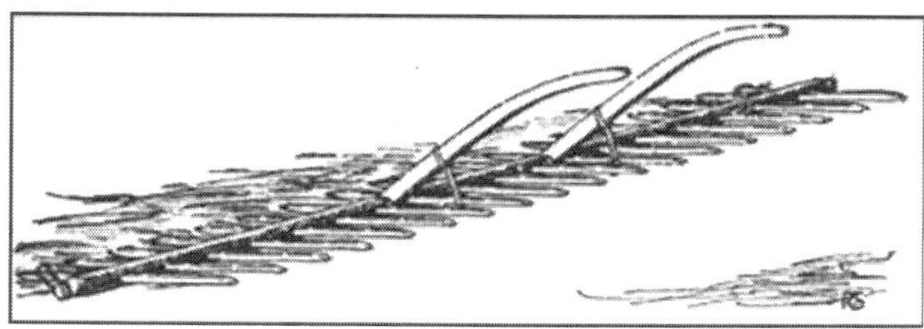

Fig. Revolving Hayrake. The center piece is 4″ x 6″ x 12′ long. The teeth are double enders 1^3D_83 square and 4′ 6″ long, which allows 24″ of rake tooth clear of the center timber. Every stick in the rake is carefully selected. It is drawn by one horse. If the center teeth stick into the ground either the horse must stop instantly, or the rake must flop over, or there will be a repair job. This invention has never been improved upon for pulling Canada peas.

Improved revolving horse rakes have a center timber of hardwood about 4 x 6 inches in diameter. The corners are rounded to facilitate sliding over the ground. A rake twelve feet long will have about eighteen double-end teeth. The teeth project about two and one-half feet each way from the center timber. Each tooth is rounded up, sled-runner fashion, at each end so it will point forward and slide along over and close to the ground without catching fast.

There is an iron pull rod, or long hook, attached to each end of the center bar by means of a bolt that screws into the center of the end of the wooden center shaft, thus forming a gudgeon pin so the shaft can revolve. Two handles are fastened by band iron straps to rounded recesses or girdles cut around the center bar.

These girdles are just far enough apart for a man to walk between and to operate the handles. Wooden, or iron lugs, reach down from the handles with pins projecting from their sides to engage the rake teeth. Two pins project from the left lug and three from the right. Sometimes notches are made in the lugs instead of pins.

Notches are better; they may be rounded up to prevent catching when the rake revolves. As the rake slides along, the driver holds the rake teeth in the proper position by means of the handles. When sufficient load has been gathered he engages the upper notch in the right hand lug, releases the left and raises the other sufficient to point the teeth into the ground.

The pull of the horse turns the rake over and the man grasps the teeth again with the handle lugs as before. Unless the driver is careful the teeth may stick in the ground and turn over before he is ready for it. It requires a little experience to use such a rake to advantage. No better or cheaper way has ever been invented for harvesting Canada peas. The only objections are that it shells some of the riper pods and it gathers up a certain amount of earth with the vines which makes dusty threshing.

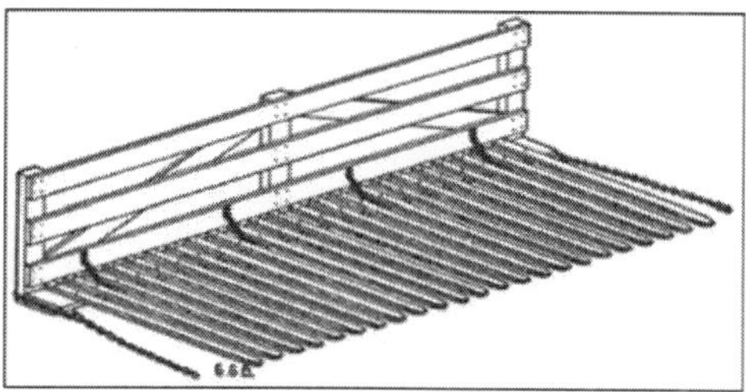

Fig. Buck Rake. When hay is stacked in the field a four-horse buck rake is the quickest way to bring the hay to the stack. The buck rake shown is 16 feet wide and the 2 x 4 teeth are 11 feet long. Two horses are hitched to each end and two drivers stand on the ends of the buck rake to operate it. The load is pushed under the horse fork, the horses are swung outward and the buck rake is dragged backward.

HAY-TEDDER

The hay-tedder is an English invention, which has been adopted by farmers in rainy sections of the United States. It is an energetic kicker that scatters the hay swaths and drops the hay loosely to dry between showers. Hay may be made quickly by starting the tedder an hour behind the mowing machine.

It is quite possible to cut timothy hay in the morning and put it in the mow in the afternoon, by shaking it up thoroughly once or twice with the hay-tedder. When clover is mixed with the timothy, it is necessary to leave it in the field until the next day, but the time between cutting and mowing is shortened materially by the use of the tedder.

Grass cut for hay may be kicked apart in the field early during the wilting process without shattering the leaves. If left too long, then the hay-tedder is a damage because it kicks the leaves loose from the stems and the most valuable feeding material is wasted. But it is a good implement if rightly used. In catchy weather it often means the difference between bright, valuable hay and black, musty stuff, that is hardly fit to feed.

Hay-tedders are expensive. Where two farmers neighbour together the expense may be shared, because the tedder does its work in two or three hours' time. Careful farmers do not cut down much grass at one time. The tedder scatters two mowing swaths at once.

In fact the mowing machine, hay-tedder and horserake should all fit together for team work so they will follow each other without skips or unnecessary laps. The dividing board of the mowing-machine marks a path for one of the horses to follow and it is difficult to keep him out of it. But two horses pulling a hay-tedder will straddle the open strip between the swaths when the tedder is twice the width of the cut.

HAY SKIDS

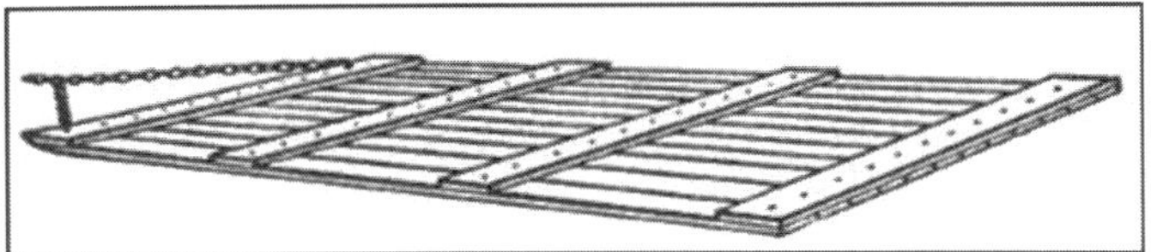

Fig. Hay Skid. This hay skid is 8 feet wide and 16 feet long. It is made of $^7/_8$″ lumber put together with 2″ carriage bolts—plenty of them. The round boltheads are countersunk into the bottom of the skid and the nuts are drawn down tight on the cleats. It makes a low-down, easy-pitching, hay-hauling device.

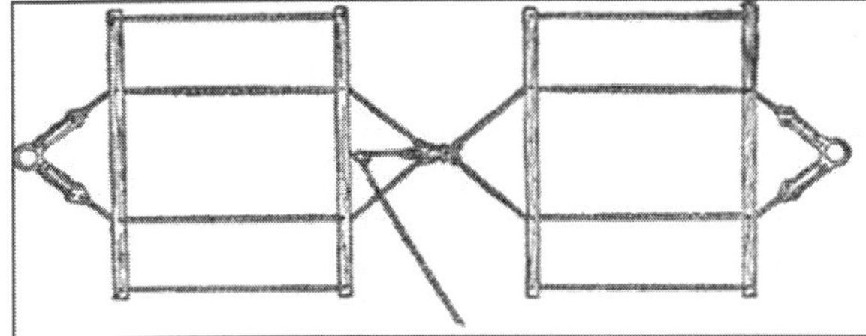

Fig. Hay Sling. It takes no longer to hoist 500 pounds of hay than 100 pounds if the rig is large and strong enough. Four feet wide by ten feet in length is about right for handling hay quickly. But the toggle must reach to the ends of the rack if used on a wagon.

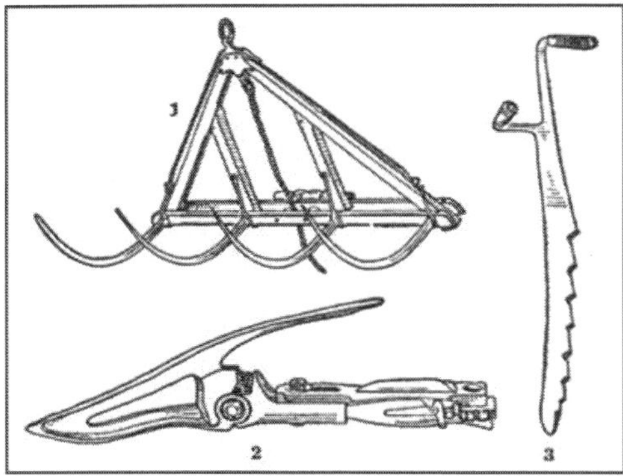

Fig. (1) Four-Tined Derrick Fork. (2) Pea Guard. An extension guard to lift pea-vines high enough for the sickle is the cleanest way to harvest Canada peas. The old-fashioned way of pulling peas with a dull scythe has gone into oblivion. But the heavy bearing varieties still persist in crawling on the ground. If the vines are lifted and cut clean they can be raked into windrows with a spring tooth hayrake. (3) Haystack Knife. This style of hay-cutting knife is used almost universally on stacks and in hay-mows. There is less use for hay-knives since farmers adopted power hayforks to lift hay out of a mow as well as to put it in.

Hay slips, or hay skids, are used on the old smooth fields in the eastern states. They are usually made of seven-eighths-inch boards dressed preferably on one side only. They are used smooth side to the ground to slip along easily. Rough side is up to better hold the hay from slipping. The long runner boards are held together by cross pieces made of inch boards twelve inches wide and well nailed at each intersection with nails well clinched. Small carriage bolts are better than nails but the heads should be countersunk into the bottom with the points up. They should be used without washers and the ends of the bolts cut close to the sunken nuts.

The front end of the skid is rounded up slightly, sled runner fashion, as much as the boards will bear, to avoid digging into the sod to destroy either the grass roots or crowns of the plants. Hay usually is forked by hand from the windrows on to the skids.

Sometimes hay slings are placed on the skids and the hay is forked on to the slings carefully in layers lapped over each other in such a way as to hoist on to the stack without spilling out at the sides. Four hundred to eight hundred pounds makes a good load for one of these skids, according to horse power and unevenness of the ground. They save labour, as compared to wagons, because there is no pitching up. All hoisting is supposed to be done by horse power with the aid of a hay derrick.

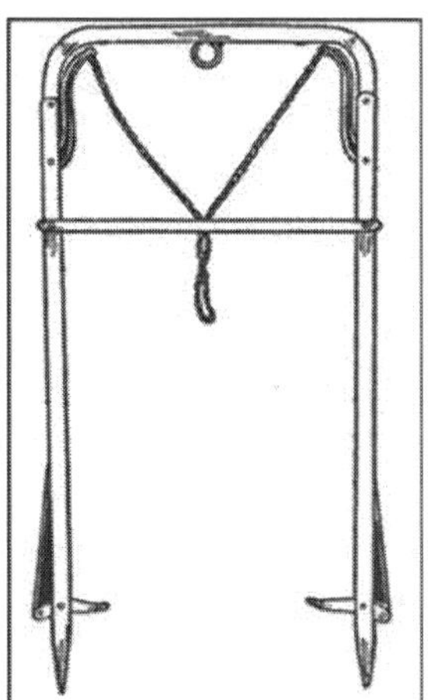

Fig. Double Harpoon Hayfork. This is a large size fork with extra long legs. For handling long hay that hangs together well this fork is a great success. It may be handled as quickly as a smaller fork and it carries a heavy load.

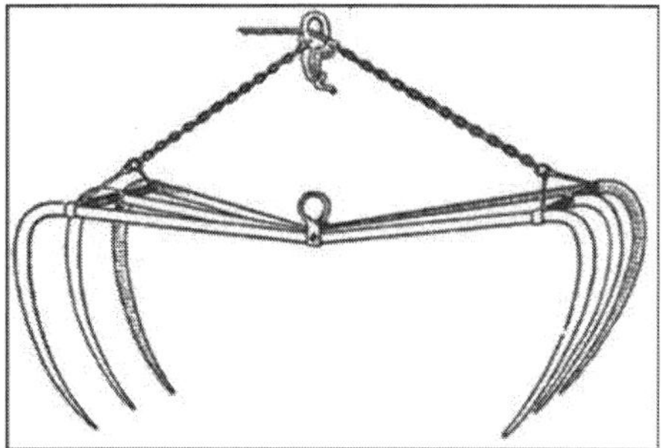

Fig. Six-Tined Grapple Hayfork. It is balanced to hang as shown in the drawing when empty. It sinks into the hay easily and dumps quickly when the clutch is released.

WESTERN HAY DERRICKS

Two derricks for stacking hay, that are used extensively in the alfalfa districts of Idaho. The derrick to the left is made with a square base of timbers which supports an upright mast and a horizontal boom. The timber base is sixteen feet square, made of five sticks of timber, each piece being 8 x 8 inches square by 16 feet in length.

Two of the timbers rest flat on the ground and are rounded up at the ends to facilitate moving the derrick across the stubble ground or along the road to the next hayfield. These sleigh runner timbers are notched on the upper side near each end and at the middle to receive the three cross timbers. The cross timbers also are notched or recessed about a half inch deep to make a sort of double mortise.

The timbers are bound together at the intersections by iron U-clamps that pass around both timbers and fasten through a flat iron plate on top of the upper timbers. These flat plates or bars have holes near the ends and the threaded ends of the U-irons pass through these holes and the nuts are screwed down tight. The sleigh runner timbers are recessed diagonally across the bottom to fit the round U-irons which are let into the bottoms of the timbers just enough to prevent scraping the earth when the derrick is being moved. These iron U-clamp fasteners are much stronger and better than bolts through the timbers.

There are timber braces fitted across the corners which are bolted through the outside timbers to brace the frame against a diamond tendency when moving the derrick. There is considerable strain when passing over uneven ground. It is better to make the frame so solid that it cannot get out of square. The mast is a stick of timber 8 inches square and 20 or 24 feet long.

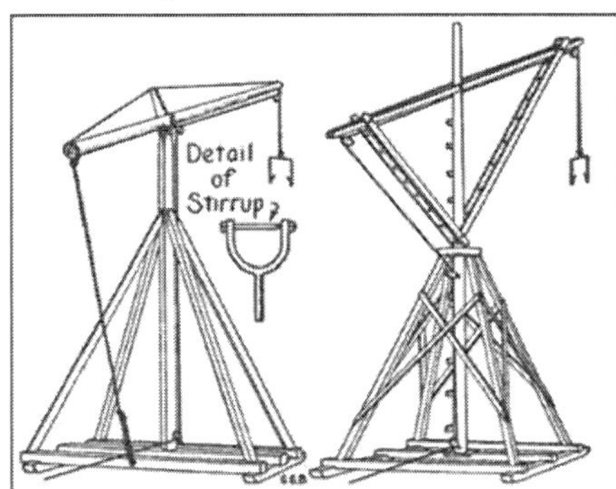

Fig. Idaho Hay Derricks. Two styles of hay derricks are used to stack alfalfa hay in Idaho. The drawing to the left shows the one most in use because it is easier made and easier to move. The derrick to the right usually is made larger and more powerful. Wire cable is generally used with both derricks because rope wears out quickly. They are similar in operation but different in construction. The base of each is 16 feet square and the high ends of the booms reach up nearly 40 feet. A single hayfork rope, or wire cable, is used; it is about 65 feet long. The reach is sufficient to drop the hay in the center of a stack 24 feet wide.

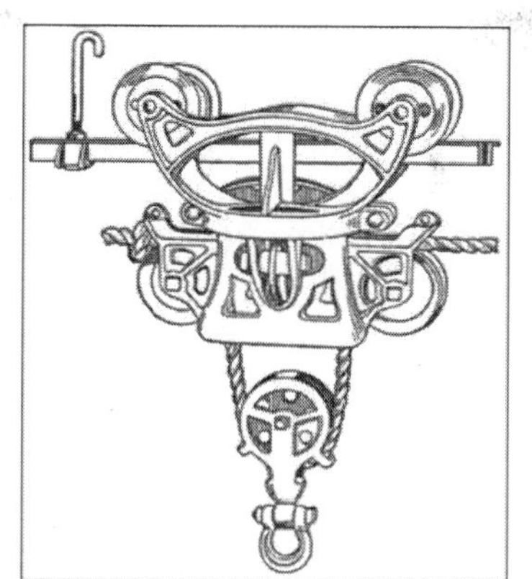

Fig. Hay Carrier Carriage. Powerful carriers are part of the new barn. The track is double and the wheels run on both tracks to stand a side pull and to start quickly and run steadily when the clutch is released.

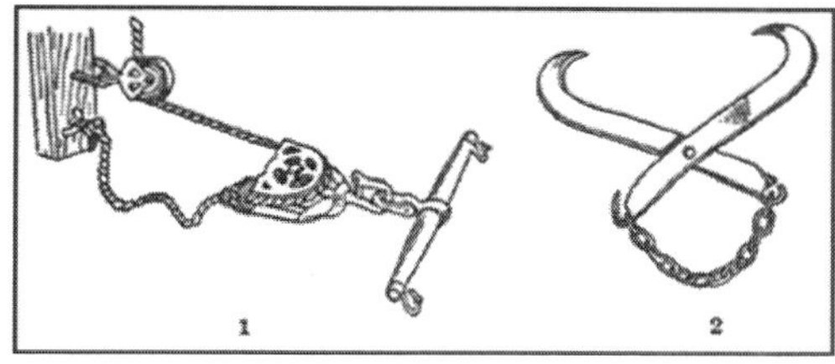

Fig. (1) Hayfork Hitch. A whiffletree pulley doubles the speed of the fork. The knot in the rope gives double power to start the load. (2) Rafter Grapple, for attaching an extra pulley to any part of the barn roof.

This mast is securely fastened solid to the center of the frame by having the bottom end mortised into the center cross timber at the middle and it is braced solid and held perpendicular to the framework by 43 x 43 wooden braces at the corners. These braces are notched at the top ends to fit the corners of the mast and are beveled at the bottom ends to fit flat on top of the timbers. They are held in place by bolts and by strap iron or band iron bands.

These bands are drilled with holes and are spiked through into the timbers with four-inch or five-inch wire nails. Holes are drilled through the band iron the right size and at the proper places for the nails. The mast is made round at the top and is fitted with a heavy welded iron ring or band to prevent splitting.

The boom is usually about 30 feet long. Farmers prefer a round pole when they can get it. It is attached to the top of the mast by an iron stirrup made by a blacksmith. This stirrup is made to fit loosely half way around the boom one-third of the way up from the big end, which makes the small end of the boom project 20 feet out from the upper end of the mast. The iron stirrup is made heavy and strong. It has a round iron gudgeon $1^{1}D_{2}3$ in diameter that reaches down into the top of the mast about 18 inches. The shoulder of the stirrup is supported by a square, flat iron plate which rests on and covers the top of the mast and has the corners turned down. It is made large to shed water and protect the top of the mast. This plate has a hole one and a half inches in diameter in the center through which the stirrup gudgeon passes as it enters the top of the mast. A farm chain, or logging chain, is fastened to the large end of the boom by passing the chain around the boom and engaging the round hook. The grab hook end of the chain is passed around the timber below and is hooked back to give it the right length, which doubles the part of the chain within reach of the man in charge. This double end of the chain is lengthened or shortened to elevate the outer end of the boom to fit the stack. The small outer end of the boom is thus raised as the stack goes up.

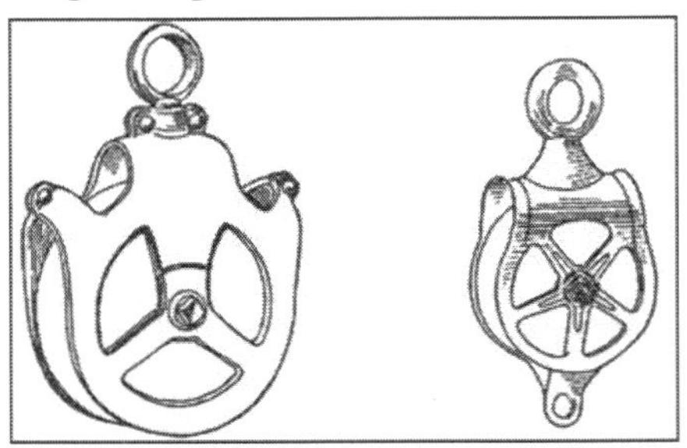

Fig. Hay Rope Pulleys. The housing of the pulley to the left prevents the rope from running off the sheaves.

An ordinary horse fork and tackle is used to hoist the hay. Three single pulleys are attached, one to the outer end of the boom, one near the top of the mast, and the other at the bottom of the mast so that the rope passes easily and freely through the three pulleys and at the same time permits the boom to swing around as the fork goes up from the wagon rack over the stack. This swinging movement is regulated by tilting the derrick towards the stack so that the boom swings over the stack by its own weight or by the weight of the hay on the horse fork. Usually a wire truss is rigged over the boom to stiffen it. The wire is attached to the boom at both ends and the middle of the wire is sprung up to rest on a bridge placed over the stirrup.

Fig. Gambrel Whiffletree, for use in hoisting hay to prevent entanglements. It is also handy when cultivating around fruit-trees.

Farmers like this simple form of hay derrick because it is cheaply made and it may be easily moved because it is not heavy. It is automatic and it is about as cheap as any good derrick and it is the most satisfactory for ordinary use. The base is large enough to make it solid and steady when in use. Before moving the point of the boom is lowered to a level position so that the derrick is not top-heavy. There is little danger of upsetting upon ordinary farm lands. Also the width of 16 feet will pass along country roads without meeting serious obstacles. Hay slings usually are made too narrow and too short. The ordinary little hay sling is prone to tip sideways and spill the hay. It is responsible for a great deal of profanity. The hay derrick shown to the right is somewhat different in construction, but is quite similar in action. The base is the same but the mast turns on a gudgeon stepped into an iron socket mortised into the center timber.

Fig. Cable Hay Stacker. The wire cable is supported by the two bipods and is secured at each end by snubbing stakes. Two single-cable collars are clamped to the cable to prevent the bipods from slipping in at the top. Two double-cable clamps hold the ends of the cables to form stake loops.

The wire hoisting cable is threaded differently, as shown in the drawing. This style of derrick is made larger, sometimes the peak reaches up 402 above the base. The extra large ones are awkward to move but they build fine big stacks.

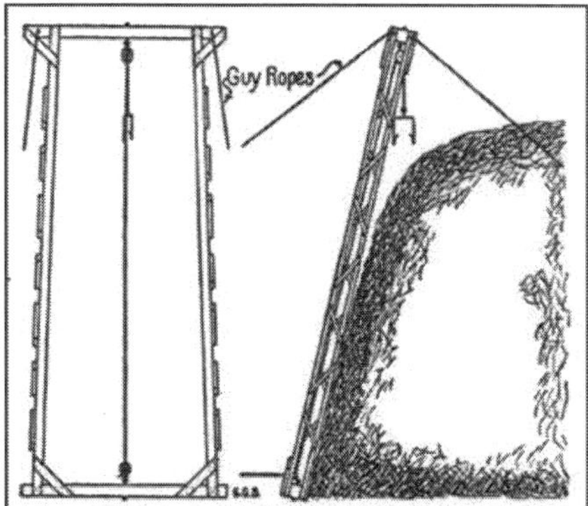

Fig. California Hay Ricker, for putting either wild hay or alfalfa quickly in ricks. It is used in connection with home-made buck rakes. This ricker works against the end of the rick and is backed away each time to start a new bench. The upright is made of light poles or 2 x 4s braced as shown. It should be 28 or 30 feet high. Iron stakes hold the bottom, while guy wires steady the top.

MECHANICAL COMMAND TO DRIVE MODERN FARM MACHINERY

At one time ninety-seven per cent of the population of the United States got their living directly from tilling the soil, and the power used was oxen and manual labour. At the present time probably not more than thirty-five per cent of our people are actively engaged in agricultural pursuits. And the power problem has been transferred to horses, steam, gasoline, kerosene and water power, with electricity as a power conveyor.

Fifty years ago a farmer was lucky if he owned a single moldboard cast-iron plow that he could follow all day on foot and turn over one, or at most, two acres. The new traction engines are so powerful that it is possible to plow sixty feet in width, and other machines have been invented to follow the tractor throughout the planting and growing seasons to the end of the harvest. The tractor is supplemented by numerous smaller powers. All of which combine to make it possible for one-third of the people to grow enough to feed the whole American family and to export a surplus to Europe.

At the same time, the standard of living is very much higher than it was when practically everyone worked in the fields to grow and to harvest the food necessary to live.

Farm machinery is expensive, but it is more expensive to do without. Farmers who make the most money are the ones who use the greatest power and the best machinery. Farmers who have a hard time of it are the ones who use the old wheezy hand pump, the eight-foot harrow and the walking plow. The few horses they keep are small and the work worries them. The owner sympathizes with his team and that worries him. Worry is the commonest form of insanity.

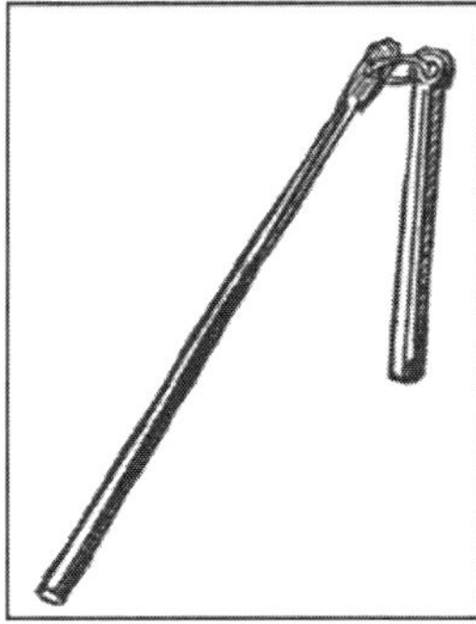

Fig. Flail, the oldest threshing machine, still used for threshing pedigreed seeds to prevent mixing. The staff is seven or eight feet long and the swiple is about three feet long by two and one-half inches thick in the middle, tapering to one and one-half inches at the ends. The staff and swiple are fastened together by rawhide thongs.

Fig. Bucket Yoke. It fits around the neck and over the shoulders. Such human yokes have been used for ages to carry two buckets of water, milk or other liquids. The buckets or pails should nearly balance each other. They are steadied by hand to prevent slopping.

At a famous plowing match held at Wheatland, Illinois, two interesting facts were brought out. Boys are not competing for furrow prizes and the walking plow has gone out of fashion. The plowing at the Wheatland plowing match was done by men with riding plows. Only one boy under eighteen years was ready to measure his ability against competition. The attendance of farmers and visitors numbered about three thousand, which shows that general interest in the old-fashioned plowing match is as keen as ever. A jumbo tractor on the grounds proved its ability to draw a big crowd and eighteen plows at the same time.

It did its work well and without vulgar ostentation. Lack of sufficient land to keep it busy was the tractor's only disappointment, but it reached out a strong right arm and harrowed the furrows down fine, just to show that it "wasn't mad at nobody."

Fig. Well Sweep. The length of the sweep is sufficient to lower the bucket into the water and to raise it to the coping at the top of the brickwork. The rock on the short end of the sweep is just heavy enough to balance the bucket full of water.

Modern farm methods are continually demanding more power. Larger implements are being used and heavier horses are required to pull them. A great deal of farm work is done by engine power. Farm power is profitable when it is employed to its full capacity in manufacturing high-priced products. It may be profitable also in preventing waste by working up cheap materials into valuable by-products. The modern, well-managed farm is a factory and it should be managed along progressive factory methods. In a good dairy stable hay, straw, grains and other feeds are manufactured into high-priced cream and butter.

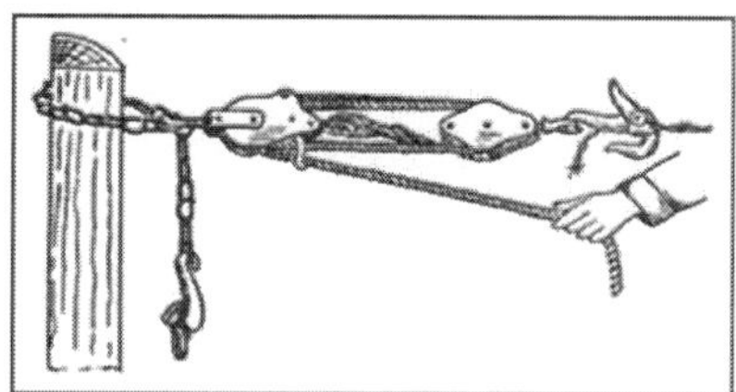

Fig. Wire Stretcher. A small block and tackle will stretch a single barb-wire tight enough for a fence. By using two wire snatches the ends of two wires may be strained together for splicing.

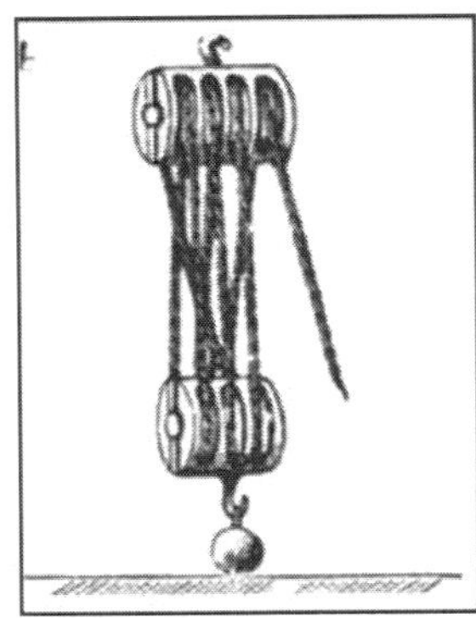

Fig. Block and Tackle. The rope is threaded into two double blocks. There is a safety stop that holds the load at any height.

Farming pays in proportion to the amount of work intelligently applied to this manner of increasing values. It is difficult to make a profit growing and selling grain. Grain may sell for more than the labour and seed, but it takes so much vitality from the land that depreciation of capital often is greater than the margin of apparent profit. When grains are grown and fed to live-stock on the farm, business methods demand better buildings and more power, which means that the farmer is employing auxiliary machinery and other modern methods to enhance values.

In other manufacturing establishments raw material is worked over into commercial products which bring several times the amount of money paid for the raw material.

Fig. Farm Hoists. Two styles of farm elevating hoists are shown in this illustration. Two very different lifting jobs are also shown.

The principle is the same on the farm except that when a farmer raises the raw material he sells it to himself at a profit. When he feeds it to live-stock and sells the live-stock he makes another profit. When the manure is properly handled and returned to the soil he is making another profit on a by-product.

Farming carried on in this way is a complicated business which requires superior knowledge of business methods and principles. In order to conduct the business of farming profitably the labour problem has to be met. Good farm help is expensive. Poor farm help is more expensive. While farm machinery also is expensive, it is cheaper than hand labour when the farmer has sufficient work to justify the outlay. It is tiresome to have agricultural writers ding at us about the superior acre returns of German farms. German hand-made returns may be greater per acre, but one American farmhand, by the use of proper machinery, will produce more food than a whole German family.

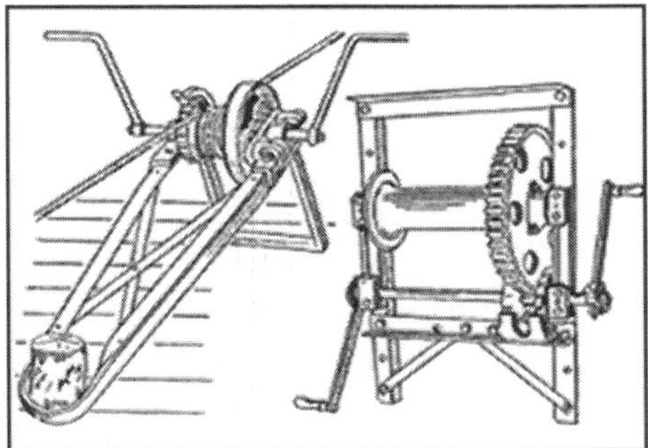

Fig. Two Powerful Winches. The one to the left is used for pulling small stumps or roots in the process of clearing land. The rope runs on and off the drum to maintain three or four laps or turns. The winch to the right is used for hoisting well drilling tools or to hang a beef animal. The rope winds on the drum in two layers if necessary.

7

Agri-Imports and Exports

INDIAN AGRO PRODUCTS

Agricultural sector is the mainstay of the rural Indian economy around, which the socio-economic privileges and deprivations revolve, and any change in its structure is expected to have a corresponding impact on the existing pattern of social equality. The growth of India's agriculture sector during the 50 years of independence remain impressive at 2.7 per cent per annum. About two-third of this production growth is aided by gains in crop productivity. The need based strategies adopted since independence and intensified after mid – sixties primarily focused on feeding the growing population and making the country self reliant in food production.

Indian agriculture has attained an impressive growth in the production of food grains that has increased around four times during the planned area of development from 51 million tons in 1950-51 to 199.1 million tonnes in 1997-98. The growth has been really striking since sixties after the production and wide spread usage of high yielding varieties of seed, fertilization, pesticides, especially in assured irrigated areas.

HISTORY

Over the 10,000 years since agriculture began to be developed, peoples across the world have discovered the food value of wild plants and animals and domesticated and bred them. Primary importance of these are cereals such as rice, wheat, barley, corn, and rye; sugarcane and sugar beets; meat animals such as sheep, cattle, goats, and pigs or swine; poultry such as chickens, ducks, and turkeys; and products like milk, cheese, eggs, nuts, and oils. Fruits, vegetables, and olives are also an important category of agriculture products; feed grains for animals include field corn, soybeans, and sorghum.

Modern agriculture in India primarily depends on engineering and technology and on the physical and biological sciences. Irrigation, drainage, conservation and sanitation, each of these stages are essential in successful farming, and require specialized knowledge and expert skills of agricultural engineers. Mechanization, the spectacular characteristic of late 19th and 20th-century agriculture, has eased much of the backbreaking toil of the farmers. More importantly, mechanization has considerably increased the farm efficiency and productivity.

OVERVIEW OF INDIAN AGRICULTURE MARKET HISTORY

In several agricultural sectors, India is the world's leading or one of the largest producers. For example, the country is second largest milk producing country in the world. The agricultural sector in the country is known for its high degree of product diversity. The complementary nature of a number of important Indian agricultural products, in comparison to those produced in west and other countries, provide India considerable export opportunities to these markets. At present, the Indian agriculture industry is on the brink of a revolution, which will modernize the entire food chain, as the total food production in the country is likely to double in the next ten years.

According to recent studies, the total turnover of Indian food market is approximately ₹.250000 crores (US $ 69.4 billion), out of which, the share of value-added food products is around ₹.80000 crores (US $ 22.2 billion). The Government of India has also sanctioned proposals for joint ventures, foreign collaborations, industrial licenses and 100 per cent export oriented units conceiving of an investment of ₹.19100 crores (US $ 4.80 billion) out of which foreign investment is over ₹ 9100 crores (US $ 18.2 Billion).

The Indian agricultural food industry also assumes significance owing to country's sizable agrarian economy that accounts for over 35 per cent of GDP and employs around 65 per cent of the population. Both in terms of number of joint- ventures/ foreign collaborations and foreign investment, the consumer food segment has the top priority. The other salient features of the Indian agro industry, which have the capacity to lure foreigners with assuring benefits are the aqua culture, deep sea fishing, milk and milk products, meat and poultry segments.

EXPORTS

Agricultural exports were 44 per cent of total exports in FY 1960; they decreased to 32 per cent in FY 1970, to 31 per cent in FY 1980, to 18.5 per cent in FY 1988, and to 15.3 per cent in FY 1993. This drop in share of agriculture in total exports was somewhat misleading because agricultural products, such as jute and cotton, which were exported in the raw form in the 1950s, have been exported as cotton yarn, fabrics, ready-made garments, coir yarn, and jute manufactures since the 1960s.

The composition of agricultural and allied products for export changed primarily due to the continuing increase of demand in the domestic market. This demand cut into the excess available for export in spite of a continuing desire, on the part of government, to shore up the invariant foreign-exchange shortage. In FY 1960, tea was the major export by value. Oil cakes, cashew kernels, tobacco, raw cotton and spices were about equal in value but were only one-eighth of the value of tea exports. By FY 1980, tea was still a major export commodity, however rice, coffee, fish, and fish products came close, followed by oil cakes, cashew kernels, and cotton. In 1992-93 fish and fish products became the main agricultural export, followed by oil meals, then cereals, and then tea. The share of fish products rose steadily from less than 2 per cent of all agricultural exports in FY 1960, to 10 per cent in FY 1980, to around 15 per cent for the 3-year period ending in FY 1990, and to 23 per cent in FY 1992. The contribution of tea in agricultural exports fell from 40 per cent in FY 1960 to around per cent per cent in the FY 1988-FY 1990 period, and to only 13 per cent by FY 1992.

Excellent export prospects, competitive pricing of agricultural products and standards, which are internationally comparable have created enormous trade opportunities in the Indian agro industry.

Exports of Agricultural products (2004-05)

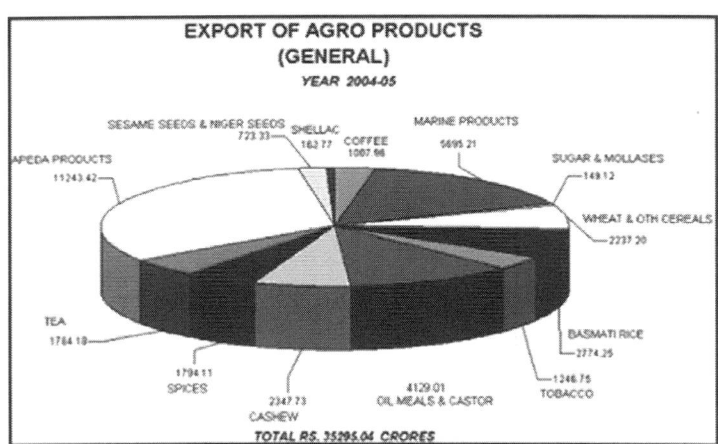

Major Export Markets

Major destinations for export of Indian agricultural products (2006-07), include -

Product	Major Markets
Floriculture	USA, Japan, UK, Netherlands and Germany
Fruits and Vegetable Seeds	Pakistan, Bangladesh, USA, Japan and Netherlands
Fresh Onions	Bangladesh, Malaysia, Sri Lanka, UAE, Pakistan and Nepal
Other Fresh Vegetables	UAE, Bangladesh, Pakistan, Nepal and Sri Lanka
Walnuts	Spain, Egypt, Germany, UK and Netherlands
Fresh Mangoes	UAE, Bangladesh, UK, Saudi Arabia and Nepal

Fresh Grapes	Netherlands, UK, UAE, Bangladesh, Belgium
Other Fresh Fruits	Bangladesh, UAE, Netherlands, Nepal, Saudi Arabia
Dried and Preserved Vegetables	Russia, France, USA, Germany and Spain
Mango Pulp	Saudi Arabia, Netherlands, UAE, Yamen, Arab Republic and Kuwait
Pickles and Chutneys	Russia, USA, Belgium, Netherlands and France
Other Processed Fruits	USA, Netherlands, UK, UAE and Saudi Arabia
Buffalo Meat	Malaysia, Philippines, Saudi Arabia, Jordan and Angola
Sheep/ Goat Meat	Saudi Arabia, UAE, Qatar, Oman and Kuwait
Poultry Products	UAE, Kuwait, Oman, Germany and Japan
Dairy Products	Bangladesh, Algeria, UAE, Yamen, Arab Republic and Egypt
Animal Casings	Germany, Portugal, France, Spain and Italy
Processed Meat	Seychelles, UAE, Hong Kong, Germany and USA
Groundnuts	Indonesia, Malaysia, Philippines, UK and Singapore
Guar Gum	USA, China, Germany, Italy and Netherlands
Jaggery and Confectionery	Portugal, USA, Bangladesh, Pakistan and Nepal
Cocoa Products	Nepal, Netherlands, Malaysia, Yamen Arab Republic and UAE
Cereal Preparations	USA, UK, Nepal, Sri Lanka and UAE
Alcoholic Beverages	Jamaica, Thailand, UAE, Angola and Bhutan
Miscellaneous Preparations	UAE, Iran, USA, UK and Indonesia
Milled Products	USA, UK, Indonesia, Maldives and UAE
Basmati Rice	Saudi Arabia, Kuwait, UK, UAE and Yamen Arab Rep.
Non Basmati Rice	Nigeria, Bangladesh, South Africa, UAE and Ivory Coast
Wheat	Bangladesh, Philippines, UAE, Sudan and Myanmar
Other Cereals	Bangladesh, Sri Lanka, Sudan, Benin, Thailand
Natural Honey	USA, Germany, Saudi Arabia, UK and UAE
Pulses	Bangladesh, Sri Lanka, Pakistan, UAE and Nepal

Product-wise Export Data (2002-03 to 2006-07)

FUTURE FORECASTS

According to experts, India has to play a bigger role in the global markets in agriculture products in the future. The country is expected to strengthen its position among the worlds leading exporters of rice. Presently it is the 2nd largest rice producer after China and the 3rd largest net-exporter after Thailand and Vietnam.

However, recent reports states that agriculture plays an important, though declining role in Indian economy. Its contribution in overall GDP fell from 30 per cent in the early nineties, to below 17.5 per cent in 2006. The country is a world leader in specialist products, such as buffalo milk, spices and bananas, mangoes, chickpeas etc., which are considered as important in the Indian diet and are also exported. India is the 5th largest cultivator of biotech crops across the world, ahead of China. In the year 2006, around 3.8 million hectares of land were cultivated with genetically modified crops, by about 2.3 million farmers. The primary GM crop is Bt Cotton that was introduced in 2002. The future growth in agriculture sector must come from -

- Advanced technologies that are not only "cost effective" but also "in conformity" with natural climatic regime of the country
- Technologies applicable to rain-fed areas particularly
- Continued genetic improvements for improved seeds and yields
- Improvements in data for superior research, results, and sustainable planning
- Bridging the gap between knowledge and practice; and
- Judicious land use resource surveys, effective management practices and sustainable use of natural resources.

IMPORT AND EXPORT

Since Independence, India has made a lot of progress in agriculture in terms of growth in output, yields and area under crops. It has gone through a Green Revolution (food grains), a White Revolution (milk), a Yellow Revolution (oilseeds) and a Blue Revolution (aquaculture). Today, India is one of the largest producers of milk, fruits, cashew nuts, coconuts and tea in the world. It is also well known for the production of wheat, vegetables, sugar, fish, tobacco and rice.

Certain types of agriculture such as horticulture, organic farming, floriculture, genetic engineering, packaging and food processing have the potential to see a surge in revenues through exports. Over the past few years, the government has stressed on the development of horticulture and floriculture by creating vital infrastructure for cold storage, refrigerated transportation, packaging, processing and quality control. If India wishes to optimize the production and export potential of these commodities, then it is essential to improve these facilities, marketing and export networks much further.

In recent years, the Central Government has offered different fiscal incentives for bettering storage facilities in rural areas. It also provides financial assistance to the State Governments for acquiring and distributing food grains at subsidized rates, especially to families with annual income below the poverty line. Today, the improved availability of bank credit through priority lending, favourable terms of trade and liberalized domestic and external trade for agricultural commodities have also encouraged private players to invest in agriculture.

The major thrust of the policies and programmes of the Government of India relating to livestock and fisheries is in the areas of rapid genetic upgradation of milch animals, improvement in the delivery mechanism of breeding inputs, control of animal diseases, creation of disease free zones, increased availability of nutritious feed, development of dairy activities and backyard poultry, development of processing and marketing facilities and enhancement of production and profitability of livestock.

Agricultural Exports

Agricultural exports have shown an increase from around ₹.60 billion in 1990 - 91 to ₹.398 billion in 2005-06. The Government's special efforts to encourage export of food grains in recent years through grant of World Trade Organization or WTO compatible subsidies has lead to India becoming one of the leading exporters of food grains in the international market

Agricultural Imports

The imports of agricultural products improved from ₹.12 billion in 1990 - 91 to ₹.220 billion in 2005- 06. The share of agri-imports to total merchandise imports in 2005-06 was 4.59 per cent. Edible oil is the single largest agricultural product imported into the country and accounts for around two-thirds of the total agricultural imports.

EXPORT PROCEDURE

A successful exporter should thoroughly research the markets. The fewer intermediaries a person has the better as every intermediary needs a percentage for his share in his business. This leads to fewer profits. The first thing you need to do is set up a business organization depending on your export needs. If you decide to incorporate a private limited company, then you have to register the same with the Registrar of Companies. There are two types of exporters - merchant exporters who buy goods from the market or from manufacturers and sell them to foreign buyers and a manufacturer exporter who manufactures the goods he exports.

The second thing you need to do is to open a current account in the name of the organization in whose name you intend to export, at a bank that is authorized to deal in foreign exchange. Carefully select the products you wish to export and study current export trends. Also, find out about the import regulations of the commodity in the importing countries. Send export letters to targeted companies with information about your company, product, pricing and other services offered. Negotiate with buyers for a good deal. Once an export order is received it must be processed and a contract that lists item specifications, payment conditions, marketing requirements, arbitration, shipping and insurance must be drawn out.

Exporters face risks such as credit risk, currency risk, carriage risk and country risk. These can be countered through steps like insisting upon an irrevocable letter of credit from the overseas buyer and taking out a marine insurance policy. All exporters have to register with a regional licensing authority that provides them with an Import Export Code (IEC) number. To get benefits and concessions under the export-import policy, exporters should register with an appropriate export promotion agency by obtaining a registration-cum-membership certificate. Goods that are exported are eligible for exemption from both Sales Tax and Central Sales Tax. For this purpose, you should get yourself registered with the Sales Tax Authority of your State.

Agricultural products have to go through quality control and pre-shipment inspections before export. Under the Export (Quality Control and Inspection) Act, 1963, about 1,000 commodities under the major groups of food and agriculture, fishery, minerals, organic and inorganic chemicals, rubber products, jute products and coir are subject to compulsory pre-shipment inspection. However, products that have an ISI Certification Mark or Agmark do not need to be inspected by any agency. All goods should be labelled, packaged, packed and marked before export.

IMPORT PROCEDURE

Imports to India are governed by the Foreign Trade (Development and Regulation) Act 1992. Under this Act, imports of all goods are free except for the items regulated by the policy or any other law in force. The present, foreign trade arrangements for different commodities are stated in the EXIM Policy of 2004-2009. This policy is announced once every five years with annual supplements coming out every year. It is also known as the Foreign Trade Policy or Export Import Policy.

Items on the 'Prohibited' list like tallow, fat or oils of any animal origin, animal rennet and wild animals including their parts and products and ivory cannot be imported. For import of items that appear in the 'Restricted' list you need secure an import licence. Import of items that are enumerated in the canalised list of items are permitted to be imported through canalising Agencies. All other products can be freely imported.

Registration with a regional licensing authority is a precondition for the import of goods. Customs officials will not permit clearance of goods unless the importer gets an Import Export Code (IEC) number from the regional licensing authority.

Import Procedure for Livestock Products

Livestock products include meat and meat products of different types that comprise fresh, chilled and frozen meat as well as tissue or organs of poultry, pig, sheep and goat. It also consists of egg and egg powder; milk and milk goods; pet foods of animal origin and embryos, ova or semen of cows, sheep and goats. No livestock product may be imported into India without a valid sanitary import permit. All livestock products with valid sanitary import permits may be brought into India only through seaports or airports where Animal Quarantine and Certification Services Stations are situated. These stations are located in the cities of Delhi, Mumbai, Kolkata and Chennai. When livestock products arrive at the checkpoint, they will be checked by the Officer-in-charge of the Animal Quarantine and Certification Services Station or any other veterinary officer duly approved by the Department Of Animal Husbandry and Dairying.

Import Procedure for the Fisheries Sector

License under EXIM policy is not required for the import of 125 species/ groups of fish, crustaceans, molluscs and other aquatic invertebrates covered under FREE policy under the EXIM policy. Import of five groups of live fish is permitted under Restricted Policy. Import of Whale Shark (Rhincodon types) and parts and products of the species is restricted.

Import Procedure for Horticulture - According to the Foreign Trade Policy, there is no quantitative restriction on import of spices into the country except for items such as 'seed quality' spices and garlic. The tariffs for import have also been steadily brought down. Under a bilateral agreement with Sri Lanka, duty free import of spices is permitted. This is useful for value addition and re-exportation.

NATIONAL PRICES

The government's price policy for agricultural products seeks to provide remunerative prices to farmers by encouraging them to invest and produce more for the agricultural sector. Minimum support prices for main agricultural products are declared every year. They are fixed after taking into account the recommendations of the Commission for Agricultural Costs and Prices (CACP).

The CACP was founded in January 1965 to advise the Government on price policy of major agricultural commodities. Assurance of a remunerative and stable price environment is considered very important for increasing agricultural production and productivity since the market place for agricultural produce tends to be inherently unstable, which often inflict undue losses on the growers, even when they adopt the best available technology package and produce efficiently.

CACP, while recommending prices takes into account important factors, such as:

- Cost of Production
- Changes in Input Prices
- Input/Output Price Parity
- Trends in Market Prices
- Inter-crop Price Parity
- Demand and Supply Situation
- Effect on Industrial Cost Structure
- Effect on General Price Level
- Effect on Cost of Living
- National Market Price Situation

- Parity between Prices Paid and Prices Received by farmers (Terms of Trade)

International Prices

The international rates of agricultural commodities constantly change due to various economic factors.

Prices of most agricultural commodities are highly volatile. The international prices of a majority of agricultural goods have been declining over the years. This is happening because of a situation of higher production and stagnant demand. Other than demand and supply, some of the other factors that affect international prices of agricultural commodities are -

- Growth of the world economy
- Conditions such as drought, floods and natural disasters.
- Value of the US dollar
- Government policies with regard to import tariffs, subsidies and other rules.
- Stability of governments
- Inflation or Depression
- Competitiveness of the product with regard to quality and price compared to those offered by other countries.

It is estimated that typically four per cent or less of commodities that are traded are actually delivered, since the basic purpose of a futures contract is to provide price-change protection.

Commodity Exchange

National Multi-Commodity Exchange of India Ltd. (NMCE)- External web site that opens in a new window is the first de-mutualised electronic multi-commodity exchange of India. It was granted national status on a permanent basis by the Government of India and started operations on 26th November 2002 with futures trading in 24 commodities. This basket of commodities has grown substantially since then to include -

- Oil Seeds
- Oil Cakes
- Oils
- Spices
- Pulses
- Metals
- Cash Crops

It is estimated that typically four per cent or less of commodities that are traded are actually delivered, since the basic purpose of a futures contract is to provide price-change protection. India's farmers and downstream industrial users of agricultural output are exposed to extremely high risks and the creation of commodities derivatives markets gives them the choice of hedging themselves against price fluctuations.

EXIM POLICY

EXIM Policy is the export import policy of the government that is announced every five years. It is also known as the Foreign Trade Policy. This policy consists of general provisions regarding exports and imports, promotional measures, duty exemption schemes, export promotion schemes, special economic zone programmes and other details for different sectors. Every year the government announces a supplement to this policy.

The EXIM Policy of 2002-2007 emphasized the importance of agricultural exports and announced measures like the setting up of agri export zones, removal of procedural restrictions and marketing cost assistance. Agri Export Zones are considered the most important creation of this policy -

Agri Export Zones

Agri Export Zones were formed as a result of this policy. These zones are meant to promote agricultural exports from the country and provide remunerative returns to the farming community regularly. They are to be identified by the State Government, which would evolve a comprehensive package of services to be provided by all State Government agencies, State Agriculture Universities and all institutions and agencies of the Union Government for intensive delivery in these zones. Corporate sector companies with proven credentials would be encouraged to sponsor new agri export zones or take over already notified agri export zones.

Services that would be managed and coordinated through this scheme include the provision of pre/post harvest operations, plant protection, processing, packaging, storage and related research and development. APEDA will supplement, within its schemes and provisions, the efforts of State Governments for facilitating exports. Click here for a list of the Agri Export Zones.

After, a change of government at the centre, a new EXIM Policy of 2004 - 2009 was announced. This policy came up with export promotional measures such as Towns of Export Excellence, Target Plus, Free Trade and Warehousing Zones and the Vishesh Krishi Upaj Yojana.

Here are details on these schemes:

a. Towns of Export Excellence - Here, towns in specific areas that produce goods of ₹.250 crores and above in the handloom, agriculture, handicraft and fisheries sector will be notified as Towns of Exports Excellence on the basis of their potential for growth in exports. They will be granted this recognition to maximize their potential, enable them to move higher in the value chain and tap new markets.

b. Target Plus - In this scheme, exporters who have attained a large increase in growth of exports would be allowed duty free credit based on incremental exports substantially higher than the general actual export target fixed. Rewards will be granted according to a tiered approach. For incremental growth of over 20, 25 and 100 per cent, the duty free credits would be 5, 10 and 15 per cent of Free on Board (FOB) value of incremental exports.
c. Vishesh Krishi Gram Udyog Yojana - It aims to promote exports of fruits, vegetables, flowers, fruits, and other value-added products. This year it has been expanded to include soybean and coconut oil as well as food preparations such as soups. Plus, the benefit of the scheme has been extended to 100 per cent export-oriented units.

The Annual Supplement of 2007 to the Foreign Trade Policy of 2004-2009(File referring to external site opens in a new window) - The government has exempted service tax on services rendered abroad and charged on exports from India. Similarly, service tax on services rendered in India but utilised by exporters would be exempted or remitted. For diversifying exports to tap unexplored markets, the focus market scheme has been expanded to include 16 new countries.

New items such as barley, oats, soybean, cigar/cheroots, bovine fats and copra have been included as focus products. To encourage agro-processing, status holders would be rewarded with duty credit scripts equal to 10 per cent of the value of agricultural exports, provided they use them for duty redemption on imports of cold storage and reefer vans.

MARKETING POLICY

The Agricultural Marketing Policy is governed by the Agricultural Produce (Grading and Marketing) Act of 1937. The Directorate of Marketing and Inspection of the Ministry of Agriculture is responsible for administering federal statutes concerned with the marketing of agricultural produce. Other Central Government organizations that are involved in agricultural marketing and promotion of exports are the Commission for Agricultural Costs and Prices, the Food Corporation of India, the Cotton Corporation of India and the Jute Corporation of India. There are also special marketing boards for rubber, coffee, tea, tobacco, spices, coconut, oilseeds, vegetable oil and horticulture. A network of cooperatives societies at the local, State and national levels and the National Agricultural Cooperative Marketing Federation of India are also involved in marketing and export promotion functions.

These boards have come up with various schemes to better marketing and boost exports. This includes the establishment of regulated markets, storage facilities, grading and standardizing agricultural products and providing standard measuring instruments.

Here is some information on major agricultural export marketing agencies in India:

a. The Agricultural and Processed Food Products Export Development Authority or APEDA. This organization was established in 1986 to develop the agricultural commodities and processed foods industry in India and to promote export. It spreads information about agri exports by organizing or participating in product promotions, seminars, buyer seller meets and international trade fairs.
b. Central Silk Board -is involved in creating publicity and promoting the export of natural silk goods.
c. The Marine Products Exports Development Authority - This organization offers subsidies to buy equipment to increase the production of fish at capture and culture facilities. This includes free installation of Turtle Excluder Devices in fishing trawlers and financial assistance for installation of other gear like fish finders, refrigerated fish holds and ice making machines.
d. The Federation of Indian Export Organizations - This is the apex body of different export promotion organizations.
e. NAFED - is the National Agricultural Cooperative Marketing Federation of India.

UREA IMPORT RULE EASED FOR FERTILIZER USE

The government last week allowed private companies to import urea for preparation of complex fertilizers used in agriculture. Till now, such companies could only import urea for industrial use, in preparation of chemicals.

For agricultural purposes, private companies used to source urea from imports made by government canalising agents, such as Indian Potash Ltd and state trading houses MMTCand STC. Two Indian companies, Coromondal Internationaland Zuari Industries, are allowed to import urea for agricultural purposes. Now, these companies can directly import without involving canalising agents, official sources said.

However, the permission is given with a rider that the urea cannot be sold directly in the market. The imported urea is to be used only to make the NPK complex fertilizer, which these companies can then sell.

Further, say the rules, strict monitoring will be done for usage of imported urea in manufacturing of the fertilizer. A company does not get a subsidy if its uses indigenously manufactured urea in preparation of complex fertilizers.

India produces about 22 million tonnes (mt) of urea in a year and consumes a little more than 30 mt. In 2010, the government had increased the retail price of urea by 10 per cent to ₹ 5,310 per tonne. This is still the current price.

Meanwhile, the ministry of chemicals and fertilizers has sought to exempt urea from the proposed Goods and Services Tax (GST), and to remove customs duty on import of plant and machinery for fertilizer projects. A ministry report on the duty structure on fertilizer and its inputs says the government distributes urea much below the cost of import or production. Taxes and duties are levied on the maximum retail price fixed by the government, which covers only 25-40 per cent of the cost. The inputs for urea production are, however, taxed at full cost, resulting in tax incidence on the inputs far in excess of that on the finished fertilizer.

Thus, under the proposed GST, the input tax credit will far exceed the tax payable on fertilizer, meaning the input credits will be far more than what could ever be availed on the outputs. This would block large amounts of input tax credit of fertilizer companies with the government on a recurring basis, even if there was provision of periodic cash refund. Currently, there are no provisions for refund of unadjusted credits in GST, except for refunds on exports.

Hence, goes the argument, the sector (meaning, urea) should be exempt from GST.

Also, it notes, a number of crucial inputs for urea manufacturing like natural gas, electricity generation and petroleum products are out of the GST ambit. Besides, the GST model seeks to exempt the food, health and educational sectors.

AGRI EXPORTS OUTPACE OTHER COMMODITIES'

Growth in India's agricultural exports has exceeded the rise in exports of other products. Through the past few years, these products have consistently seen a rise in their share in the export basket, primarily due to the huge stocks resulting from bumper output, as well as favourable government policies.

According to data from the commerce ministry, in 2010-11, agricultural exports stood at $17.35 billion, in 2011-12 $27.43 billion, in 2012-13 $31.86 billion and in the first 11 months of 2013-14, it stood at $29.3 billion.

During this four-year period, overall exports recorded 93 per cent growth. The share of agricultural commodities in India's overall export basked rose to 10.66 per cent in 2012-13 from 7.06 per cent in 2009-10.

"Agricultural exports growth will continue in the future, too, with improved prospects and favourable long-term policy support," Commerce Minister Anand Sharma had said on the sidelines of a Business Standard awards function here on Saturday.

Sharma said from now, basmati rice would incorporate the new Pusa 1121 variety. Various promotional efforts in major importing countries made the Pusa 1121 variety a preferred choice. Also, the Centre had signed free trade agreements with a number of agricultural product-deficient countries. The government also allowed exports of foodgrains and other agricultural commodities on a quota basis, with a positive response. Recently, India had allowed limited exports of pulses to Maldives. It has already been announced subsidy for sugar had resulted in a spurt in exports. Buffalo meat and guar gum are other major products seeing significant growth in the export basket. Though guar gum prices have fallen in the past year, its exports have risen in volume terms.

"The growth momentum in India's agricultural exports is expected to continue in the next few years, with an increased share of processed food, including mango pulp, dried and preserved vegetables, meat and poultry items. Factors such as reduced transaction costs, time, better port gate management and fiscal incentives contributed to this upward trend. With continued focus on issues such as food safety and compliance with international standards, we can surely reach new heights," said Piruz Khambatta, chairman and managing director, Rasna, and chairman, Confederation of Indian Industry's national committee on food processing.

According to the World Trade Organization, global export and import of agricultural and food products stands at $1.66 trillion and $1.82 trillion, respectively, of which India's shares are 2.07 per cent and 1.24 per cent', respectively. This indicates India is a net exporter of agricultural products. The country ranks 10th in terms of global agricultural and food exports.

In recent years, the government's policy impetus has provided stability to agricultural exports. Given sufficient stocks of foodgrains in the central pool, the government has allowed exports of wheat. Also, efforts have been taken to promote horticulture exports. "Though these measures are in the right direction, a consistent long-term trade policy, with tariff in a narrow band, might be required to acquire international presence in commodities, wherein it has comparative advantage," Economic Survey 2012-13 had stated.

Sharma said the government was working on a long-term policy for sustainable growth in agricultural commodities. Currently, India is the world's largest rice exporter and second, in terms of wheat exports. Horticulture exports have also seen good growth. To achieve the desired growth, "India needs to change the cropping pattern, with a larger focus on north India", Sharma had said.

R S Rawat, secretary-general, Associated Chambers of Commerce and Industry of India, said, "The government must take policy reforms to support growth in agricultural commodities. To achieve the $70-billion export target for 2017 will not be too ambitious, with the possibility of policy implementation increasing productivity and promoting diversity of crops and specialised items to meet specific demands abroad."

Among agricultural commodities, exports of basmati rice have risen 46 per cent to $3.47 billion in the first nine months of this financial year, compared with $2.37 billion in the year-ago period. Exports of non-basmati rice rose seven per cent to $2.13 billion in the April-December 2013 period from $1.99 billion in the year-ago period. Exports of dairy products recorded 138 per cent growth in April-December 2013 at $435.93 million, against $183.24 million in the corresponding period last year. Due to declining global wheat prices, India's realisation from wheat exports fell 5.24 per cent — from $1.24 billion to $1.17 billion.

AGRI EXPORT ZONES GET GOING

The government has so far sanctioned 10 Agri Export Zones (AEZs) which are being set up in various states with an estimated investment of more than ₹.200 crore. Out of the total investment, around ₹.80 crore will flow from various central government agencies like the Agriculture and Processed Food Export Development Authority (APEDA), National Horticulture Board, Ministry of Agriculture and the Ministry of Food Processing Industries. It is estimated that these Agri Export Zones will generate exports of around ₹.1700 crore over the next 5 years, besides generating a substantial amount of direct and indirect employment. While introducing the Agri Export Zones concept in the Exim Policy announced on 31st March, 2001, Shri Murasoli Maran, Union Minister of Commerce and Industry, had said that these zones would be part of the effort to provide improved access for India's agricultural and allied products in the international market with a view to providing remunerative returns to the farming community in a sustained manner. In view of the popularity of the scheme, proposals with regard to setting up of similar Agri Export Zones are being considered by a large number of states and it is expected that 10 more AEZs will be set up during the financial current year. The 10 AEZs already sanctioned include those for pineapple in West Bengal; litchi in Uttaranchal; gherkins in Karnataka; vegetables in Punjab; mangoes and potatoes in Uttar Pradesh; grapes and wines in Maharashtra; flowers in Tamil Nadu; and potatoes in Punjab.

The emphasis in all the Agri Export Zones is on convergence. Thus, the objective is to utilise the on going schemes of various Central and State Government Agencies in a co-ordinated manner to cover the entire value chain from farm to the consumer. In all the projects, necessary commitments have been given by the State Governments and the Central Government Agencies have also agreed to the project in principle.

The concept of Agri Export Zone attempts to take a comprehensive look at a particular produce/products located in a contiguous area for the purpose of developing and sourcing the raw materials, their processing/packaging, leading to final exports. Thus, the entire effort is centered around a cluster approach of identifying the potential products, the geographical region in which these are grown and adopting an end to end approach of integrating the entire process, right from the stage of production till it reaches the market. The anticipated benefit to accrue as a consequence of setting up of such zones are as follows: (i) Strengthening of backward linkages with a market oriented approach; (ii)

Product acceptability and its competitiveness abroad as well as in the domestic market; (iii) Value addition to basic agricultural produce; (iv) Bring down cost of production through economy of scale; (v) Better price for agricultural produce; (vi) Improvement in product quality and packaging; (vii) Promote trade-related research and development; and (viii) Increase employment opportunities.

Agricultural and Processed Food Products Development Authority (APEDA) was appointed as the nodal agency by the Central Government to promote setting up of Agri Export Zone.

A lot of headway has already been made in the first six Agri Export Zones that were sanctioned. After the signing of MOU's between the representative bodies of the Central and State Governments, it is expected that export of potatoes will commence from Punjab and Uttar Pradesh from January 2002 itself. The export of litchi from Uttaranchal will commence from the next season, *i.e.*, June 2002, as necessary preparations are already under way. Similarly, in case of West Bengal, the State Government is in the process of finalising the names of entrepreneurs with whom the MOU's will be signed soon. In view of the nature of the project, export from this green field Agri Export Zone is likely to commence in the year 2003.

However, in case of Agri Export Zones for mangoes in Uttar Pradesh, the exports are likely to commence during the next season, as the necessary ground work is already under way and it is expected that the pack houses would come up before the commencement of the season. An over-all marketing strategy has already been worked out for promoting export of mangoes in specific destinations. Gherkins from Karnataka and vegetables from Punjab are already been exported but this process is being further strengthened through a variety of interventions under Agri Export Zones.

GREEN OUTPUTS MARKET TRENDS AND POTENTIALS IN INDIA

Demand for green agricultural products is a stimulant for growth for input market. In other words if there is demand in market for organically produced farm products will encourage farmers to implement the organic farming practices and also to use organic input like bio-fertilizers, bio-pesticides, varmi-compost, green manure and FYM.

Estimating area under organic agriculture in India is a difficult task as there is no central agency that collects and compiles this information. Different agencies have estimated the area under organic agriculture differently for instance the study undertaken by FIBL and ORG-MARG (Garibay S V and Jyoti K, 2003) the area under organic agriculture to be 2,775 hectares (0.0015 per cent of gross cultivated area in India). But other estimation undertaken by SOEL-Survey shows that the land area under organic cropping is 41000 hectare. The total numbers of organic farms in the country as per SOEL-Survey are

5661 but FIBL and ORG-MARG survey puts it as 1426. Some of the major organically produced agricultural crops in India include crops like plantation, spices, pulses, fruits, vegetables and oil seeds etc.

Table. Major products produced in India by organic farming

Major products produced in India by organic farming

Type of Product	Products
Commodity	Tea, Coffee, Rice, Wheat
Spices	Cardamom, Black pepper, white pepper, Ginger, Turmeric, Vanilla, Tamarind, Clove, Cinnamon, Nutmeg, Mace, Chili
Pulses	Red gram, Black gram
Fruits	Mango, Banana, Pineapple, Passion fruit, Sugarcane,
Orange, Cashew nut, Walnut	
Vegetables	Okra, Brinjal, Garlic, Onion, Tomato, Potato
Oil seeds	Mustard, Sesame, Castor, Sunflower
Others	Cotton, Herbal extracts

India is best known as an exporter of organic tea and also has great export potential for many other products. Other organic products for which India has a niche market are spices and fruits. Org-Marg's survey also proves this fact as around 30 per cent of respondents that includes producer, exporters and traders has responded that organic tea is produced in India and this is highest response for any single crop, next are spices, fruits, vegetables, rice and coffee (Garibay S V and Jyoti K, 2003). There is small response for cashew, oil seed, wheat and pulses. Among the fruit crops Bananas, Mangos and oranges are the most preferred organic products.

Export Market: Organic agricultural export market is one of the major drivers of greening of agriculture in India. The current production of organic crops is around 14,000 tons (Garibay S V and Jyoti K, 2003). Out of this production, tea and rice contributes around 24 per cent each, fruits and vegetables combine makes 17 per cent of this total production. From India around 11,925 tons of organic product is exported, that makes around 85 per cent of total organic crop production. Major export market for Indian producers are Australia, Belgium, Canada, France, Germany, Italy, Japan, Netherlands, Sweden, Singapore, South Africa, Saudi Arabia, UAE, UK, and USA. Estimated quantity of various products that are exported from India in 2002.This shows that around 3000 tons of tea was exported and in quantity term it was the highest, next major exports are rice (2500 tons), fruits and vegetables (1800 tons), cotton (1200 tons) and wheat (1150 tons) (Garibay S V and Jyoti K, 2003).

Table. Major organic products exported from India

Product	Sales (Tons)
Tea	3000
Coffee	550
Spices	700

Rice	2500
Wheat	1150
Pulses	300
Oil Seeds	100
Fruits & Vegetables	1800
Cashew Nut	375
Cotton	1200
Herbal Products	250
Total	11,925

The burgeoning US and European green markets provides enormous scope for Indian exporters. International Trade Centre's (ITC) overview of estimation of organic food world-wide shows that there is strong growth in retail sales from US $ 10 billion in 1997 to US$ 17.5 billion in 2000 and about US$ 21 billion in 2001 (in 16 European countries, USA and Japan).

If the demand of so called 'green product' in Japan that is not certified as organic product is excluded from total estimation, than also it is US$ 16 billion for 2999 and reached around US$ 19 billion in 2001. Though the current market share for organic produced is estimated between one to two per cent of total food products market, but the forecasting by experts show that this market is likely to grow at a higher pace. According to experts in medium term around five per cent of the market is expected to be organic market share (Minou Yussefi and Heldge Willer, 2003).

Europe is largest market of organic produces in world and consumes around half of the world produce of organic production (Minou Yussefi and Helga Willer, 2003). EU is a net importer of cereals, oilseeds, potatoes and vegetables. For 2001, European market for organic food was estimated to be around US$ 9 billion but with expected annual growth rate of around 20 per cent depending upon the market, and for 2003 the retail sales for organic food in this market is expected to US$ 10-11billion (Rudy Korbech-Olesen, ITC, UNCTAD/WTO). Within Europe, Germany is largest market for organic products with sales value of around 2.5 billion Euros ($2.3 billion (US)). On an average the per capita spending on organic produces in Europe was euro 23 for 2000. In terms of per capita consumption of organic products countries like Denmark (72 Euros per head), Switzerland (Euros 68), Austria (Euros 40) and Germany (Euros 31) fared much better than others. While some of the European countries like Italy not only meet their internal demand for organic products but they also cater to the demands of other neighbouring countries. There are other countries like United Kingdom, which are highly dependent on imports to meet their domestic organic product demand.

In North America, retail sales of organic products for 2002 was estimated nearly $12 billion (US) out of this US alone is contributing for $11 billion (US). US retail sale for organic product has grown 20-24 per cent per year for the past 12 years and the same growth trends expected to continue for future.

Current retail sales for organic food is around 2 per cent of total retail food sale in US (Minou Yussefi and Heldge Willer, 2003).

In Asia largest market for organic food product is Japan and it is estimated to have the retail sales of organic food and beverage of around $(US) 2.5-3.0 billion (Minou Yussefi and Heldge Willer, 2003). Of this total value of organic market, imports are estimated to be to the tune of around $(US) 360 million. Though the organic food market is not more than 0.5 per cent of total food market of Japan but according to the Japanese Integrated Market Institute, import of organic products is likely to grow by 40 per cent. Other global markets for organic products in Middle East are Saudi Arabia and UAE. Within Africa, South Africa is the only country that has organic market potential. As seen from the global market growth trends for organic foods there are enormous potentials for India to exploit.

Table. Percentage of organic food and medium term growth expected in selected markets

Overview for World Market for organic food and beverages in 2000 (estimates)

Markets	% of total food sales	% Expected growth - Medium term
Germany	1.6-1.8	10-15
U.K.	1.0-2.5	15-20
Italy	0.9-1.1	10-20
France	0.8-1.0	10-15
Switzerland	2.0-2.5	10-15
Denmark	2.5-3.0	10-15
Austria	1.8-2.0	10-15
Netherlands	0.9-1.2	10-20
Sweden	1.0-1.2	15-20
Belgium	0.9-1.1	10-15
U.S.A.	1.5-2.0	20

Total commodity wise demand (in volume terms) has been estimated in some selected export markets (Germany, Holland, UK, Switzerland, USA, and Japan) by the FIBL and ORG- MARG survey which shows that for Banana it is 6,410 tons, for wheat and soy bean 1,000 tons, for pineapple around 900 tons and for mango this is around 650 tons (Garibay S V and Jyoti K, 2003).

The attractiveness of organic market gets enhanced also because of the price premium that these products have over the conventionally produced products. Price premium for various organic produces vary in different countries depending upon the distribution channels and market conditions. This premium varies from 30-50 per cent (trader level) for different products.

As seen from above there are immense opportunities for organic agricultural exports for Indian to exploit. Some of the prerequisites for exploiting this potential include:

- Farmers capacity to produce the organically the agricultural products which have global market and
- Prior experience of exporters and traders in exporting agricultural commodities to these markets

An attempt has been made to develop a matrix by depicting the conventional agricultural commodities, which India has been exporting to various countries across the world as well as the existence of organic market for these commodities in these countries. Depicting the current conventional agricultural market indicates the capabilities of India in exporting the specific agricultural commodities to different countries and similarly depicting the existence of organic agriculture market for specific commodities in these countries reveal the existence of opportunities for exporting organic agricultural commodities. In developing this matrix we have used annual exports of agricultural commodities as published in CMIE agricultural sector reports and for exploring organic market in different countries for different commodities we have used the internet resources mentioned at the end of this article.

The matrix reveals that India has demonstrated capabilities of exporting agricultural commodities like rice, wheat, tea, coffee, spices, oil meals, sugar, fruits and vegetables etc to countries like USA, U. K, Germany, Japan, France, Saudi Arabia, South Africa, CIS Countries, Poland, Netherlands, Italy etc. It also shows that in most of these countries there is a demand for organically produced commodities, which attract price premiums ranging from 10 per cent to 100 per cent. This showcases a window of opportunity that is yet to be exploited to its maximum potentials by Indian exporters and producers of agricultural commodities.

8

Agricultural Credit and Foreign Trade

'AGRICULTURAL CREDIT'

Any of several credit vehicles used to finance agricultural transactions, including loans, notes, bills of exchange and banker's acceptances. These types of financing are adapted to the specific financial needs of farmers, which are determined by planting, harvesting and marketing cycles.

Short-term credit finances operating expenses, intermediate-term credit is used for farm machinery, and long-term credit is used for real-estate financing.

IMPACT OF AGRICULTURAL CREDIT ON AGRICULTURE PRODUCTION

A large proportion of the population in India is rural based and depends on agriculture for a living. Enhanced and stable growth of the agriculture sector is important as it plays a vital role not only in generating purchasing power among the rural population by creating on-farm and off-farm employment opportunities but also through its contribution to price stability. In India, although the share of agriculture in real GDP has declined below one-fi fth, it continues to be an important sector as it employs 52 per cent of the workforce. The growing adult population in India demand large and incessant rise in agricultural production. But per capita availability of food, particularly cereals and pulses, in recent years has fallen significantly. As a result, slackening growth of agriculture during last decade has been a major policy concern.

Three main factors that contribute to agricultural growth are increased use of agricultural inputs, technological change and technical efficiency. With savings being negligible among the small farmers, agricultural credit appears to be an essential input along with modern technology for higher productivity. An important aspect that has emerged in last three decades is that the credit is not only obtained by the small and marginal farmers for survival but also by the large farmers for enhancing their income. Hence, since independence, credit has been occupying an important place in the strategy for development of agriculture. The agricultural credit system of India consists of informal and formal sources of credit supply. The informal sources include friends, relatives, commission agents, traders, private moneylenders, etc. Three major channels for disbursement of formal credit include commercial banks, cooperatives and micro-finance institutions (MFI) covering the whole length and breadth of the country. The overall thrust of the current policy regime assumes that credit is a critical input that affects agricultural/ rural productivity and is important enough to establish causality with productivity. Therefore, impulses in the agricultural operations are sought through intervention in credit.

In order to improve the flow of credit to the agricultural sector, the Reserve Bank had advised public sector banks to prepare Special Agricultural Credit Plans (SACP) in 1994-95. Under the SACP, the banks are required to fix self-set targets for achievement during the year. The targets are generally fixed by the banks about 20 to 25 per cent higher over the disbursements made in the previous year. With the introduction of SACP, the flow of credit to agricultural sector has increased significantly from ₹.8,255 crore in 1994-95 to ₹.1,22,443 crore in 2006-07 which were higher than the projection of ₹.1,18,160 crore. As against the target of ₹.1,52,133 crore for the financial year 2007-08, disbursements to agriculture by public sector banks under the plan were ₹.1,11,543 crore (provisional). As recommended by the Advisory Committee on Flow of Credit to Agriculture and Related Activities from the Banking System (Chairman: Shri V.S. Vyas), the Mid-Term Review of Annual Policy of RBI for 2004-05 made the SACP mechanism applicable to private sector banks from the year 2005-06. Disbursements to agriculture by private sector banks under SACP during 2006-07 aggregated to ₹.44, 093 crore against the target of ₹.40,656 crore. As against the target of ₹.41,427 crore for the financial year 2007-08, disbursements to agriculture by private sector banks aggregated to ₹.45,905 crore (provisional).

From the Government side, with a view to doubling credit flow to agriculture within a period of three years and to provide some relief to farmers affected by natural calamities within the limits of financial prudence, the Union Finance Minister announced several measures on June 18, 2004. Accordingly, the Reserve Bank and NABARD issued necessary operational guidelines to banks. From the very beginning, the actual disbursements exceeded the targets for each of the last four years. As against the target of ₹.2,25,000 crore for 2007-08, all banks disbursed ₹.2,25,348 crore (provisional). During 2007-08, 7.29 million new farmers were financed by commercial banks and RRBs as against the target of 5 million farmers fixed by the Union Finance Minister for the year. The Finance Minister, in his Budget Speech for the year 2008-09, urged the banks to increase the level of credit to ₹.2,80,000 crore during the year 2008-09.

Such efforts have, however, not been transmitted to the growth in agriculture output. Since the mid-1990s, the growth of the agricultural sector has been low as well as volatile; the growth decelerated from an annual average of 4.7 per cent per annum during 1980s to 3.1 per cent during the 1990s and further to 2.2 per cent during the Tenth Plan period. Growth in agricultural production has decelerated during 2006-07 with the agriculture sector characterised by stagnation in output of major food grains. Per capita annual production of cereals declined from 192 kilogram (kg) during 1991-95 to 174 kg during 2004-07 and that of pulses from 15 kg to 12 kg over the same period. Per capita availability of food grains has, thus, fallen close to the levels prevailing during the 1970s.

Volatility in agricultural production has not only affected overall growth but also exerted persistence pressure on maintaining low and stable infl ation. Demand-supply gaps were refl ected in higher domestic food prices in recent years. All these evidences apparently point to the fact that higher credit to agriculture is not translated into commensurate increase in agricultural output. In this paper, we examine this basic premise empirically using both macro- and micro level data.

LITERATURE REVIEW

India has systematically pursued a supply leading approach to increase agricultural credit. The objectives have been to replace moneylenders, relieve farmers of indebtedness and to achieve higher levels of agricultural credit, investment and agricultural output. Among earlier studies, Binswanger and Khandker (1992) found that the output and employment effect of expanded rural finance has been much smaller than in the non-farm sector. The effect on crop output is not large, despite the fact that credit to agriculture has strongly increased fertilizer use and private investment in machines and livestock. High impact on inputs and modest impact on output clearly mean that the additional capital investment has been more important in substituting for agricultural labour than in increasing crop output.

Between bank nationalization in 1969 and the onset of financial liberalization in 1990 bank branches were opened in over 30,000 rural locations which had no prior presence of commercial banks (called un-banked locations). Alongside, the share of bank credit and savings which was accounted for by rural branches raised from 1.5 and 3 per cent respectively to 15 per cent each (Burgess and Pande, 2005). This branch expansion was an integral part of India's social banking experiment which sought to improve the access of the rural poor to cheap formal credit. The estimates suggested that a one per cent increase in the number of rural banked locations reduced rural poverty by roughly 0.4 per cent and increased total output by 0.30 per cent. The output effects are solely accounted for by increases in non-agricultural output – a finding which suggests that increased financial intermediation in rural India aided output and employment diversifi cation out of agriculture.

In a detailed paper, Mohan (2006) examined the overall growth of agriculture and the role of institutional credit. Agreeing that the overall supply of credit to agriculture as a percentage of total disbursal of credit is going down, he argued that this should not be a cause for worry as the share of formal credit as a part of the agricultural GDP is growing.

This establishes that while credit is increasing, it has not really made an impact on value of output figures which points out the limitations of credit. In another study, Golait (2007) attempted to analyse the issues in agricultural credit in India. The analysis revealed that the credit delivery to the agriculture sector continues to be inadequate. It appeared that the banking system is still hesitant on various grounds to purvey credit to small and marginal farmers. It was suggested that concerted efforts were required to augment the flow of credit to agriculture, alongside exploring new innovations in product design and methods of delivery, through better use of technology and related processes. Facilitating credit through processors, input dealers, NGOs, etc., that were vertically integrated with the farmers, including through contract farming, for providing them critical inputs or processing their produce, could increase the credit flow to agriculture significantly.

In general, it is difficult to establish a causal relationship between agriculture credit and production due to the existence of critical endogeneity problem. However, Sreeram (2007) concluded that increased supply and administered pricing of credit help in the increase in agricultural productivity and the well being of agriculturists as credit is a sub-component of the total investments made in agriculture. Borrowings could in fact be from multiple sources in the formal and informal space. Borrowing from formal sources is a part of this sub-component. With data being available largely from the formal sources of credit disbursal and indications that the formal credit as a proportion of total indebtedness is going down, it becomes much more difficult to establish the causality.

He also stated that the diversity in cropping patterns, holding sizes, productivity, regional variations make it difficult to establish such a causality for agriculture or rural sector as a whole, even if one had data. Finally, he argued that mere increase in supply of credit is not going to address the problem of productivity, unless it is accompanied by investments in other support services. In the present study, we take a re-look at the problem by quantitatively assessing the impact of institutional credit expansion on agriculture.

ASSESSMENT OF PROGRESS IN AGRICULTURAL CREDIT IN INDIA

In India, the share of agriculture in the gross domestic product has registered a steady decline from 36.4 per cent in 1982-83 to 18.5 per cent in 2006-07. Agriculture growth has remained lower than the growth rates witnessed in the industrial and services sectors. The gap between the growth of agriculture and non-agriculture sector began to widen since 1981-82, and more particularly since 1996-97, because of acceleration in the growth of industry and services sectors.

Even though the share of agriculture in GDP has declined over the years, the number of people dependent on agriculture for their food and livelihood has remained unchanged. Therefore, a number of measures were taken by the Reserve Bank and the Government of India for facilitating increased credit flows to the agriculture sector. With a view to doubling credit flow to agriculture within a period of three years, several measures have been announced. From the very beginning, the actual disbursements exceeded the targets for each of the last four years.

Table. Average GDP growth rates of agriculture and other sectors at 1999-2000 prices

(Per cent)					
Period		**Total Economy**	**Agriculture & allied**	**Crops & livestock**	**Non-agriculture**
1	2	3	4	5	6
Pre-Green Revolution	1951-52 to 1967-68	3.7	2.5	2.7	4.9
Green Revolution period	1968-69 to 1980-81	3.5	2.4	2.7	4.4
Wider technology dissemination period	1981-82-1990-91	5.4	3.5	3.7	6.4
Early Reform Period	1991-92 to 1996-97	5.7	3.7	3.7	6.6
Ninth and Tenth	1997-98 to 2006-07	6.6	2.5	2.5	7.9
Plan	2005-06 to 2006-07	9.5	4.8	5.0	10.7

The increased credit flow to agriculture has not resulted in the commensurate increase in production. The average rate of growth of foodgrains production decelerated to 1.2 per cent during 1990-2007, lower than annual rate of growth of population, averaging 1.9 per cent. The per capita availability of cereals and pulses has witnessed a decline during this period. The consumption of cereals declined from a peak of 468 grams per capita per day in 1990-91 to 412 grams per capita per day in 2005-06, indicating a decline of 13 per cent during this period.

The consumption of pulses declined from 42 grams per capita per day (72 grams in 1956-57) to 33 grams per capita per day during the same period. The overall production of food grains was estimated at 217.3 million tonnes in 2006-07, an increase of 4.2 per cent over 2005-06. Compared to the target set for 2006-07, it was, however, lower by 2.7 million tonnes (1.2 per cent). Over a medium term, there has generally been a shortfall in the achievement of target of foodgrains, pulses and oilseeds during 2000-01 to 2006-07. The actual production of foodgrains on an average was 93 per cent of the target. Actual production, however, was only 87.7 per cent of target for pulses and 85.3 per cent of target for oilseeds.

Table. Targets and actual disbursement to agriculture by banks

(₹ crore)								
Agency	**(2004-05)**		**(2005-06)**		**(2006-07)**		**(2007-08)***	
	Target	**Disbursement**	**Target**	**Disbursement**	**Target**	**Disbursement**	**Target**	**Disbursement**
1	2	3	4	5	6	7	8	9
Comm. Banks	57,000	81,481	87,200	1,25,477	1,19,000	16,64,486	1,50,000	1,56,850
Coop. Banks	39,000	31,231	38,600	39,786	41,000	42,480	52,000	43,684
RRBs	8,500	12,404	15,200	15,223	15,000	20,435	23,000	24,814
Other Agencies		193						
Total	**1,05,000**	**1,25,309**	**1,41,000**	**1,80,486**	**1,75,000**	**2,29,401**	**2,25,000**	**2,25,348**

Table. Actual production relative to targets

(Per cent)							
Crop	**2001-02**	**2002-03**	**2003-04**	**2004-05**	**2005-06**	**2006-07**	**Average (2001-07)**
1	**2**	**3**	**4**	**5**	**6**	**7**	**8**
Rice	101.5	77.2	95.2	88.9	104.5	100.6	94.7
Wheat	93.3	84.3	92.5	86.3	91.8	100.4	91.4
Coarse cereals	101.1	79	110.6	90.9	93.3	92.9	94.6
Pulses	89.1	69.6	99.4	85.8	88.4	93.7	87.7
Foodgrains	**97.6**	**79.4**	**96.9**	**88.1**	**97**	**98.8**	**93**
Oilseeds	**37.8**	**55**	**102**	**93**	**105.2**	**82.6**	**85.3**
Sugarcane	91.4	89.8	73.1	87.8	118.4	131.7	98.7
Cotton	68.9	57.5	91.5	109.5	112.1	122.3	93.7
Jute & Mesta	106.2	94	93.1	87.1	96.1	99.9	96.1

Table. Rate of growth of production for major crops

(Per cent)							
Year	**Rice**	**Wheat**	**Pulses**	**Food grains**	**Cotton**	**Oilseeds**	**Sugarcane**
1	**2**	**3**	**4**	**5**	**6**	**7**	**8**
1989-90 to 2006-07	1.17	1.9	-0.03	1.18	2.04	1.25	1.13
1992-93 to 1996-97	1.73	3.6	0.66	1.88	4.88	3.57	3.74
1997-98 to 2001-02	1.13	1.26	-2.52	0.67	5.79	-4.68	1.23
2002-03 to 2005-06	1.75	0.42	3.27	1.61	20.22	9.81	-1.23

The production of agriculture crops, besides the weather-induced fluctuations, significantly depends on the availability of inputs like fertilizers, irrigation, certified seeds, credit support and appropriate price signals. Minimum support prices indicated upfront and before the sowing seasons act as effective incentives for acreage response of the agricultural crops. Deviations in foodgrains and agricultural output from their long-term trends are determined, among other factors, by variations in monsoon around its long-term trend and the area under irrigation

On the productivity front, India is not only low relative to other countries in crop production, there are considerable inter-state variations. The productivity of wheat in 2005-06 varied from a low of 1,393 kg per ha in Maharashtra to a high of 4,179 kg in Punjab. The Steering Committee on Agriculture for the Eleventh Five Year Plan has observed that not only the yields differed across the States, there was a significant gap between the performance and potential as revealed by actual yield and yield with improved practices adopted by farmers.

Table. Interstate Variation in Actual yield and yield with improved practices

State	Improved practice(I)	Farmer practice(F)	Actual yield 2003-04(A)	Gap (per cent)	
				I and F	I and A
1	2	3	4	5	6
Wheat (Yield : Kg/ha – 2002-03 to 2004-05)					
Bihar	3651	2905	1783	25.7	104.8
Madhya Pradesh	3297	2472	1789	33.4	84.3
Uttar Pradesh	4206	3324	2794	26.5	50.5
Rice (irrigated) (Yield : Kg/ha - 2003-04 to 2004-05)					
Uttar Pradesh	7050	5200	2187	35.6	222.4
Bihar	4883	4158	1516	17.4	222.1
Chhattisgarh	3919	3137	1455	24.9	169.4
Sugarcane					
Maharashtra	127440	99520	51297	28.1	148.4
Karnataka	147390	128000	66667	15.1	121.2
Bihar	74420	49440	40990	50.5	81.6

The need to analyze the effect of institutionalized agriculture credit on agriculture output in an econometric framework. Considering the diverse nature of agriculture activities and agriculture credit distribution across the country, the Arellano-Bond (1991) dynamic panel estimation at district level is most suitable as it is designed for situations with 1) "small T, large N" panels, meaning few time periods and many individuals; 2) a linear functional relationship; 3) a single left-hand-side variable that is dynamic, depending on its own past realizations; 4) independent variables that are not strictly exogenous, meaning correlated with past and possibly current realizations of the error; 5) fixed individual effects; and 6) heteroskedasticity and autocorrelation within individuals, but not across them. Arellano-Bond estimation starts by transforming all regressors, usually by differencing, and uses the Generalised Method of Moments, and so is called "difference GMM".

METHODOLOGY FOR ESTIMATING THE EFFECT OF AGRICULTURAL CREDIT ON PRODUCTION: A DYNAMIC PANEL REGRESSION APPROACH

ECONOMIC FRAMEWORK

When the farmer faces a credit constraint, additional credit supply can raise input use, investment, and hence output. This is the liquidity effect of credit. But credit has another role to play. In most developing countries where agriculture still remains a risky activity, better credit facilities can help farmers smooth out consumption and, therefore, increase the willingness of risk-averse farmers to take risks and make agricultural investments. This is the consumption smoothing effect of credit. Therefore, specifi cation of an appropriate model of agricultural credit and output is fraught with several econometric difficulties. First, time series data on informal credit do not exist.

If expansion of formal credit causes a reduction in informal credit, a regression of output on formal credit will measure the effect of expansion of credit net of the effect of reduced informal credit. The second econometric problem is the joint dependence of output and credit on other variables such as the weather, prices, or technology. Credit advanced by formal lending agencies such as banks is an outcome of both the supply of and demand for formal credit.

The amount of formal credit available to the farmer, his credit ration, enters into his decision to make investments, and to finance and use variable inputs such as fertilizer and labour. There is, therefore, a joint dependence between the observed levels of credit used and aggregate output. A twostage instrument variable (IV) procedure can solve this identifi cation problem. The third econometric problem arises because formal agriculture lending is not exogenously given or randomly distributed across space. That means, the banks will lend more in areas where agricultural opportunities are better, risk is lower, and hence, chances for loan recovery are higher. An unobserved variable problem thus arises for the econometric estimation and is associated with unmeasured or immeasurable region, say district characteristics. This problem can be overcome by the use of district-level panel data. The credit delivery to agriculture will have differential impacts over different regions in India.

Even various districts in same state may be responding in a different manner to the change in credit delivery. Therefore, any methodology where data are pooled together will not be appropriate. Again, assuming exogeneity in the independent variables may lead to wrong results as the variables like area under cultivation may depend on last periods' output. For example, an increase (decrease) in output in a particular district at any particular year leads to the chances of more (less) area of showing in the next year which increases (decreases) the likelihood of higher production in the subsequent year. This led us analyzing the data using a dynamic panel data analysis with instrumental variables using Arellano-Bond Regression. Using some select states, district level panel data is obtained and analyzed allowing district level unobserved heterogeneity.

PANEL DATA METHODOLOGY

Panel data is widely used to estimate dynamic econometric models. Its disadvantage over cross-section data is that we cannot estimate dynamic models from observations at a single point in time, and it is rare for single cross section surveys to provide suffi cient information about earlier time periods for dynamic relationships to be investigated. Its advantages over aggregate time series data include the possibility that underlying microeconomic dynamics may be obscured by aggregation biases, and the scope that panel data offers to investigate heterogeneity in adjustment dynamics between different panels. A single equation, autoregressive-distributed lag model can be estimated from panels with a large number of cross-section units, each observed for a small number of time periods.

This situation is typical of micro panel data and calls for estimation methods that do not require the time dimension to become large in order to obtain consistent parameter estimates. Assumptions about the properties of initial conditions also play an important role in this setting, since the infl uence of the initial observations on each subsequent observation cannot safely be ignored when the time dimension is short.

In case of absence of strictly exogenous explanatory variables or instruments as strict exogeneity rules out any feedback from current or past shocks to current values of the variable, which is often not a natural restriction in the context of economic models relating several jointly determined outcomes. Identifi cation then depends on limited serial correlation in the error term of the equation, which leads to a convenient and widely used class of Generalised Method of Moments (GMM) estimators for this type of dynamic panel data model.

EMPIRICAL ANALYSIS

Before explaining the results of the regression model, association in terms of correlation between agriculture credit and output based on the district level data for 2007 is assessed.

Then, we discuss the empirical results of the dynamic panel regression using state-level data and fi nally results of the same based on districts level data of select states are presented.

ASSOCIATION BETWEEN AGRICULTURE CREDIT AND OUTPUT: SOME EMPIRICAL ASSESSMENT

Indicus Analytics, a private sector organization, came out with district GDP data, classified by agriculture, industry and others for the year 2007. With a view to analyze the association between agriculture credit and output, the data district GDP from agriculture data (DGDP_AG) of major states are juxtaposed with district-level agricultural credit data (as per place of utilization) as available in Basic Statistical Returns of all scheduled commercial banks.

In addition to the credit amount outstanding (AG_C), the number of credit accounts to agriculture (N_AG_C) is also used. The (Pearson's) correlation coefficients for districts within the states have been derived to indicate the direction and extent of relationship between GDP and credit. The elasticity of bank credit on GDP has been chosen to measure the responsiveness of the relationship to changes in bank credit to the GDP.

The correlation coefficients of GDP and bank credit in respect of agriculture for the states Andhra Pradesh, Chattisgargh, Jharkhand, Orissa, Rajasthan, Tamil Nadu, Uttar Pradesh, Uttranchal and West Bengal were positive and statistically significant at 1 per cent level.

The correlation coefficients were significant for Assam, Bihar, Maharashtra, Meghalaya, Sikkim at 5 per cent level. However, the correlation coefficients were not found significant for the states like Gujarat, Haryana, Himachal Pradesh, Jammu and Kashmir, Karnataka, Kerala, Madhya Pradesh and Punjab.

The log elasticity of bank credit to GDP of agriculture were highly significant for the states Andhra Pradesh, Assam, Bihar, Chhattisgarh, Jharkhand, Madhya Pradesh, Maharashtra, Orissa, Rajasthan, Tamil Nadu, Uttar Pradesh and Uttaranchal at one per cent level.

Extension of formal credit to agriculture may play a significant role on the agriculture output in these states. On the other hand, bank credit does not seem to play a major role on the agriculture output in Haryana, Jammu and Kashmir, Karnataka and Kerala.

The flow of credit for agriculture in the states of Uttar Pradesh, Andhra Pradesh, Maharashtra, West Bengal, Rajasthan, Tamil Nadu, Bihar, Orissa, Assam, Chhattisgarh, Jharkhand, Uttaranchal, etc., are very important as these states contributes significantly towards agriculture GDP and the elasticity of credit to GDP is also significantly high.

Interestingly, when agriculture credit is extended to more people, as demonstrated by the number of agricultural credit account, it has translated to higher output.

This is true for most of the states, except Kerala. Therefore, empirical association clearly indicates that financial inclusion of farmers in the organized financial system boosts agriculture output.

Table. Correlation of District GDP and Bank Cr edit at District Level: 2007

State	No. ofdistricts	Correlation (DGDP_ AG, AG_C)	Log-Elasticity (DGDP_ AG ,AG_C)	RankCorrelation (DGDP_ AG, N_ AG_C)	AverageDGDP_ AG (₹lakh)	AverageAG_C (₹lakh)
1	2	3	4	5	6	7
Andhra Pradesh	22	0.77*	0.66*	0.67*	248234	87346
Assam	23	0.42**	0.70*	0.52**	71483	4386
Bihar	37	0.50**	0.58*	0.60*	69559	17897
Chhattisgarh	16	0.70*	0.67*	0.79*	58961	11783
Gujarat	25	0.19	0.42**	0.50**	152787	48758
Haryana	19	0.12	0.05	0.72*	119075	51571
Himachal Pradesh	12	0.06	0.47**	0.54***	44403	10530
Jammu & Kashmir	14	0.43	0.66	0.77*	56967	7826
Jharkhand	18	0.59*	0.60	0.56**	40816	6420
Karnataka	27	0.12	0.05	0.65*	109415	75980
Kerala	14	-0.19	-0.03	0.42	127711	60785
Madhya Pradesh	45	0.21	0.47*	0.56*	72148	22951
Maharashtra	33	0.37**	0.50*	0.72*	157593	36207
Orissa	30	0.60*	0.46*	0.72*	74577	12722
Punjab	17	0.26	0.36**	0.67*	205346	74078
Rajasthan	32	0.74*	1.02 *	0.67*	106739	40342
Tamil Nadu	27	0.53*	0.64 *	0.62*	96348	64578
Uttar Pradesh	69	0.69*	0.65 *	0.83*	127330	33806
Uttaranchal	13	0.78*	0.49 *	0.70*	41327	10797
West Bengal	17	0.81*	0.55 *	0.70*	276601	28430
Total	**510**	**0.53***	**0.51 ***	**0.70***	**113195**	**35658**

Note:

*, ** and *** indicate statistical signifi cance at 1 per cent, 5 per cent and 10 per cent, respectively.

DGDP_AG= District domestic product from agriculture.

AG_C = District level bank credit to agriculture.

N_AG_C = District level bank credit account to agriculture.

With the objective identifying the role of bank credit in agriculture growth, the dynamic panel data regression with instrumental variables is performed with agriculture output as the dependant variable and total outstanding agriculture credit amount, total outstanding agriculture credit accounts, total agriculture area and rainfall as the regressors. Here analysis is done in two parts. First, an all India level analysis is undertaken involving all the variables for which data are available at state level, in a panel setup, which allows state level unobserved heterogeneity in the model. Secondly, using data of some select states, district level panel data model is estimated and analyzed allowing district level unobserved heterogeneity.

AGGREGATE ANALYSIS USING STATE-LEVEL PANEL DATA

In this chapter, 20 major states in India are included in the analysis for a period from 2001 to 2006. The period of study is confi ned to the above mentioned time period mainly due the restricted data availability. The state-wise agriculture output is obtained from the 'State wise estimates of value of output from agriculture and allied activities ', Central Statistical Organisation, Ministry of Statistics and Programme Implementation, Government of India. The total rainfall in the monsoon season (June to September) is used to obtain the excess/defi cit rainfall in different subdivisions for different years. The state level data on total credit outstanding and total number of credit accounts for the scheduled commercial banks are available in the 'Basic Statistical Returns of Scheduled Commercial Banks in India', Reserve Bank of India. This data source is utilised to obtain the credit amount outstanding and total number of agriculture accounts for direct, indirect and total agriculture credit of schedule commercial banks in India. All the variables except rainfall is standardized using population size of the state obtained from the projected data of 2001 population census. The variables used in the study are mentioned below as follows.

pcagout	Per capita agriculture output in rupees
pctacam	Per capita total agriculture credit amount outstanding in rupees
pctacn	Per capita total number of agriculture credit accounts outstanding per one lakh population
pcagar	Total agriculture area in square meter standardized by population
rain	Absolute deviation from normal rain
pcdacam	Per capita agriculture direct credit amount outstanding in rupees
pcidacam	Per capita agriculture indirect credit amount outstanding in rupees
pcdacn	Per capita number of direct agriculture credit accounts outstanding per one lakh population
pcidacn	Per capita number of indirect agriculture credit accounts outstanding per one lakh population

A dynamic panel regression model is estimated with per capita agriculture output as the dependent endogenous variable. The per capita direct agriculture credit amount outstanding, per capita direct agriculture credit number of accounts, pre capita indirect agriculture credit amount outstanding, pre capita indirect agriculture credit number of accounts and agriculture area are used as predetermined regressors, while rainfall is treated as a strictly exogenous variable.

The lagged values of the endogenous and predetermined variables; and exogenous variable are used as instrumental variables. All variables except rainfall used in this equation and subsequent equations are in their first difference form. The difference GMM is used to obtain the parameter estimates. The results obtained from one-step are reported because of downward bias in the computed standard errors in two-steps. The validity of the instruments is verified using Sargan test for over identifying restrictions and is found to be satisfactory. The difference residuals exhibited significant first order serial correlation and no second order serial correlation. Therefore it satisfi es another essential assumption for the consistency of the system GMM estimator *i.e.* no serial correlation in the error terms.

The agriculture area is significant in explaining the variation in agriculture output, even after adjustment of heteroskedasticity-consistent standard errors. Thus, the credit-output relationship of agriculture is difficult to establish at the state level. Agriculture is typically a localized regional economic activity and, therefore, its aggregation over states may hide the spatial heterogeneity. In order to establish a clear picture of the impact of agriculture credit on agriculture output.

Table. GMM estimates for the all India level regression equation

Variable	GMM estimates	Robust Std. Err.	z	Prob.	[95% Conf. Interval]	
Δpcagout (-1)	1.0724	0.1176	9.1200	0.0000	0.8420	1.3029
Δpcdacam	-0.0166	0.0276	-0.6000	0.5480	-0.0708	0.0375
Δpcdacam(-1)	0.0172	0.0335	0.5100	0.6080	-0.0485	0.0829
Δpcdacn	-0.0155	0.0102	-1.5200	0.1280	-0.0354	0.0045
Δpcdacn(-1)	0.0164	0.0132	1.2400	0.2140	-0.0095	0.0422
Δpcidacam	0.1148	0.0589	1.9500	0.0510	-0.0007	0.2302
Δpcidacam (-1)	-0.0579	0.0398	-1.4600	0.1450	-0.1359	0.0200
Δpcidacam (-2)	-0.0728	0.0582	-1.2500	0.2110	-0.1868	0.0412
Δpcidan	0.2434	0.1713	1.4200	0.1550	-0.0923	0.5792
Δpcidan (-1)	-0.2793	0.2355	-1.1900	0.2360	-0.7408	0.1822
Δpcidan (-2)	0.2371	0.1518	1.5600	0.1180	-0.0604	0.5346
Δpcagar	0.0234	0.0051	4.6000	0.0000	0.0134	0.0333
Rain	-0.1800	0.2412	-0.7500	0.4560	-0.6528	0.2928

Note:

GMM results are one-step estimates with heteroskedasticity-consistent standard errors and test statistics.

Arellaño-Bond test that average autocovariance in residuals of order 1 is 0:

H0: no autocorrelation z = -2.00 Pr > z = 0.0454

Arellano-Bond test that average autocovariance in residuals of order 2 is 0:

H0: no autocorrelation z = 0.95 Pr > z = 0.3404

Sargan test of over-identifying restrictions: chi2(25) = 6.64 Prob > chi2 = 0.99

(Sargan is a test of the over identifying restrictions for the GMM estimators, asymptotically ÷2. This test uses the minimised value of the corresponding two-step GMM estimators.)

STATE LEVEL ANALYSIS WITH DISTRICTS AS THE PANEL

As the agriculture activities tend to perform rather different across districts, a district level panel data is obtained and analyzed allowing district level unobserved heterogeneity. The data of the variables districts in India. Therefore, the analysis is performed for districts of select states namely Maharashtra, Andhra Pradesh, Punjab and West Bengal. The period of study is from 2001 to 2006. The district level agriculture output is not available as such in public domain. But district wise production and farm harvest prices of various agriculture commodities are available with Directorate of Economic and Statistics, Department of Agriculture and Corporation, Ministry of Agriculture, Government of India.

This information is utilised to estimate the district level agriculture output. For compilation of this only main agriculture commodities are selected. The cultivated area of different agriculture crops is also available from the same source and this information is utilised to obtain the area under cultivation for different districts. The subdivision wise normal and actual rainfall is published on a regular basis by Indian Metrological Department. The total rainfall in the monsoon season (June to September) is used to obtain the excess/defi cit rainfall in different subdivisions for different years.

The district level data on total credit outstanding and total number of credit accounts for the schedule commercial banks are available in Basic Statistical Returns of Scheduled Commercial Banks, Reserve Bank of India. This source of data also provide the information on credit amount outstanding and total number of agriculture accounts for direct, indirect and total agriculture credit of schedule commercial banks in India. All the variables expect rainfall is standardized using population size of the district obtained from 2001 population census.

The first model identifi es the effect of per capita direct agriculture credit amount outstanding and per capita indirect agriculture credit amount outstanding on per capita agriculture output. This model is estimated with per capita agriculture output as the dependent endogenous variable. The per capita direct agriculture credit amount outstanding, per capita indirect agriculture credit amount outstanding and agriculture area are used as predetermined regressors, while rainfall is treated as a strictly exogenous variable. The lagged values of the endogenous and predetermined variables and exogenous variables are used as instrumental variables.

All variables except rainfall used in this equation and subsequent equations are in their first difference form. The difference GMM is used to obtain the parameter estimates. Again, the results obtained from one-step are reported because of downward bias in the computed standard errors in two-step. The validity of the instruments is verified using Sargan test for over identifying restrictions and is found to be satisfactory. The difference residuals exhibited significant first order serial correlation and no second order serial correlation. Therefore it satisfi es the essential assumption for the consistency of the system

GMM estimator *i.e.* the assumption of no serial correlation in the error terms. The agriculture area is having a significant positive impact and rainfall (measured as deviation from normal) is negatively affecting the agriculture output and is significant, which is consistent with the general beliefs. The indirect agriculture credit amount and its first lag are insignificant in describing the variation in the agriculture output. On the other hand, the intervention through direct agriculture credit has significant positive impact on agriculture output. In particular, change in per capita agriculture direct credit (amount outstanding) by one per cent will lead to increase in per capita agriculture output by 0.11 per cent. This effect is, however, more stronger from the area under cultivation; an increase in per capita agriculture area by one per cent has the potential of raising the per capita agriculture output by more than 0.5 per cent. The output effect of rainfall on agriculture in India is still very severe; the deviation of rainfall from normal by one per cent could adversely affect agriculture output growth by 0.8 per cent.

Table. GMM estimates for the second regression equation

Variable	GMM estimates	Robust Std. Err.	z	Prob.	[95% Conf. Interval]	
1	2	3	4	5	6	7
Δpcagout (-1)	0.0605	0.1531	0.3900	0.6930	-0.2396	0.3605
Δpcdacam	0.1100	0.0317	3.4700	0.0010	0.0479	0.1722
Δpcidacam	0.1683	0.1739	0.9700	0.3330	-0.1726	0.5093
Δpcidacam (-1)	0.0980	0.1026	0.9600	0.3390	-0.1030	0.2990
Δpcagar	0.5268	0.1501	3.5100	0.0000	0.2326	0.8211
Rain	-0.8378	0.3039	-2.7600	0.0060	-1.4334	-0.2421

Note:

GMM results are one-step estimates with heteroskedasticity-consistent standard errors and test statistics.

Arellano-Bond test that average auto covariance in residuals of order 1 is 0:

H0: no autocorrelation z = -2.17 Pr > z = 0.0298

Arellano-Bond test that average auto covariance in residuals of order 2 is 0:

H0: no autocorrelation z = 1.09 Pr > z = 0.2753

chi2(11) = 19.52 Prob > chi2 = 0.0529

(Sargan is a test of the over identifying restrictions for the GMM estimators, asymptotically

÷2. This test uses the minimised value of the corresponding two-step GMM estimators.)

The accounts of direct agriculture credit and number of indirect agriculture credit accounts. The parameters are estimated using difference GMM. The lagged values of the endogenous and predetermined variables and exogenous variables are used as instrumental variables. The validity of the instruments and assumption of no serial correlation in the error terms are tested using Sargan test for over identifying restrictions and Arellano-Bond test and are found to be satisfactory. The direct agriculture credit amount is significant and positively explains the variation described in the agriculture output. While the number of indirect agriculture credit accounts is significant at 10 per cent level and positive at first lag. That is more people benefi ted from the indirect finance to agriculture current year may lead to higher output next year. As in the case of earlier models agriculture area is having a significant positive impact and rainfall (measured as deviation from normal) is negatively affecting the agriculture output. However, the effect of rainfall deviation from normality to agriculture output got substantially reduced once we control for the financial inclusion indicator like number of people covered under agricultural loan facilities from the formal institutional mechanism. The indirect agriculture amount outstanding and direct number of agriculture accounts is found to be insignificant in explaining the variation in agriculture output.

Table. GMM estimates for the second regression equation

Variable	GMM estimates	Robust Std. Err.	Z	Prob.	[95% Conf. Interval]	
1	2	3	4	5	6	7
Δpcagout (-1)	0.1292	0.1040	1.2400	0.2140	-0.0745	0.3329
Δpcdacam	0.1050	0.0259	4.0500	0.0000	0.0542	0.1558
Δpcdacn	-0.0121	0.0101	-1.2000	0.2320	-0.0318	0.0077
Δpcidacam	-0.1215	0.1017	-1.1900	0.2320	-0.3208	0.0779
Δpcidacam (-1)	-0.0132	0.0252	-0.5200	0.6010	-0.0625	0.0361
Δpcidan	0.1926	0.1461	1.3200	0.1870	-0.0938	0.4791
Δpcidan (-1)	0.2330	0.1251	1.8600	0.0630	-0.0122	0.4783
Δpcagar	0.6321	0.1470	4.3000	0.0000	0.3441	0.9202
Rain	-0.2447	0.2947	-0.8300	0.4060	-0.8223	0.3330

Note:

GMM results are one-step estimates with heteroskedasticity-consistent standard errors and test statistics.

Arellano-Bond test that average autocovariance in residuals of order 1 is 0:

H0: no autocorrelation z = -4.48 Pr > z = 0.0000

Arellano-Bond test that average autocovariance in residuals of order 2 is 0:

H0: no autocorrelation z = 1.61 Pr > z = 0.1082

Sargan test of over-identifying restrictions: chi2(46) = 62.74 Prob > chi2 = 0.051

(Sargan is a test of the over identifying restrictions for the GMM estimators, asymptotically

÷2. This test uses the minimised value of the corresponding two-step GMM estimators.)

INSTITUTIONAL STRUCTURE FOR AGRICULTURAL AND CREDIT POLICY

POLICY MAKING

The main authority concerned with policy design for all aspects of agriculture is the Supreme Agricultural Council (SAC), chaired by the Prime

Minister and composed of representatives of all concerned Ministries. At the administrative apex of agriculture is the Ministry of Agriculture and Agrarian Reforms (MAAR). The main functions such as statistics and planning, extension, agricultural affairs (in charge of implementation of the crop plan), plant protection, livestock, and animal health are replicated at the province and district levels. The functions finally converge at the service units at the field level.

Agriculture has been a part of the centrally planned economic system with Government organizations and agencies closely involved in all production and distribution activities. Fixed and multiple end-use oriented exchange rates, Government monopolies in procurement of all produce, fixed crop prices with balancing subsidies to neutralize production cost increases, rigidly enforced crop plans down to the individual farm level, strict control on imports through licensing and negligible use of private sector resources and energies to generate competition and efficiency were the main characteristics of the economy till the mid-1980s. Since then, there have been many major changes such as: unification of exchange rates, private sector entry into defined areas of agricultural procurement, imports of certain inputs and export of vegetables and fruits, reduced rigidities in crop planning, removal of explicit subsidies, fixation of prices according to production costs and similar measures.

The Permanent Economic Committee in the Prime Minister's Office representing Ministries of Economics and Foreign Trade, Finance, the Central Bank of Syria (CeBS) and the Banks is the principal institutional instrument to conduct monetary policy. The annual credit plan provides for the financing requirements of the public sector, which account for the bulk of bank lending. The CeBS is the major source of lending to the Government with additional resources from the Commercial Bank (CoBS) through obligatory subscription to treasury securities. Public sector undertakings borrow mainly from the CoBS.

Till recently, the chief concern of credit policy has been the need of public enterprises and finding the ways and means to meet their requirements. This situation has changed with increasing the role for the private sector in investments. Compulsions to meet the large and growing public sector credit needs and apprehension that prices might get out of control have combined to keep critical monetary determinants unchanged. Borrowing by banks are 1.44 times and 2.66 times the deposits (Demand and Time Liabilities) in the case of Industrial Bank and Commercial Bank respectively, suggesting that banks rely more on borrowings to lend and invest than on deposit mobilization.

THE FINANCIAL SYSTEM

The financial sector in its entirety is government owned and directed. With the Central Bank of Syria as the banker's bank at the apex, the system consists of five specialized banks, namely, the Commercial Bank of Syria (CoBS), the Agricultural Cooperative Bank (ACB), the Industrial Bank (IB), the People's

Credit Bank (PCB) and Real Estate Bank (REB). The CoBS is the only intermediary for foreign currency dealings and holds foreign currency deposits of companies and individuals. Banks and their allotted segments of activity are summarized.

The Public Debt Fund is a unit of the Ministry of Finance and is a source of finance for development schemes of public sector institutions, and occasionally provides assistance to overcome difficult cash situations, especially if they had been caused by policy decisions of the Government such as postponement of loans affected by drought etc. Surpluses of public institutions are transferred to PDF. Loans from CoBS and from other specialized banks to public institutions are channeled through PDF. Thus, the PDF acts as the overall pool of surpluses and deficits of public enterprises taking from each according to its capacity and giving back to each according to need.

Table. Banks and their customer segments

	ACB	CoBS	IB	PCB	REB
Number of branches	110	53	17	60	16
Farm production					
Farm development					For land, building
Input production, fixed capital					For land, building
Input production working capital					
Input trade, joint ventures of Agricultural Engineers with Syndicate					
Input and output trade working capital					
Village and town traders for farmers' other requisites and input trade, working capital				**	
Agro-processing under Law 10 i.e. with foreign investment, working capital in local currency					
Agro-processing without foreign investment, fixed capital				**	
Agro-processing without foreign investment Working capital					
Opening of Letters of Credit or foreign remittances, inward and outward					

Note:

**PCB's focus is more on small industries, artisans, producer cooperatives and small merchants.

The credit planning process takes place in two parallel streams. The Agricultural Directorate in each governorate receives from MAAR an indicative plan of expected crop production for the area. Taking local seasonal and other conditions into consideration, the Directorate, in consultation with representatives of the Farmers' Federation, prepares the plan for MAAR's approval. ACB is not involved in this process, although credit is an important input. Independently, the ACB branch prepares an annual plan of credit. The plan is largely guided by the previous year's off-take of loans and is not influenced by the production plan sponsored by the Directorate and the Farmer

Federation. At the head office of the ACB, the branch estimates are consolidated for assessing the fund requirements. The fund plan is submitted for approval of the Supreme Agricultural Council, based upon which CeBS is authorized to discount agricultural loans issued by the branches. Branches directly discount their loan documents from CeBS's local branch.

Liquidity in the Credit System: Published figures for the latest available year show that, for the banking system as a whole, against capital reserves and demand and time liabilities/deposits (DTL), after allowing for fixed assets and the statutory reserve requirements, there is a balance lendable availability of SP227 319 million, against which claims on economic sectors was SP255 056 million suggesting a strong demand for funds exerted on the system. Of this, the public sector accounted for SP179 817 million, that is, over 70 per cent.

Interest rates have remained stationary over several years and financial institutions do not have the freedom to fix deposit and lending rates. As interest rates are neither market-driven nor administratively updated to match macro-economic situations, this critical monetary instrument has remained dormant and, in times of inflation, acted as a serious disincentive to savings with consequent contraction of lendable funds for new investment.

INTEREST RATE STRUCTURE

The cost of various facilities and loans from the central bank of the country are as per schedule given below.

Table. Central bank rates of interest to specialized banks (%)

	ACB	CoBS	IB**	PCB
Rediscounts commercial transactions		5	3.5	3.25
Rediscount agricultural transactions	2.5*	3.25	1.75	
Rediscount industrial transactions		4.25	2.75	2.5
Rediscount agricultural financing	2.75*	3.5	2	
Rediscount industrial financing		4.5	3	
Loans and advances commercial		5.75	4.25	3.5-3.75
Loans and advances industrial short term		4.25	2.75	2.75-3
Loans and advances agricultural financing	2.5*	3.25	3	
Loans and advances financing export and storage		3.25	3	
Loans and advances storage of commercial wheat and barley		3.25		
Loans and advances storages of commercial commodities		4.75		

Note:

*For cooperatives the rate is 0.75 per cent lower. ** Rates to IB for rediscounting are lower than those of CoBS by 1.5 per cent.

The schedule of interest rates charged by banks is set.

Table. Banks' rates of interest (%)

Term and instruments	Sector	ACB	CoBS	IB	PCB
Discounting commercial bills	Public	-	7	7.5-8	-
	Coop	-	-	7	7-8
	Private	-	9	9	8-8.5
Loans against export	Public	-	2.25	-	-
	Coop	-	-	-	-
	Private	-	7.5	-	-
Loans against storage of agricultural products and goods to be exported	Public	-	5.5	-	-
	Coop	-	-	-	-
	Private	-	7.5	-	-
Overdrafts	Public	-	7.5	9	-
	Coop	-	-	10	-
	Private	-	9	11	-
Long term loans	Public	4	7	8	-
	Coop	4	-	7.5	7-8
	Private	5.5	8.5	9.5	7-9
Short term loans depending on whether loan is SP50 000 or more	Public	4-6	5.5	7.5	-
	Coop	4-6	-	7.25	7-8
	Private	5.5-7.5	7.5	10	7-9

The additional interest charged to the private sector is 0.5 per cent in the case of ACB, 1 per cent by PCB and 2 per cent by IB. IB's interest rates are higher across the board by 0.5 to 2 per cent for the same borrower category, compared to other banks. It is seen that cooperatives get the benefit of lowest interest and next, the public sector, with private sector subject to the highest rate because of the lower risk that banks attach to lending to public sector agencies guaranteed by the respective Ministries. In refinancing agricultural production loans, CeBS has a discriminatory margin of 25 per cent for discounting loans taken by private farmers who are not members of farmer associations.

THE AGRICULTURAL COOPERATIVE BANK

The ACB combines the functions of loan disbursement, input distribution and crop proceeds disbursement, and the last mentioned function is rendered on behalf of Government agencies for procurement of grain, cotton, seed, vegetable and sugar. ACB has a network of 108 branches distributed over all governorates. Branches operate as independent units, each regarded as a separate profit centre. Each branch reports directly to the Director General.

The ACB collects loans and conducts transactions in accordance with the Public Funds Collection Law, the Syrian Law and the Code of Procedures. The funds and rights of ACB are considered as those of the State Treasury. It has priority in claiming fixed and current assets of the debtor and those of the guarantor in respect of recoveries, regardless of whether or not such assets are mortgaged in favour of ACB, subject only to any charge prior to the date of the issuance of the loan.

Branch managers are authorized to act as registrars of documents on behalf of the Real Estate Office and mortgage endorsements made by them are legally recognized. ACB also has special powers of confiscation, under law, without having to go through elaborate legal procedures. ACB, as the lending agency, has special powers of endorsing collateral charges on ownership titles, which are legally enforceable. This is a unique feature of the Syrian system, encouraging timely repayments and acting as deterrent on willful defaulters.

All stocks of fertilizers from local production or from imports are taken over by the Bank as and when they are produced or imported. ACB is allowed an administrative charge of 2 per cent on imported fertilizers. Stocks are stored in warehouses located in different parts of the country and delivered to cooperatives and farmers at ex-warehouse prices. Seeds are delivered from ACB stores as well as from General Organization for Seed Multiplication's (GOSM) branches.

This again is according to the permitted quantities stipulated in crop licenses. In regard to agricultural chemicals, ACB has an intermediary role of collecting the value of chemicals for control of wheat bugs and herbicides distributed by the Directorate, from the crop proceeds.

It is to be noted that of the total mobilization, term deposits account only for a small proportion weakening the resource base and increasing reliance on borrowings for lending operations.

Table. ACB deposits

ACB Deposit Growth	**1996**	**1997**	**1998**	**1999**
Demand deposits & current accts	5 931	7 106	8 402	7 393
Term & savings deposits	783	916	1 089	1 223
Total	6 714	8 022	9 491	8 616
Term deposits as % of total	12	11	11	14

The sum against legal cases, indicative of likely future bad debts, has increased sharply and steadily from 8 per cent to 31 per cent. Drought conditions in 1998 and 1999 have resulted in large defaults. Prior to this, the situation was reasonably under control. Loans outstanding and to be collected are 2.54 times the annual disbursement.

Table. ACB loans outstanding

	1996	1997	1998	1999
Outstanding loans	18 452	17 957	15 651	15 029
Those under legal action	1 588	2 940	4 526	6 835
Total	20 040	20 897	20 177	21 864
Legal cases as % of total	8	14	22	31

About SP1.8 billion of the receivables are considered doubtful of recovery out of the sum of SP6.8 billion under legal action. It is possible that a significant part of the SP6.8 billion may have to be written off in stages. There being no cash flow from this "asset" there is bound to be considerable liquidity pressure especially because discounting with CeBS involves funding by ACB of 25 per cent margin and bridging of the balance 75 per cent for a few days - from disbursement to farmers and subsequent realization of discounted proceeds from CeBS. To improve liquidity, it seems necessary to infuse about 50 per cent of the debts under legal action equivalent to SP3.4 billion, partly as fresh capital and the remaining as loan.

Available information and data make it difficult to assign the assets and liabilities to the two main operations of ACB, namely, input distribution and banking, to ascertain the efficiency with which they utilize financial resources. For ACB, as a total entity, profits are declining from year to year. The loss in banking in 1999 could be due to non-recovery of interest on account of waivers/postponements to meet drought conditions in 1998 and 1999. Return on capital (ROC) is poor because of prohibitively high transaction costs at 11.59 per cent of loan disbursement.

Table. ACB profitability

	Commercial	Banking	Total	Capital	ROC
	(SP million)				(%)
1996	-7	910	903	1 916	47
1997	-185	694	509	1 982	26
1998	-57	160	103	2 017	5
1999	239	-132	107	5 153	2

THE COOPERATIVES

Although cooperatives, in the manner in which they operate in Syria, cannot be considered an intermediary financial institution at the grass root level, they play a vital part in the whole system of input supply, procurement, credit disbursement, crop proceed disbursement, dues collection and in providing farmer groups' collective guarantee for repayments. But for the cooperatives, the workload and cost per transaction for ACB, which is already high, would be even higher. This kind of intermediation limited to physical intervention, free of fiduciary involvement, has helped in steering clear of the susceptibility to mismanagement and abusive practices that often lead cooperatives to financial failures. The better run cooperatives have the potential to graduate to a more useful role of promoting and mobilizing savings, and acting as mutual help societies. It is critical to promote this concept and reduce the retailing role of credit by ACB, which results in an enormous number of transactions and paper work unduly increasing cost. This would also improve service by avoiding convergence of a large number of farmers, long queues and indefinite waiting during season.

Table. Cooperatives - Structure 1999

Multi-purpose	Others	Total	Membership	Capital ('000 SP)	Area ('000 ha)
4 145	1 250	5 395	932 639	90 564	2 488

Cooperatives are vertically organized as Peasant Unions at district and governorate levels, further integrated as the Peasants Union Federation (PUF) at the national level.

At these levels, they participate in the deliberations of the Agricultural Council. The PUF participates in the Supreme Planning Council and the Supreme Agricultural Council and has a say in matters affecting farming such as pricing, credit and marketing.

ANALYSIS OF CREDIT TO RURAL HOUSEHOLDS

Mechanisms and outcomes of the rural credit system

Farm credit accounted for 11.28 per cent of total available credit in 1990, and this declined to 9.88 per cent in 1999. There is no separate information on credit extended to input and output agencies, both in distribution and in manufacture, and to those engaged in agro-processing and exports. ACB extends assistance to private farmers, cooperative member farmers, cooperatives, Peasant Unions and federations and public sector organizations engaged in agriculture.

The private sector has access, if no society is functioning in the same area. In the case of medium- or long-term loans, the access procedure is elaborate and time-consuming. Farmers find it difficult to obtain loans for machinery like harvesters and tractors and have to depend on supplier credit at high interest rates of 20-30 per cent. According to them, lesser availability of medium- and long-term loans, affected important activities like land reclamation and fruit tree replanting.

Loans products

Short-term credit is made available for farm expenses such as plowing, harvesting, irrigation and fuel, cost of inputs, for small tools and for animal feeds and veterinary medicines. Short-term loans are for a period of 300 days and are given in cash and in kind as inputs. Medium-term credit for periods not exceeding five years is extended for greenhouses, forest tree planting, purchase of livestock, digging of canals for irrigation, equipment for poultry farms and machinery for grading, waxing and packing.

Long-term credit for periods of ten years or less is aimed at financing construction of stores, land improvement, forestry projects, fruit tree planting programmes and cold storage facilities. It is generally restricted to state farms.

Working capital is provided to agriculture graduates who are in contract with the agriculture engineers' syndicate or with the Peasant Union. No private sector dealers in inputs or output without a contract are eligible for assistance. These joint ventures, in return for the guarantee extended by the syndicate/ union to ACB for the working capital loan, should give away to the syndicate/ union 40 per cent of the profits. Working capital is given as a lump sum cash loan and not as a drawing facility subject to a limit and subject to availability of security in the form of stocks. The maximum loan under this scheme is SP300 000 against the syndicate's/union's guarantee, and is returnable in six months. As working capital is required continuously, the loan needs renewal every time it is repaid, considerably reducing its usefulness in terms of convenience, continuity and cost.

Loans for land reclamation are subject to a standard ceiling whereas the actual fund needed may be higher depending on the nature of the terrain and the soil structure. The ceilings of SP5 900 per dunnum for mountainous terrain and SP4 400 for flat land appear to have not been revised after they were fixed over ten years ago. The actual costs are estimated by farmers at SP15 000 for mountainous terrain and SP6 000 to SP10 000 for other land types; allowing for some exaggeration by complaining farmers, it seems that there is scope for review and greater flexibility. As regards finance for fruit tree planting programmes, the term of five years is clearly insufficient, as most fruit trees take longer (six-seven years for seedlings for apples, for example) to attain a commercially viable level of yields. Replanting cost for apples is about SP195 000 for irrigated farms and SP96 000 for rain-fed farms, whereas loan sizes do not often match this need.

A significant feature is the low proportion of medium- and long-term loans, and the declining percentage from year to year - from 17 per cent in 1997, to 15 per cent in 1998 and further down to 14 per cent in 1999. Medium- and long-term loans carry a margin of 25 per cent for discounting with the CeBS and it is possible that ACB's ability to enlarge the quantum of medium- and long-term loans is constricted by fund availability. Medium- and long-term loans are important for increasing productivity, improving quality and value addition and raising farmers' debt capacity

Loans disbursement, security and recovery

Loan amounts are determined strictly on the basis of input eligibilities determined in the crop license. Loan sums and inputs in kind are given to the cooperative for disbursement to individual members according to their eligibility. Private farmers who are not members of farmer associations, apply individually and make individual arrangements for drawing the cash part and taking delivery of inputs.

Table. Loans by term (SP million)

Term	1997	1998	1999
Long term 10 years	252	181	123
Medium term 5 years	1 978	1 701	1 278
Short term 300 days	4 248	4 023	3 442
Loans in kind - Short term	6 920	6 735	5 366
TOTAL	13 398	12 640	10 209

Table. Loans by sector (SP million)

Years	Public sector	Cooperative sector	Private sector	Total
1995	258	7 128	8 134	15 520
1996	210	6 920	7 932	15 062
1997	283	6 065	7 050	13 398
1998	214	5 666	6 760	12 640
1999	189	4 488	5 532	10 209
2000	216	3 703	4 839	8 758

Lending has been steadily declining. Private farmers accounted for more than 50 per cent and cooperative members about 45 per cent, the remaining going to the state farms. Many non-member farmers prefer to stay out of the association, as they do not wish to be penalized for other members' defaults. Many wished that they could buy inputs from nearby sources for cash. The loans recovery enforcement mechanism is effective and as such, repayments are generally satisfactory except in times of poor rainfall and drought, as for the last few years.

Table. Recovery situation

Loan recovery rate	1999	1998	1997	1996
Receivables*	15 810	17 919	15 509	16 118
Collections	8 960	13 386	12 569	14 530
%	57	75	81	90

Note:

*Receivables are total of loans due and overdue during year under review.

In times of natural calamities like drought, a committee, appointed by the Governor, consisting of representatives of ACB, administrative authority of the affected area, the MAAR and the Peasant Union assesses the extent of damage based upon which ACB Board is authorized to grant full or partial deferment. If damage is not more than 30 per cent of the debtors' average annual yield, 50 per cent of the sum due is deferred, and 100 per cent is deferred if the damage is more than 60 per cent. Deferment applies only to principal and interest must continue to be paid. The repayment in installments is allowed over not longer than three years. It was noted, however, that rescheduling is not allowed for medium-and long-term loans when repayment capacity is affected by drought.

In the case of crops sold in the open market such as citrus, growers may at times be financially disabled by bumper crops and precipitous fall in prices due to weak post-harvest supports. Farmers' ability to keep to the schedule of repayment is affected in such circumstances and there is no special provision to extend relief on repayment of loans to meet such market situations beyond the growers' control.

Table. Summary of collateral requirements for agricultural loans

	Cooperatives	**Public sector**	**Private sector**
Short term	Guarantee by government or crop security, collective guarantee for cooperative members.	Guarantee by controlling Ministry or crop.	Personal guarantee of 2 farmers.
	Limit of SP2.5 million.	Limit of SP2.5 million for state farms.	Limit of SP500 000 per farmer.
Medium and Long term	In-kind collateral except where the society or the association or district Peasant Union or PUF guarantees and provided individual member/member society's share is within prescribed limits.	Government guarantee up to SP2.5 million and in-kind collateral for higher amounts.	In-kind collateral.
	Limit of SP1.5 million for the society and SP300 000 per member for medium term and SP1.5 million and SP100 000 respectively for long term.	Limit of SP1.5 million.	Limit of SP1 million.

Real estate collateral is the most widely accepted form of security and value cover for loans ranging between 80-100 per cent.

The real cost of credit

The general structure of interest rates. In addition to interest, association members have to pay the association 1 per cent of the loan for their services. Although the term is 300 days, farmers do not find any advantage in delaying the sale of the harvest, as the official prices for the crop are the same for all months. Farmers who are in debt sell the crop the very day of harvest, particularly cereals. Sometimes lack of proper storage at village level also compels immediate disposal. Consequently, the duration of the loan is only about

180 days. As such, the commission of 1 per cent is equivalent to 2 per cent annualized. It is estimated that other incidental expenses incurred in getting through the formalities and visiting the bank to submit the application and to collect the cash/input would be about SP200. If the average loan is reckoned as SP20 000, this is equal to 1 per cent and again annualizes as 2 per cent. However, this expense is incurred by private farmers and not by association members. The commission, on the other hand, is incurred by association members and not by private members. Thus, it can be said that the effective cost is about 2 per cent more than the official rate. To this must be added 0.1 per cent administration charge, 0.15 per cent stamp duty and 0.25 per cent on each bag of fertilizer - totalling 0.5 per cent which, again, annualizes as 1 per cent. The total effective rate is 3 per cent more than the apparent rate.

Fruit growing farmers mentioned that they had to bear "quite a lot of expenses" in getting the loan. Fees for the inspection team, collection agent's fee and fee for registering collateral, besides the incidental expenses for completion of formalities constitute the additional charges. However, because of the extended term of the loan, these may not annualize at more than 1 per cent even if the total expenses are 5 per cent of the loan value. The real cost may lie in the additional penal charges of 1 per cent per month or 12 per cent per annum, payable if the loan is not discharged in five years. In situations where a waiting of, two years beyond the official loan term is involved to discharge the loan, the additional 12 per cent per year for the two years is 24 per cent.

Spread over a seven-year term, this additional incidence is, arithmetically, about 3.5 per cent per annum. Thus, the effective additional cost would be about 4.5 per cent (inclusive of 1 per cent mentioned earlier) per year on medium-term loans.

Alternative sources of finance for rural households

Fruit growers take medium- and long-term loans from friends or relatives living outside the country, mostly Arab countries and Lebanon. Machinery suppliers are also a common source of finance for purchase of equipment. The procedures are short and simple although interest rates are high. For production expenses, the alternative sources are the input supplier and the output dealers, exporters' agents and cold storage units.

Input dealers are generally small traders and do not have the capacity to extend credit covering the whole crop duration. In the case of agro-chemicals, however, profit margins being good, because of the cost-plus pricing system and withdrawal of control, large agro-chemical marketing firms are able to extend prolonged credit to their wholesalers enabling the latter, in turn, to sell on credit to farmers. Output dealers, exporters' agents and cold storage units are active in fruit and vegetable growing areas. Financing by them takes several forms. Direct advances ahead of the season are given with, and sometimes without, an agreement on the unit price at which the harvest would be sold. The farmer is thus under obligation to sell the crop to the dealer at a price to be negotiated after the money has been taken and finds himself under pressure to accept any price offered by the lender. Another form of financing by the output buyer is to agree on a lump sum to be paid to the grower for the entire output. The lump sum is paid in suitable installments to enable the grower to meet production expenses. The expected yield is estimated by the contractor in such a way that he recovers interest at a high rate. Sometimes to meet production expenses farmers liquidate their assets. When the need is large they are forced to sell the house and equipment. Smaller farmers sell their animals.

ASSESSMENT OF ACB

There is little flexibility in loan structures to suit different crop and cash flow situations. There is no distinction between a good borrower and a bad one, between a borrower who keeps his value addition in the bank and one who either does not produce the value addition or squanders it. The author hardly heard from any of the farmer group reports of high interest rates. On the other hand, their readiness to resort to more convenient and costlier alternative sources is indicative of the higher value they place upon better service and easy access than on cost. The low percentage of medium- and long-term loans is not conducive to promote long-term investment to improve farm productivity.

Subsidy dependence

The subsidy dependence of the rural financial institution is the percentage by which its average on-lending rate would have to increase to make it sustainable (Jacob Yaron and Others, Rural Finance, The World Bank, 1997). In the present case, the present average lending rate of 7.44 per cent needs to be increased, according to the mission's estimate, by 3.26 per cent points towards interest, plus 9.59 per cent points towards additional transaction cost incurred by ACB - that is, by a total of 12.85 per cent. This gives a subsidy dependence index of 1.73 (12.85/7.44), which is extremely high and untenable. Of this 1.73, untenably low interest rates account for 0.44 and the very high transaction cost for the balance 1.29. This stresses the urgency of restructuring ACB and lowering its transaction cost per loan.

Effectiveness

ACB has partially succeeded in meeting the needs of the farm sector for cash and in-kind loans. Loan volumes have been declining year after year. An important objective of ACB is to encourage cooperative societies. While cooperatives have been supported through concessional terms, not much effort has been made to promote them as second tier institutions for savings mobilization and retail lending. This would have shifted ACB's focus from retail lending to wholesale lending, leaving to grassroot financial institutions the responsibility of credit rating of borrowers and ensuring timely collections. Encouragement of mechanization, micro-irrigation and similar activities are part of ACB's objectives, but these have not made satisfactory progress as indicated by the low ratio of medium-long term loans.

Outreach is normally measured by several indicators, such as number of borrowers, average loan size as proxy for income level, percentage lent to zones with less rainfall, percentage of women borrowers and number of branches. It is observed that the number of beneficiaries in 1999 was only 54 per cent of the number in 1994. Even allowing for the poor rainfall in 1999 and 1998, it is seen that contraction had already set in from 1995. The number of borrowers was 749 703 in 1989, nearly three times the client base in 1999. This trend is indeed cause for concern. The possible reason could be either that those loans are not reaching farmers or that farmers are unwilling to utilize the facility from the Bank resorting to alternative avenues, or that farmers are becoming self-sufficient for financing production activity. The last mentioned possibility seems most unlikely, going by the impression one gains from meeting farmer groups in different parts of the country. Loans advanced have also been going down - in 2000 to 61 per cent of the 1994 level despite increase of area during the same period. Not surprisingly, therefore, the average size of loans has been increasing and is presently 1.32 times the size of six years ago. The higher average size is suggestive of a movement towards larger farmers and/or towards better-endowed zones. Statistical information is not available on lending by zones or lending to women farmers, to ascertain the proportion of credit extended to disadvantaged segments.

There has been no major expansion of ACB's branch network in recent years. It is recognized that each branch has to be self-sustaining in terms of earnings as far as possible. When overall profitability is satisfactory, there is room to cross-subsidize branches in remote areas. Another possibility is providing skeleton service or weekly rotating branch with the same staff, taking turns over two or three small branches in nearby locations. No innovative work of this nature is in evidence.

The very low ratio of medium- and long-term lending is hardly conducive to farm productivity and enhancement of its debt capacity. ACB is unfortunately under constant liquidity stress and is not able to act beyond the role of a passive credit conduit.

Delegation of powers

The Credit Committee at the branch, consisting of the branch manager and heads of credit, collection and information divisions, is authorized to sanction to a maximum of SP50 000 per farmer for short-term loans, SP200 000 for medium- term loans and SP100 000 for long-term loans. For societies, federations, governorate unions and state farms the corresponding limits are SP2.5 million, 1.5 million and 1.5 million respectively. Sums in excess are sent to the Board of Directors, consisting of the Director General and Directors of divisions. The Board has authority to sanction short-term loans without any ceiling on their authority, but medium- and long-term loans subject to limits of SP500 000 and 250 000 respectively.

The Bank management does not have the authority to design and introduce new loan or savings products suited to different crops, clientele and conditions. Insufficient delegation is not conducive to placing upon bank managements the responsibility for mobilizing savings, operating within prudence norms, controlling costs, designing products and finally, producing an acceptable return. There are no goals to score and no competitor to outplay.

CREDIT TO SERVICES PROVIDERS

FOOD INDUSTRY

Food industry's share of borrowing from the banking system is steadily on the increase, from 20 per cent in 1990 to 39 per cent in 1999. In absolute terms it has grown from SP298 million to SP976 million, by well over three times. Significantly, while sums borrowed have increased, the number of borrowers has declined from 1 129 to 976 suggesting that average size of loans has increased. This might imply a use of better technology, or higher degree of automation or scale increase, or a combination of any of these. CoBS does not have figures classified according to industry and nature of activity.

Against a theoretical potential of SP73 252 million, calculated on the basis of approved projects under Law 10, IB has met cumulatively SP5 844 million from 1990 to 1999 - less than 10 per cent of the potential. To the extent of working capital funding from CoBS, the gap would narrow. The low percentage is also due to a poor rate of maturation of approved investments, as well as to low participation by the private sector in bank credit.

INPUT DISTRIBUTION

With regard to fertilizer, ACB is the sole distributor for the whole country. ACB finances its purchases from the GFC and from GEZA, with original Government capital and additional infusions, if any, of monies from depositors on which an interest of 8 per cent is payable and by rediscounting its sale bills with the Central Bank of Syria at rates ranging from 2-3 per cent. The rediscounting rate is dependent upon the term of loans under rediscount and whether the loans are to cooperative or to other sectors, the former getting a preferential rate. Liquidity gaps arising out of defaults are covered by commercial borrowings from other banks or from the public debt fund. Whenever loans are rescheduled at the behest of the Government, under drought and similar natural calamities, the Government funds the delayed cash inflow.

GOSM, the seed producing organization, meets its working capital requirements through commercial borrowing from CoBS.

The CoBS provides working capital to private traders, importers, exporters and agro-processing companies. Working capital is extended as a drawing facility and not as a loan in the sense of a cash disbursement. Private importers and distributors of seeds and agricultural chemicals met by the mission were not found to be overly enthusiastic about the facilities available from the CoBS. Many of them rely on owners' funds and borrowings or capital contributions. The reason given is that borrowing from the Commercial Bank of Syria is fraught with time-involving procedural hurdles eventually resulting in credit limits that are too small for the scale of operation. The interest on borrowings from the Commercial Bank is 9 per cent for private sector and 7.5 per cent for the public sector, and thus places a handicap on the private sector.

INPUT RETAIL

ACB gives working capital loans to joint ventures (JV) at 7.5 per cent. The JV, however, has to pay a 4 per cent commission to the syndicate or farmer union guaranteeing the loan, this represents 40 per cent of the profit share. At a distribution margin of 10 per cent on inputs, the JV has to give 4 per cent towards this share. As this is an absolute sum and not an annual rate, the 4 per cent translates to 8 per cent on an annual basis on the six-month duration allowed for the loan.

While the CoBS finances medium and large agricultural trading companies, the PCB focuses on small traders, professionals and vocational categories. Both long-term and short-term credit are available depending on the purpose. Ceiling for credit granted by PCB has been increased to SP500 000 for short-term loans and to SP one million for medium-term loans. In respect of medium and long-term loans, invariably a fee of 2.5 to 3 per cent is charged from the principal at the time of disbursement, which has the effect of increasing the effective rate by 1 per cent. PCB deducts the interest in advance in the case of loans of less than SP1 million and borrowers of higher values have the option of paying interest in advance or along with each installment. Security for long-term loans is as high as five times the value of the loan, and it is surprising that such stringent terms should be imposed upon those whose resources are limited and whose economic status the PCB is aiming to improve.

POST-HARVEST ACTIVITIES

For agro-processing companies, fixed capital for acquiring plant and machinery, land and buildings as well as working capital are available from IB. It finances all activities except requirements of foreign currency, including the opening of Letters of Credit in favour of foreign parties. For these, the borrower has to go to the CoBS. IB also extends working capital in local currency. Generally, IB contributes 30 per cent of the project cost with the promoters contributing the balance. Olive processing, cold storage, manufacture of veterinary and agricultural chemicals, fabrication of agricultural machinery, and export oriented activities are classified as of higher priority eligible for 50 per cent contribution from IB towards fixed capital for initial establishment and for expansion, the promoter group contributing the balance of 50 per cent. Security for the long-term loan is in the form of property collateral in addition to the assets of the establishment. For short-term loans, two guarantors are needed, each of them with property to cover 100 per cent of the loan, the valuation taken at 75 per cent of market value. Availability of working capital seems a major problem for many agro-processing units.

LOW PARTICIPATION BY PRIVATE SECTOR

The reluctance of the private sector to avail itself of bank credit, assuming that it is available without much procedure and red tape is due to several reasons. Religious considerations preclude lending and borrowing against interest. The private sector is still unfamiliar with legitimate procedural and security requirements of the banking industry. Many promoters do not appreciate the criticality of working capital and do not tie up working capital along with financial arrangements for capital for plant and machinery. They look for working capital after commencement of production, and quite often find themselves having to grapple with an acute cash crunch. Slow progress in lending to the private sector is also due to the fact that banks still look upon the public sector as their chief customer. Public sector lending is the softer part of the market, needing less effort in evaluation, securitizing, monitoring and recovery. The preferential treatment to public sector borrowers (example: lower interest) not only places private sector at a competitive handicap, but sends out a wrong signal from the government to banks that the private sector is lower priority.

ALTERNATIVE SOURCES FOR SERVICE PROVIDERS

Inadequate support from the banking system leads private entrepreneurs to resort to many ways of funding new projects and expansion of existing ones. Many invest their own savings and those of close relations and friends, forming the promoter group. Promoter exposure is necessary up to 50 or 60 per cent of project cost when a lending institution extends financial assistance. However, in the current situation in Syria, the promoter group is obliged to "own" 100 per cent of the capital to meet the project cost instead of 50-60 per cent. In the absence of any leverage through borrowing, the cost of capital is very high and many projects that are worthwhile with partial financing by the bank may not be attractive for investment entirely with owners' capital.

Private funding is apparently taking place on a fairly large scale outside the banking system. Return expectations in the informal market are anywhere from 24 to 36 per cent, and sometimes even higher. Private flow of funds between lenders and borrowers is not the most efficient way of establishing a fair equilibrium price for capital, as it is determined very often by the mutual bargaining strength of the two parties without either of them having any information on alternative availability and terms. The same funds flowing through the banking system would result in time and demand deposits, greater lending ability with banks, more quasi money, multiplier effect of deposits creating loans and loans creating deposits and a generally higher liquidity. Viewed from this angle, private flow of funds only infuses money into the market, and that too inefficiently. The informal flow denies the economic system the benefits of the multiplier effect and market efficiency.

An important source of funding is from external commercial borrowing. Although this is illegal, some promoters do resort to this source where interest is low but exposure to adverse changes in the exchange rate is high. Allowing for the exchange rate change, the cost is still attractive because of the high interest charges in the informal market.

The informal funding of projects leads to a situation where only projects with very high returns would pass the test. As a result, many projects with attractive returns, by normal standards, would get neglected, arresting economic growth. It also affects expansion and modernization necessary to acquire global competitiveness. Lack of competitiveness forces manufacturers to confine themselves to the domestic market and exerts pressure on the Government to protect local industry for its low efficiency and poor product design and quality.

AGRICULTURAL POLICIES AND THE RURAL FINANCIAL SYSTEM

SAVINGS

Savings has been appropriately called the "forgotten half of rural finance", as provision of financial services often focuses more on extending credit, neglecting other services like savings, family budgeting and insurance. Earning rates on savings having remained constant for over ten years now. The attractiveness of the rate has varied depending on the inflation rate and opportunity cost of capital. The following picture emerges deflating the retail price index change from the interest rate of 8 per cent. Savings increased when "real" interest rate improved and more so when it turned positive. Savings in 1990 were a bare SP66 291 million. Since then, it has increased more than four-fold. This also goes to show that opportunity cost of capital in the informal market is not such a heavy counter-force as to dampen the effect of improvement in positive rates of return in the formal system. Although informal lending rates are cited as varying from 24 to 36 per cent, such markets not being so fluid and well organized, access to opportunities may not be easy, apart from the higher risk involved in such investments compared to keeping money in an institution having the backing of the Government.

Table. "Real" interest on savings and savings growth

	1994	1995	1996	1997	1998	1999
General Index	154	170	185	189	188	184
Increase over previous year	-	10.4%	8.8%	2.2%	- 0.5%	-2.1%
Savings Interest of 8 % minus inflation	-	- 2.4%	- 0.8%	+5.8%	+8.5%	+10.1%
Savings - Banks + Post Office + Investment Bond - SP million	141 719	159 535	179 003	204 041	228 675	278 437
Increase in Savings SP million	-	17 816	19 468	25 038	24 634	49 762

Several factors inhibit savings mobilization. Banks do not have the freedom to design savings products carrying different rates of return and cash flow features to meet varying saving characteristics. Rates and other terms being standard, there is a sterile uniformity among the banks. This lack of variation dampens any semblance of competition to attract savers. As lending rates cannot be varied by the management, there is no way for the bank being able to distinguish a good customer with large deposits and maintaining a good account from, say, a one-time borrower. Any disciplinary minimum ratio of banks' own funds and mobilization as a condition of refinancing and discounting, if it exists, is weakly enforced. So, there is no pressure to mobilize savings and deposits. Where agricultural credit is concerned, the low rate of interest based on 2.5 per cent discounting by the central bank to meet all lending needs, has taken away the purpose of mobilizing deposits costing 8 per cent. More the deposits, the higher would be the loss to ACB.

Total demand for funds has grown 3.2 times and loans to sectors other than the public sector have grown faster than that for the public sector apparently stimulated by the policy reforms. The public sector, however, continues to dominate the capital market, with its share dropping a little bit from 74.72 to 70.46 per cent. In absolute terms also, public sector absorption of available resources is still very substantial indeed.

THE FINANCIAL SECTOR IN NEED OF ENLARGEMENT

Priority areas enumerated in the agricultural policy have not fared well. Loans for irrigation declined in 1999 to a little more than a third of 1990. Greenhouses, which form the thrust for improved quality and competitive costs for export, have limped from SP301 million to SP475 million in 1999. The share of these special purpose loans declined from 20 per cent of total loans to 12 per cent in 1999, besides registering a fall in absolute terms from SP1 695 million to SP1 271 million.

Table. Loans for agriculture priority purposes (SP million)

Purpose	1990	1995	1996	1998	1999
Irrigation	1329	1292	1076	531	513
Greenhouse	301	865	742	510	475
Machinery	65	423	414	368	283
Sub-total	1695	2580	2232	1409	1271
Total loans	8607	15440	15060	12599	10222
Percentage	20	17	15	11	12

Funding of food related industries by IB increased from SP298 million in 1990 to SP976 million in 1999, registering a growth of 3.3 times. The share of this sector grew from 20 per cent in 1990 to 39 per cent in 1999 of the total lending by IB.

Syrian agriculture has responded well over the years to rapidly increasing population, over 3 per cent annually till the 1980s and close to that number thereafter, by providing adequate supply of calories. Wheat production increased from1.2 million tonnes in 1991 to 2.5 million tonnes in 1998, ignoring the steep fall in 1999 because of acute drought conditions - an increase of 101 per cent. Similarly rain-fed wheat, barley, lentils and chickpeas, the principal food items, registered increases over the same period of 46 per cent, 67 per cent (but in 1996 after which there has been a decline), 213 per cent and 216 per cent respectively, presenting, on the whole a very good performance. Loans classified by crops show that 71 per cent of the loans-in-kind went to wheat and 19 per cent for cotton. All other crops took up the remaining 10 per cent of loans issued. This is indicative of the narrow base of the formal rural credit system. This, together with the trend of average loan size having increased by 32 per cent in the last few years, shows that formal credit has enormous potential for reaching out to other crops and smaller farmers to increase overall productivity.

TYPES OF AGRICULTURAL LOANS

In this guide, agricultural loans are categorized as short-term, intermediate-term or long-term, depending on their maturity. Lenders often describe loans by the purpose or terms of the loan. For example short-termloans are often used for operating expenses. Loan maturity usually matches the length of the agricultural production cycle (*e.g.*, 3 to 18 months), hence a short-term loan. However, this may be described as line-of-credit financing under a credit commitment, which specifies the amount and timing of the disbursements and payments of the loan. The line-of-credit may be a single disbursement due at a specified future date or arevolving line-of-credit in which the borrower may borrow and repay as needed during a specified time period, usually subject to a maximum borrowing level. On a non-revolving line-of-credit, a borrower is entitled to a specified amount of funds, and repayment does not allow the borrower to draw those funds again. A non-revolving line-of-credit is sometimes referred to as a draw note.

Intermediate-term loans are used to finance depreciable assets such as machinery, equipment, breeding livestock and improvements. In addition, intermediate-term loans are sometimes used to restructure a borrower's balance sheet to provided additional working capital. Lenders often describe them as capital, orinstallment, loans. Loans usually range from 18 months to 10 years.

Long-term loans are used to acquire, construct and develop land and buildings, and usually are amortized over periods longer than 10 years. Lenders may describe them as real estate mortgages because they are usually secured by real estate. Long-term loans are sometimes referred to as contract financing, in which case a seller provides financing directly to a buyer.

LOAN DOCUMENTS

Loan transactions typically include several documents for the borrower to sign, depending upon the type of loan. The note or promissory note is a document in which the borrower agrees to repay a loan at a stipulated interest rate within a specified period of time. The note may specify a variable, fixed or adjustable rate, and whether line-of-credit financing is being used. A loan agreement is a written agreement between a lender and a borrower stipulating the terms and conditions associated with a financing transaction, and the expectations and rights of the parties involved. The loan agreement may indicate reporting requirements, possible sanctions for lack of borrower performance and any restrictions placed on a borrower.

A security agreement is a legal document signed by a borrower granting a security interest to a lender in specified personal property pledged as collateral to secure a loan. Essentially, a security agreement states what happens to the collateral if a borrower fails to perform as promised. A financing statement is a document filed by a lender with public official. The statement reports the security interest or lien on the borrower's non-real estate assets. The mortgage serves the same purpose in financing real estate.

TERMS AND CONDITIONS OF THE LOAN

Disbursement of Funds

Disbursements for intermediate- and long-term loans are usually a single payment advanced at a specified time. Some short-term operating loans may be single disbursements, but the trend in the lending industry is to establish lines-of-credit. This feature allows the borrower to reduce interest costs by using funds when needed and repaying funds as surplus cash is available. Disbursement of funds on lines-of-credit is handled many ways. Many commercial banks allow the customer to phone or electronically submit a request for a specified amount to be deposited into the borrower's checking account. The borrower's loan balance is increased and funds are added to the borrower's account. Or, the lender may provide the borrower a book of drafts. A draft can be used instead of a check to pay bills. The borrower's loan balance increases when the draft clears the financial system and returns to the borrower's financial institution. Lenders usually restrict drafts to business-related expenses.

Payment Type

Payment type refers to the method of repayment. Payments on line-of-credit financing generally occur when the borrower has surplus funds. The lender usually establishes a payment schedule for intermediate- and long-term loans. A borrower should ask the lender to produce a copy of a payment schedule that specifies principal and interest payments over the life of the loan. The borrower can then compare payment patterns on different loans.

A borrower should be aware of any demand clauses in a note or loan agreement. A demand clause is a provision that allows the lender to demand payment at *any* time. Even though the demand provisions are seldom carried out, a borrower should be *comfortable* with paying the loan upon demand, especially in times of economic uncertainty.

There are three common payment types. One payment type for intermediate- or long-term loans is the fixed payment method. This method requires a fixed payment (interest plus principal), which repays a loan over a specified period of time at a specified interest rate. This repayment process if often referred to as equal amortization. Part of each payment is allocated to principal and part to interest, with successive payments retiring more and more principal.

Another way to calculate the payment on an intermediate- or long-term loan is fixed principal payment with interest due on the unpaid balance. The fixed principal amount is usually calculated by dividing the loan amount by the total number of payments. Under this method, the initial payments of principal and interest are the largest, and the ability to cash flow these payments must be considered. This method of payment requires less total interest over the life of the loan because more of the principal is repaid earlier in the loan.

A comparison between the fixed payment method and the fixed principal method. The loan is for $100,000, to be repaid over five years at 10 per cent interest. The payment remains constant ($26,380) with the fixed payment method. With the fixed principal method, the annual payment ranges from $30,000 in year 1 to $22,000 in year 5. Total interest payments are $1,898 higher with the fixed payment method. A third payment type is a balloon payment loan. Balloon payment loans are relatively shorter-term loans (*e.g.*, five years). At the end of the period, the entire unpaid balance of the loan is due; the principal must either be paid in full or new loan terms must be negotiated. The initial payments are usually based on a longer amortization period (*e.g.*, 10 to 30 years) under the assumption that the loan will be paid off, renewed or financed at maturity. If interest rates fall and credit conditions improve, a borrower could negotiate more favorable loan terms at renewal. On the other hand, if interest rates rise or credit tightens, the loan terms may become less favorable. In addition, the borrower's risk is considerably higher since the lender may decide not to renew the loan at maturity. Borrowers considering balloon payment loans need to enquire about the fees added each time the loan is renewed.

Table. Fixed payment vs. fixed principal.

Loan Terms: $100,000, five years, 10% interest, one payment per year.										
	Fixed Payment Method					Fixed Principal Method				
Year	Beginning Balance	Principal Payment	Interest Payment	Total Payment	Ending Balance	Beginning Balance	Principal Balance	Interest Payment	Total Payment	Ending Balance
1	100,000	16,380	10,000	26,380	83,620	100,000	20,000	10,000	30,000	80,000
2	83,620	18,018	8,362	26,380	65,602	80,000	20,000	8,000	28,000	60,000
3	65,602	19,820	6,560	26,380	45,782	60,000	20,000	6,000	26,000	40,000
4	45,782	21,802	4,578	26,380	23,980	40,000	20,000	4,000	24,000	20,000
5	23,980	23,980	2,398	26,378	0	20,000	20,000	2,000	22,000	0
Total		100,000	31,898	131,898	NA	NA	100,000	30,000	130,000	NA

A balloon payment loan. Payments in the first four years are identical to a 20-year amortized loan. After the fifth payment, $10,660 of the loan has been paid off, leaving an outstanding loan balance of $89,340. This amount must be paid in full or refinanced at interest rates prevailing in year 5. Although the lender may refinance the balloon payment loan, there is *no* legal obligation to do so.

Table. Payment pattern of a balloon payment loan.

Loan Terms: $100,000, five years, 10% interest, one payment per year based on a 20-year amortization.					
Year	**Beginning Balance**	**Principal Payment**	**Interest Payment**	**Total Payment**	**Ending Balance**
1	100,000	1,746	10,000	11,746	98,254
2	98,254	1,921	9,825	11,746	96,333
3	96,333	2,113	9,633	11,746	94,220
4	94,220	2,324	9,422	11,746	91,898
5	91,896	2,556	9,190	11,746	89,340
5 (Balloon Payment)	89,340	89,340	0	89,340	0

The decision to renew will be based on the lender's consideration of credit and economic factors as they apply at the time of renewal. A borrower selecting a balloon payment loan should be comfortable with risks associated with balloon payment loans. If a borrower is considering refinancing with a different lender upon maturity, additional administrative and closing costs may be incurred.

Interest Rate

Since it is the visible "price tag" of a loan, the interest rate is often used to compare loans. Loans carry fixed, adjustable or variable interest rates. A fixed rate loan carries the same interest rate until the loan is paid off. A variable or adjustable rate loan has provisions to change the interest rate based on changes in market rates of interest, a specified index or other factors determined by a lender. Interest rates on adjustable rate loans or mortgages can *only* change at intervals specified in a not or loan agreement. For example, the interest rate on a five-year adjustable rate mortgage can change once every five years.

A variable rate loan may also designate intervals in which interest rates may change, but in some variable rate loans a change in the interest may be at the *discretion* of the lender. If a borrower has a variable or adjustable rate loan, he or she should know how often and how much the interest rate may change. The borrower should also be able to calculate how changes in interest rates affect the loan payment. A borrower should ask the lender to estimate the scheduled payment at various rates of interest. The borrower should be comfortable with the uncertainty involved with potential interest rate changes. If not, the borrower may request loan terms that reduce the interest rate risk.

If the interest rate on a variable or adjustable rate loan is linked to a specified index rate, a lender typically adds a margin above the index rate to determine the interest rate. For example, if the index rate is 9 per cent and the margin is 2 per cent, the interest rate on a variable rate or adjustable rate loan is 11 per cent. If the index rate changes to 11 per cent in the next adjustment period, the interest rate charged will be 13 per cent.

Many characteristics of variable and adjustable rate loan differ among lenders. The interest rate index (if any), margin, length of adjustment period and caps (upper limits) are the major distinguishing features. These features may be negotiable.

Interest rate index: The variable of adjustable interest rate is sometimes linked to an interest rate index. Many lending institutions use their average cost of funds or another internal rate as the basis to price loans. Other common indices include 1-year Treasury securities rates, 90-day Treasury bills, prime rate charged at money center banks, federal funds rate and the London Interbank Offer Rate (LIBOR). Differences between the indices can be substantial. Federal funds rates and 90-day Treasury bill rates can change every day, while the prime rate changes less frequently. A borrower should ask the lender about historical patterns of the index rate. In addition, if the lender is using the institution's internal rate, a borrower should ask how often the lender changes this rate.

Margin: The margin refers to the percentage points that the lender adds to the rate index to determine the rate charged to the borrower. The margin covers the costs of administering the loan, a risk premium, and a profit margin for the lender. The note or loan agreement will state if the margin is to remain constant over the maturity of the loan.

Length of adjustment period: The adjustment period is the length of time before the lender can change the borrower's interest rate. At the end of each adjustment period, the interest rate may be adjusted to reflect changes in the index (if an index is used). The note *may* allow for other terms of the loan to change at each adjustment period.

Caps: Rate caps may be associated with variable or adjustable rate loans. They limit how much the interest rate can change at each adjustment period. Many loans also have life-of-loan rate caps which limit interest rate movements over the entire life of the loan. Often a cap may be purchased as an optimal feature of the loan.

A borrower should be aware of each of these factors affecting a variable or adjustable rate loan. Moreover, the combination of the factors and the resulting implications must be considered. For example, if a lender has a volatile interest rate index, a borrower should consider some type of cap. Lenders will negotiate on the different variable and adjustable rate features. For example, a lender may lengthen the adjustment period in exchange for a higher margin. A borrower should feel comfortable with the variable or adjustable rate features and be willing to discuss changes in a loan package.

Fees and Service Charges

As a general rule, loan fees or "points" are charged at a the time the loan is made. A point is 1 per cent of the amount loaned. In addition, there may be other service charges for which the lender will require reimbursement. Service charges and fees are typically charged for:

- Real estate appraisals,
- Credit searches,
- Legal costs,
- Recording mortgages and deeds,
- Mortgage title insurance premiums and
- Title searches.

Fees and service charges increase the borrower's cost. The estimated increased cost over the life of a loan for one point paid at origination on a 10 per cent loan with annual payments. For example, the effective interest rate on a 7-year loan increases about 0.30 per cent (30 basis points) for one point paid at origination. In other words, if the stated interest rate on a 7-year loan is 10 per cent and the lender charges a one-point origination fee, the effective interest rate to the borrower would be approximately 10.30 per cent. Using this technique, a borrower could compare the effective interest rates on various loan products. A loan comparison FAST tool can also be used to estimate more precisely the impact of fees on the costs of loans.

Some lenders will reduce the interest rate in exchange for a fee at origination. This is called a rate buydown. A borrower could also use to estimate the potential benefit from a rate buy down. For example, suppose a lender offers to reduce a borrower's interest rate on a fixed-rate loan by 0.20 per cent (20 basis point) for 1 point at origination. The cost is the same if the loan is expected to be paid off in 12 to 13 years (intersection of 0.20 per cent and line). If the loan is expected to be paid off in more than 13 years, a borrower should benefit by therate buydown, while a borrower's anticipated return would be less if the loan is expected to be paid off in less than 12 years. This graph is only an approximation of a 10 per cent fixed payment loan. The break-even point would be different if the loan were to be paid before maturity. A borrower should ask the lender to estimate the break-even points for each loan alternative.

Stock Purchases and Compensating Deposit Balances

Farm Credit System lenders usually require a stock purchase. The stock requirement may be as low as 2 per cent of the loan amount or a maximum of $1,000. A stock purchase increases the effective interest rate on the loan. The effect of the stock purchase on the effective interest rate diminishes as the loan maturity lengthens and/or loan size increases. The purchase of stock is a financial investment in the issuing institution which is typically paid back at loan maturity, but the lender is not obligated to do so.

A compensating deposit balance is aminimum deposit balance that is sometimes required by a bank from a borrower. The balance is usually expressed as a percentage of the total loan commitment and/or a stipulated percentage of the amount of commitment actually used by the customer. In some cases, compensating balances can be used as a negotiating device by the borrower.

Payment Frequency

The frequency of payments differs among loans. Typically, intermediate- and long-term loans are structured with monthly, quarterly, semiannual or annual payments. More frequent principal payments generally reduce the total interest paid over the life of the loan. A similar factor to consider is the timing of the payments. Obviously, it is preferable to have payments which correspond with high cash inflows. A borrower should establish a payment pattern with a lender that coincides with his or her cash flow.

Note: The shift upward if interest rates are higher than 10 per cent, while the graph will be slightly lower if interest rates are less than 10 per cent. Account for the time value of the tax difference between points paid at origination and interest paid over the life of the loan.

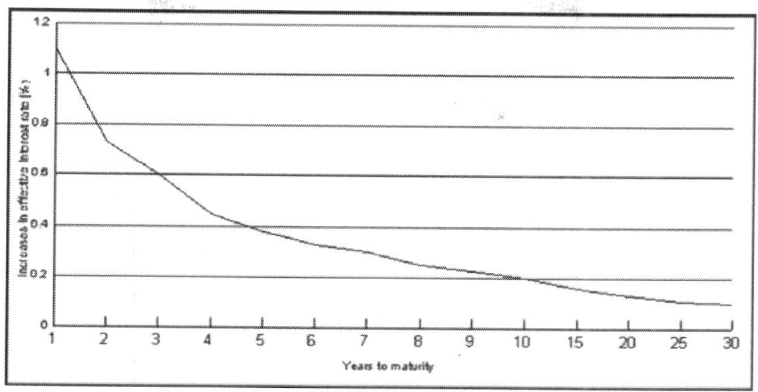

Fig. Increased cost of loan per point origination fee.

Maturity

Loan maturity is simply the time until the loan is fully due and payable. A borrower should evaluate his or her ability to generate cash to repay debt when comparing loans with different maturities. As a rule of thumb, a borrower should not select a loan maturity that is longer than the anticipated life of the asset being financed. Shorter maturities result in lower total interest payments over the life of the loan and more rapid accumulation of equity in the asset being financed. In contrast, loans with longer maturities will have lower loan payments and, therefore, free up cash for other uses. Thus, tradeoffs between shorter and longer loan maturities should be carefully evaluated.

Annual payments among loans of different maturities and interest rates. For example, the annual payments on a $100,000, 20-year loan at 10 per cent would be $11,746. The payments on a similar 30-year loan would be $10,608. A borrower could also use Table 3 to estimate the change in annual payments as interest rates changed.

Table. Annual loan payments for $100,000 loan, fixed-payment method.

Annual Payments						
	Interest Rate					
Years to Maturity	8%	10%	12%	14%	16%	18%
5	25,046	36,380	27,741	29,128	30,541	31,978
10	14,903	16,275	17,698	19,171	20,690	22,251
20	10,185	11,746	13,388	15,099	16,867	18,682
30	8,883	10,608	12,414	14,280	16,189	18,126

Collateral Requirements

Collateral refers to the assets pledged as security in a loan transaction. The legal documents representing a lender's interest in collateral include a mortgage or deed of trust in the case of farm real estate loans and asecurity agreement for operating and intermediate-term loans. Nearly all farm real estate loans are secured by a mortgage or deed of trust on a tract of land. Operating loans and intermediate-term loans may be secured or unsecured, although secured loans are more common. Unsecured loans generally involve smaller loans to financially strong borrowers who usually are long-term customers of the lending institution.

Intermediate-term loans generally are secured by the asset being purchased. Examples are tractors, combines, equipment, facilities and breeding livestock. Operating loans usually are secured by current assets and sometimes by intermediate assets as well. Examples of collateral for operating loans are farm supplies, crop and livestock inventories, growing crops, government payments and deposit accounts. A blanket filing may be used on a line-of-credit financing so that the security agreement applies to essentially all of the current and intermediate assets and, if stipulated, to property acquired in the future as well. A security agreement usually includes covenants about selling, insuring and/or maintaining collateral. Many security agreements for real estate purposes now include provisions regarding the storage and disposal of hazardous wastes. A borrower should be aware of the procedures and notifications that need to be made upon selling or modifying assets used as collateral.

A borrower should also be aware of the lender's right to collateral upon default. A security agreement or mortgage will specifically outline the lender's and borrower's rights upon default.

Prepayment Penalties

A borrower should be aware of prepayment penalties. A prepayment penalty is a fee charged by a lender when a loan is paid prior to its maturity. Prepayment penalties vary significantly among lenders. Prepayment penalties will *increase* the cost of refinancing a mortgage and *reduce* the flexibility of changing loan alternatives, such as refinancing if interest rates decline.

Refinancing

As interest rates fall, refinancing may become more attractive. Better service, lengthening of maturity, more favorable non-interest lending terms, and customer dissatisfaction may also motivate a borrower to refinance. In addition to the interest rate reduction, a borrower should estimate fees and penalties that would result from refinancing the loan. These fees and penalties may overcome any interest rate savings. Furthermore, if the borrower is switching from a fixed-rate loan to a variable- or adjustable-rate loan, he or she should evaluate the differences in risk associated with these loans.

Loan Conversion

A conversion option provides the ability to convert from one type of loan to another (*e.g.*, from fixed rate to variable, and visa versa) at any time or at the end of an adjustment period. This option may require a fee.

Reporting Requirements

The detail and frequency of financial reports required from a borrower will differ from lender to lender and by type and size of loan. Many real estate lenders require annual financial statements. Others only require statements when originating or renegotiating a loan. The borrower should feel comfortable with the financial statement requirements, but choosing a lender that requires the fewest reports may not be in the borrower's best interest. Most borrowers are already preparing periodic financial reports for management purposes. Thus, reasonable reporting requirement should not be a significant factor in the borrower's choice of lender.

The financial statements include a balance sheet formulated at the same time each year and an income statement. A cash flow budget may also be helpful. Using accurate and verifiable data is important.

Credit Evaluation Procedures

Many agricultural lenders analyze the creditworthiness of agricultural borrowers using a combination of judgement and formal credit scoring models or risk assessment worksheets. This approach seeks to combine various measures of business performance (*e.g.*, profitability, solvency, liquidity, repayment history and collateral) with other information about the borrower to reach an overall credit score. This credit score may be used in making loan decisions, determining the borrower's interest rate, deciding about loan supervision and monitoring business performance. Borrowers can aid in the evaluation process by keeping comprehensive financial records (*e.g.*, balance sheets, income statements, flow of funds summaries) and presenting this information to the lender in a timely, well-organized and thoroughly documented fashion. Annual statements should be prepared at the same time each year to allow for more appropriate comparisons over time. In turn, the borrower can ask a lender to review the results of the credit evaluation process so that both parties clearly understand the strengths and weaknesses of the farm business, and thus develop more effective financial plans in the future.

Risk Management

Practices by borrowers that help to stabilize farm income and improve the liklihood of successful loan repayment. Examples include enterprise diversifiication, resistant seed varieties, crop insurance, holding financial reserves, limits on borrowing, crop share leases, off farm work, and marketing practices sucgh as forward contracting, frequency of sales, futures and option contracts and others. Lenders may respond to these practices with more favorable fnancing terms, lower interest rates, and access to credit.

Late-Payment Penalties and Foreclosure Provisions

Even though most borrowers plan to make timely loan payments, unforeseen circumstances sometimes results in late payments. A borrower should compare the penalties associated with late payments and grace periods that are specifically written in loan documents. A borrower should also be aware of the conditions under which the loan is considered in default.

INTERNATIONAL TRADE AND INDIAN AGRICULTURE SECTOR

The backbone of Indian economy is considered as 'agriculture'. It has an important role in Indian economy though the share of agriculture in GDP is declining, it still remains a significant contributor nearly 30 to 35 per cent coming from this sector. It provides livelihood for nearly two-thirds of our population. Agriculture provides raw material for industrial growth and use *e.g.* sugar, jute, cotton, agro-based industries such as food processing are gaining in importance.

NATURE OF INDIAN AGRICULTURE: THE BACKDROP

Though more than five decades have passed since independence and agricultural production has increased several folds.

Agriculture growth rate in India:

1950-1960	1961-1970	1971-1980	1981-1990	1991-2002
3.1%	2.5%	1.9%	3.8%	2.8%

But certain features of Indian agriculture hamper the balanced growth and development.

Finance is the major constraint here. Despite the introduction of rural banking, credit societies, kisan credit cards, the small and marginal farmers still go to the same old pawn broker or money lender and become their victims. Low wages for agricultural labour due to excessive population is another major issue, which is ailing this sector. Lack of opportunities, low skills of agriculture laborers denies them to think something else other than agriculture and cheapness of labour leads to adoption of labour-intensive methods of production instead of modern technology. Added to all these, the Indian agriculture depends heavily on monsoon. Nearly 70 per cent of the areas continue to depend on monsoon rather than irrigation. Agricultural inputs, seeds, irrigation, electricity, plant protection, fertilizers and agricultural investment are seen as other major problems characterizing Indian agriculture.

INTERNATIONAL TRADE AND ITS IMPACT ON INDIAN AGRICULTURE

Agriculture trade contributes 15 per cent of total foreign exchange earnings. Broadly, agricultural and agri-based products can be divided into three categories, they are raw products, semi-raw products and processed and ready to yield products.

The major agri-exports of India are cereals, rice, basmati rice and non-basmati rice, spices, oilcake, tobacco un-manufactured, tea, coffee and marine products. Lack of market access in the developed market economy countries due to high tariffs and pronounced Non-tariff barriers has bees acting as a deterrent for the exports. As against agricultural exports, agri-imports constitute only a small proportion of the countries total imports 5 per cent.

Subsequent to the economic reforms initiated in June 1991, removing the restrictions and protective licensing regime, free trade in a large number of items has become the order of the day. With the removal of QRs on agricultural items and urea the Indian farmer community has been placed to face stiff competition from the developed nations.

Two-thirds of Indian farmers are in the small and marginal farmers cultivating two hectares of land contrary to the popular perceptions given a level playing field.

These farmers are not able to compete with their western counterparts in agricultural production and exports.

Each farmer in the developed countries gets on an average a subsidy of US $ 29,000 a year. The US domestic support for farmers is US $ 25.5 billion in 1996 while the European Union it was US $ 85 billion. In the US and EU farmers constitute less than 3 per cent of the population in contrast India's domestic support to its farmers works out to a negative US $ 23.75 billion in 1996 even after providing fertilizer, electricity, irrigation and seed subsidies.

Added to all these things, the emergence of World Trade Organization (WTO)

In 1995 laid a specific set of rules in agriculture trade popularly known as AoA, which came into force on 1stJan. 1995.

According to WTO norms, member countries are required to fulfill the following commitments:

1. All non-tariff barriers are to be replaced by tariff barriers and tariffs will have to be reduced by 36 per cent by industrialized countries and 24 per cent by developing nations.
2. Countries with closed farm market will have to import at least 3 per cent of domestic consumption of the product, raising to 5 per cent over a period of 6 years.
3. Trade support to farmers will have to cut by 20 per cent over a period of 6 years by developed and by 13.3 per cent by developed countries.
4. The value of direct export subsidy will have to be cut by 36.1 per cent and the volume of subsidized exports by 21 per cent over a period of 6 years, while in the case of developing countries direct export subsidies will have to reduce by 24 per cent and the quantity

of subsidized exports will have to reduce by 14 per cent over a period of 10 years.

WTO norms on Agriculture sector for developing and developed nations

Developed countries (1995-00) (1995-04)	Developing countries	
Tariffs (avg. cut for all agricultural products)	36%	24%
Minimum cut per product line	15%	10%
Domestic support	20%	13%
Export subsidies	36%	24%
Subsidized quantities	21%	14%

IMPLICATIONS

It is feared that the agreements reached is not favorable to many developing nations including India for the following reasons:

(a) The government will be forced to withdraw/reduce subsidies to its farmers.
(b) The country will be compelled to import at least 3 per cent of the domestic demand for agricultural products
(c) Policies like Public Procurement and PDS may have to be abandoned.
(d) Patenting of seeds will force Indian farmers to buy seeds from Multi nations foreign enterprises.

TRADE AND EXPORT POLICY AND EXTERNAL SECTOR

The trade and external sector of a country plays an important role in its economic development, as it is responsible for integrating the domestic economy through the twin channels of trade and capital flows. Indian markets have been the best performing markets among emerging market peers so far in the year 2014, according to a recent study. In the past two years, the Indian economy has grown from US$ 500 billion to US$ 2 trillion and the trade and external sector has had a big role to play in this development. The Indian government is keen to improve on exports and provide more jobs for the young, talented, well-educated and even the semi-skilled and unskilled people of the country.

CAPITAL INFLOWS

India's foreign exchange (Forex) reserves stood at US$ 318,579.8 million as on August 29, 2014. Foreign currency assets aggregated to US$ 291,318.2 million and the value of gold reserves stood at US$ 21,173.8 million, as on August 29, 2014, according to the weekly statistical data released by Reserve Bank of India (RBI).

FOREIGN DIRECT INVESTMENTS (FDI)

India received total foreign investment (including equity inflows, 're-invested earnings' and 'other capital') worth US$ 341,357 million in the period

April 2000 - August 2014 and a cumulative amount of US$ 229,595 million during the same period. The country was one of the top destinations for FDI inflows from Asian countries, with Mauritius contributing 36 per cent and Singapore 12 per cent; the UK contributed 9 per cent of the total foreign inflows in the mentioned period.

FOREIGN INSTITUTIONAL INVESTORS (FIIS)

The total number of FIIs registered in India was 1,710 in FY14. The net investments by overseas investors into India so far this year has reached US$ 30 billion, while their cumulative total inflows into the country has crossed the US$ 200 billion mark.

The net investments by foreign investors into Indian debt markets since the beginning of 2014 have reached US$ 17 billion, while the same for equities stand around US$ 13 billion.

Buoyed by the steps taken by the central government, FIIs have started investing more in India in the past few months. Foreign investment inflows are expected to more than double to touch US$ 60 billion this fiscal, as per an industry study.

EXPORTS

According to data released by Ministry of Commerce, exports from India during July 2014 were valued at US$ 27727.60 million which was 7.33 per cent higher than the US$ 25835.08 million achieved during July, 2013.

The top five countries receiving Indian exports in the period April-July 2014 are the USA, UAE, Saudi Arab, Hong Kong and China. The value of exports to USA amounted to US$ 14,261.85 million, which is an increase of 8.16 per cent over the same period in 2013.

India's sugars and sugar confectionery exports for the month of July 2014 has grown to US$ 70.79 million, which is an increase of 15.6 per cent over the previous month, according to industry data. This growth can be attributed to the developments in the field of agriculture in the country in the recent past.

According to Engineering Export Promotion Council (EEPC), India's engineering exports have shown remarkable year-on-year (y-o-y) growth, ranging from 100-150 per cent in July, 2014. Similarly, for the April-July period, the engineering goods exports to Turkey have grown by 106 per cent to US$ 638 million from US$ 309 million.

EXTERNAL SECTOR

The Prime Ministers of India and Japan after their Annual Summit meeting issued a joint statement to further enhance cooperation in several fields such as infrastructure, railways and agriculture, among others. Japan also plans to double its investments in India in the next five years from around US$ 2 billion last year. China is seeking to prop up its investments in India and has planned to invest over US$ 5 billion in two industrial parks in Gujarat and Maharashtra with a focus on automobile manufacturing and power transmission and generation equipment manufacturing respectively.

In a boost to India's nuclear power industry, Mr Tony Abbott, the Prime Minister of Australia, is expected to sign an agreement to sell uranium to India. This move is slated to help both the countries as Australia has high uranium reserves which it exports, while India depends on imported fuel for its 20 nuclear reactors.

Argentina has given full access of its US$ 6 billion drug market to Indian companies. This has increased the scope of exports to finished pharmaceuticals formulations from just raw materials earlier.

The Ministry of Railways, Government of India, signed a memorandum of understanding (MoU) with Czech Railways (Ceske Drahy), Czech Republic, for technical cooperation in the field of railways. The MoU is valid for a period of three years which can be extended further for successive periods of one year at a time.

FOREIGN TRADE POLICY

India has signed a Comprehensive Economic Partnership Agreement with South Korea, which will give enhanced market access to Indian exports as part of the Foreign Trade Policy's (FTP) strategy of market expansion. These trade agreements are in line with India's Look East Policy.

To help exporters get into long-term contracts, the RBI has simplified the rules for credit to exporters. This will now enable them to get long-term advance from banks for up to 10 years to service their contracts. It will also aid the overall export performance and help the rupee in the short to medium term.

All exports and import-related activities are governed by the FTP, which is mainly aimed at enhancing the country's exports and use trade expansion as an effective instrument of economic growth and employment generation.

The Government of India is expected to announce the new five-year FTP (2014-19) as it seeks to boost manufacturing and exports, among other things. The Federation of Indian Export Organisation (FIEO) has called for more focus on export of services and hi-tech products in the new policy.

ROAD AHEAD

Boosted by the forthcoming FTP, India's exports are expected to cross the US$ 350 billion mark in 2014-15 and reach US$ 750 billion by 2018-19 according to FIEO. Also, with the Government of India striking important deals with the governments of countries such as Japan, Australia and China, the external affairs sector is contributing heavily to the economic development of the country and growth in the global markets. Moreover, by implementing the

FTP 2014-19, India's share in world trade is expected to double from the present level of three per cent by the year 2020.

FOREIGN TRADE POLICY

In India, the main legislation concerning foreign trade is the Foreign Trade (Development and Regulation) Act, 1992. The Act provides for the development and regulation of foreign trade by facilitating imports into, and augmenting exports from, India and for matters connected therewith or incidental thereto. As per the provisions of the Act, the Government:- (i) may make provisions for facilitating and controlling foreign trade; (ii) may prohibit, restrict and regulate exports and imports, in all or specified cases as well as subject them to exemptions; (iii) is authorised to formulate and announce an export and import policy and also amend the same from time to time, by notification in the Official Gazette; (iv) is also authorised to appoint a 'Director General of Foreign Trade' for the purpose of the Act, including formulation and implementation of the export-import policy.

Accordingly, the Ministry of Commerce and Industry has been set up as the most important organ concerned with the promotion and regulation of foreign trade in India. In exercise of the powers conferred by the Act, the Ministry notifies a trade policy on a regular basis with certain underlined objectives. The earlier trade policies were based on the objectives of self-reliance and self-sufficiency. While, the later policies were driven by factors like export led growth, improving efficiency and competitiveness of the Indian industries, etc.

With economic reforms, globalisation of the Indian economy has been the guiding factor in formulating the trade policies. The reform measures introduced in the subsequent policies have focused on liberalization, openness and transparency. They have provided an export friendly environment by simplifying the procedures for trade facilitation. The announcement of a new Foreign Trade Policy for a five year period of 2004-09, replacing the hitherto nomenclature of EXIM Policy by Foreign Trade Policy (FTP) is another step in this direction. It takes an integrated view of the overall development of India's foreign trade and provides a roadmap for the development of this sector. A vigorous export-led growth strategy of doubling India's share in global merchandise trade (in the next five years), with a focus on the sectors having prospects for export expansion and potential for employment generation, constitute the main plank of the policy. All such measures are expected to enhance India's international competitiveness and aid in further increasing the acceptability of Indian exports. The policy sets out the core objectives, identifies key strategies, spells out focus initiatives, outlines export incentives, and also addresses issues concerning institutional support including simplification of procedures relating to export activities.

The key strategies for achieving its objectives include:-

- Unshackling of controls and creating an atmosphere of trust and transparency;
- Simplifying procedures and bringing down transaction costs;
- Neutralizing incidence of all levies on inputs used in export products;
- Facilitating development of India as a global hub for manufacturing, trading and services;
- Identifying and nurturing special focus areas to generate additional employment opportunities, particularly in semi-urban and rural areas;
- Facilitating technological and infrastructural upgradation of the Indian economy, especially through import of capital goods and equipment;
- Avoiding inverted duty structure and ensuring that domestic sectors are not disadvantaged in trade agreements;
- Upgrading the infrastructure network related to the entire foreign trade chain to international standards;
- Revitalizing the Board of Trade by redefining its role and inducting into it experts on trade policy; and
- Activating Indian Embassies as key players in the export strategy.

The FTP has identified certain thrust sectors having prospects for export expansion and potential for employment generation. These thrust sectors include: (i) Agriculture; (ii) Handlooms and Handicrafts; (iii) Gems and Jewellery; and (iv) Leather and Footwear. Accordingly, specific policy initiative for these sectors has been announced.

For the agriculture sector:-

- A new scheme called "Vishesh Krishi Upaj Yojana (Special Agricultural Produce Scheme)" to boost exports of fruits, vegetables, flowers, minor forest produce and their value added products has been introduced. Under the scheme, exports of these products qualify for duty free credit entitlement (5 per cent of Free On Board (f.o.b) value of exports) for importing inputs and other goods;
- Duty free import of capital goods under Export Promotion Capital Goods (EPCG) scheme, permitting the installation of capital goods imported under EPCG for agriculture anywhere in the Agri- Export Zone (AEZ);
- Utilizing funds from the 'Assistance to States for Infrastructure Development of Exports (ASIDE) scheme' for development of AEZs;
- Liberalization of import of seeds, bulbs, tubers and planting material, and liberalization of the export of plant portions, derivatives and extracts to promote export of medicinal plants and herbal products.

For the handlooms and handicraft sector:-

- Enhancing to 5 per cent of Free On Board (f.o.b) value of exports duty free import of trimmings and embellishments for handlooms and handicrafts;

- Exemption of samples from countervailing duty (CVD);
- Authorizing Handicraft Export Promotion Council to import trimmings, embellishments and samples for small manufacturers; and
- Establishment of a new Handicraft Special Economic Zone.

For the gems and jewellery sector:-

- Permission for duty free import of consumables for metals other than gold and platinum up to 2 per cent of Free On Board (f.o.b) value of exports;
- Duty free re-import entitlement for rejected jewellery allowed up to 2 per cent of f.o.b value of exports;
- Increase in duty free import of commercial samples of jewellery to ₹ 1 lakh; and
- Permission to import of gold of 18 carat and above under the replenishment scheme.

For the leather and footwear sector, the specific policy initiatives are mainly in the form of reduction in the incidence of customs duties on the inputs and plants and machinery. These include:-

- Increase in the limit for duty free entitlements of import trimmings, embellishments and footwear components for leather industry to 3 per cent of Free On Board (f.o.b) value of exports and that for duty free import of specified items for leather sector to 5 per cent of f.o.b value of exports;
- Import of machinery and equipment for Effluent Treatment Plants for leather industry exempted from customs duty; and
- Re-export of unsuitable imported materials (such as raw hides and skin and wet blue leathers) has been permitted.

In order to review the progress and policy measures, each year, "Annual Supplements" to the five year Foreign Trade Policy (FTP) have been announced by the Ministry:-

- The Annual Supplement announced in April, 2005 incorporated additional policy initiatives and further simplified the procedures. It provided for an active involvement of the State Governments in creating an enabling environment for boosting international trade, by setting up an Inter-State Trade Council. Also, different categories of advance licences were merged into a single category for procedural facilitation and easy monitoring. The supplement provided renewed thrust to agricultural exports by extension of 'Vishesh Krish Upaj Yojna' to poultry and dairy products and removal of cess on exports of all agricultural and plantation commodities.
- The Annual Supplement put forward in April 2006, announced the twin schemes of 'Focus Product' and 'Focus Market'. To further meet the objective of employment generation in rural and semi urban areas,

export of village and cottage industry products were included in the 'Vishesh Krishi Upaj Yojana', which was renamed as "Vishesh Krishi and Gram Udyog Yojana". Also, a number of measures were introduced in order to achieve the objective of making India a gems and jewellery hub of the world. These include:- (i) allowing import of precious metal scrap and used jewellery for melting, refining and re-export; (ii) permission for export of jewellery on consignment basis; (iii) permission to export polished precious and semi precious stones for treatment abroad and re-import in order to enhance the quality and afford higher value in the international market.

- Likewise, the third Annual Supplement to the Foreign Trade Policy was announced on 19 April,2007 (effective from 1 st April, 2007). Some of the important measures introduced by it are:- (i) exemption from service tax on services (related to exports) rendered abroad; (ii) service tax on services rendered in India and utilized by exporters would be exempted/remitted; (iii) categorization of exporters as 'One to Five Star Export Houses' has been changed to 'Export Houses and Trading Houses', with rationalization and change in export performance parameters; (iv) expansion of ceiling, scope and coverage under the 'Focus Market Scheme (FMS)' and 'Focus Product Scheme (FPS)'.
- The final annual supplement to the Foreign Trade Policy for 2004-2009 was announced in April 2008 in which several innovative steps were proposed. They included the following:
  1. Import duty under the EPCG scheme is being reduced from 5 per cent to 3 per cent, in order to promote modernization of manufacturing and services exports.
  2. Income tax benefit to 100 per cent EOUs available under Section 10B of Income Tax Act is being extended for one more year, beyond 2009.
  3. To promote export of sports and toys and also to compensate disadvantages suffered by them, an additional duty credit of 5 per cent over and above the credit under 'Focus Product Scheme' is being provided.
  4. Our export of fresh fruits and vegetables and floriculture suffers from high incidence of freight cost. To neutralize this disadvantage, an additional credit of 2.5 per cent over and above the credit available under Visesh Krishi and Gram Udyog Yojana (VKGUY) is proposed.
  5. Interest relief already granted for sectors affected adversely by the appreciation of the rupee is being extended for one more year.
  6. DEPB scheme is being continued till May 2009.

Trade Facilitation Measures (Supplement To Foreign Trade Policy 2004-09) Announced On 26th February 2009,

- DUTY CREDIT SCRIPS under DEPB scheme to be issued without waiting for realization of export proceeds;
- Special package of ₹ 325 crore for leather and textiles sector;
- STCL, DIAMOND INDIA, MSTC, GEM and JEWELLERY EPC and STAR TRADING HOUSES added as nominated agencies for import of precious metals;
- Gem and Jewellery export: import restrictions on worked corals removed;
- Bhilwara and Surat recognized as towns of export excellence for textiles and diamonds;
- Threshold limit for recognition as premier trading houses reduced to ₹ 7500 crore;
- Under EPCG scheme, export obligation extended till 2009-10 for exports during 2008-09;
- DEPB/DUTY CREDIT SCRIP utilization extended for payment of duty for import of restricted items also;
- Procedure for claiming duty drawback refund and refund of terminal excise duty further simplified;
- Re-credit of 4 per cent SAD for VKGUY, FPS and FMS allowed;
- A new office of DGFT to be opened at Srinagar;
- Value cap under DEPB revised for two products;
- Electronic message transfer facility for advance authorization and EPCG to be established;
- Gem and Jewellery units in EOU to be allowed – personal carriage of gold up to 10 kg;
- Advance licenses issued prior to 1.4.2002 requiring MODVAT/ CENVAT certificate dispensed with;
- Export obligation period against advance authorizations extended up to 36 months;
- Reimbursement of additional duty of excise levied on fuel to be admissible for EOUS;
- Early refund of service tax claims and further simplification of refund procedures on the anvil;

In accordance with the provisions of the Act, a “Directorate General of Foreign Trade (DGFT)” has been set up as an attached office of the Ministry of Commerce and Industry.

It is headed by the ‘Director General of Foreign Trade’ and is responsible for formulating and executing the Foreign Trade Policy/Exim Policy with the main objective of promoting Indian exports. The DGFT also issues licences to exporters and monitors their corresponding obligations through a net work of

32 regional offices located at the following places:- Ahmedabad; Amritsar; Bangalore; Baroda (Vadodara); Bhopal; Kolkata; Chandigarh; Chennai; Coimbatore; Cuttack; Ernakulam; Guwahati; Hyderabad; Jaipur; Kanpur; Ludhiana; Madurai; Moradabad; Mumbai; New Delhi; Panaji; Panipat; Patna; Pondicherry; Pune; Rajkot; Shillong; Srinagar(Functioning at Jammu); Surat; Thiruvananthapuram; Varanasi; and Vishakhapatnam.

9

E-Commerce and Information Technology for Agribusiness

ELECTRONIC COMMERCE FOR AGRIBUSINESS

Electronic commerce (e-commerce) is relatively new to the agricultural industry, and affects such aspects of the organization as its strategy, processes, customer relationships, information technology, and business culture.

IMPACTS OF E-COMMERCE AND INFORMATION TECHNOLOGY ON GLOBAL AGRICULTURAL MARKETS

E-commerce and information technology have had an important impact on productivity in the United States and on general trade flows. In an attempt to understand specific impacts on agricultural trade and on the agricultural and agribusiness sectors, the Farm Foundation, USDA's Economic Research Service and the University of Minnesota sponsored a symposium in Washington, D.C., May 29 - 30, 2001, to define researchable issues, data needs and policy issues.

Over 60 people from agribusiness, government agencies and universities attended presentations by industry representatives, academicians and government researchers. The workshop explored the impact of the Internet on transaction costs in agricultural markets, transportation and storage costs, and trading patterns. It also examined the use of the Internet to target niche food markets, to identify preservation of food products, and to develop product standards in the private sector. There was also discussion of the business experiences of agricultural Internet firms.

E-COMMERCE HELP AGRICULTURAL MARKETS

E-commerce refers to the use of the Internet to market, buy and sell goods and services, exchange information, and create and maintain web-based relationships between participant entities (Fruhling and Digman, 2000). Based on its demonstrated impact in industrial retail markets (Elia, Lefebvre, and Lefebvre 2007), e-commerce is believed to have the potential to increase profitability in agricultural markets by increasing sales and decreasing search and transactions costs. The creation of electronic markets that are expected to be more transparent and competitive than physical markets may attract more consumers by increasing demand and improving the firm's strategic position with customers seeking specific niche products or having geographical restrictions (Batte and Ernst, 2007; and Montealegre, Thompson and Eales, 2007).

However, due to the relatively new state of e-commerce in agriculture, its impact has not been widely measured and documented. We developed an evaluation framework and applied it to measuring the performance of the agricultural e-commerce platform MarketMaker. The analysis focuses on the impact of MarketMaker on producers and farmers' markets and consists of both the perceived impacts based on survey responses and a willingness–to-pay analysis, as well as the examination of factors that affect the impacts of the web site. Our findings provide guidance for future development of agricultural e-commerce-enabling platforms like MarketMaker, as well as future evaluation efforts of these platforms.

WHAT IS MARKETMAKER?

MarketMaker is an interactive e-commerce platform that provides food marketing information to food entrepreneurs—agricultural producers, buyers, processors, wholesalers, food retailers, restaurants—and their customers. The site was created in 2000 by a team of University of Illinois Extension personnel with the goal of building an electronic infrastructure that would easily connect Illinois food producing farmers with economically viable new markets. In 2005, a multi-state partnership of land grant institutions and agriculturally focused organizations was formed to build a national network of interconnected MarketMaker sites. By December 2012, 19 states and the District of Columbia became part of the national network. The site currently includes nearly 660,000 profiles of food-related enterprises including 8,618 agricultural producers and experiences about 1 million hits per month from over 85,000 users. The original MarketMaker project was funded by the Illinois Council on Food and Agricultural Research and the Illinois Department of Agriculture. As other states have joined the MarketMaker network, funding has typically come from state departments of agriculture and land grant universities along with other sources such as Sea Grant.

MarketMaker provides information about product availability by geographic location and market orientation to help inform decisions of both producers and consumers. As an electronic farm directory/food marketing tool, MarketMaker directly competes with other web sites such as Local Harvest, Eat Well Guide, Rural Bounty, Local Farm Link, Chef Collaborative, Agricultural Business, Green People, Pick Your Own, various state locally grown promotion web sites,

Farm Bureau, local food directories supported by a host of local organizations, and directories provided by state departments of agriculture. Different from some food marketing web sites, such as Local Harvest, MarketMaker does not have a selling feature, meaning that users cannot purchase products directly through the web site. In contrast to farm directory web sites, such as Rural Bounty, Chef Collaborative, Agricultural Business, and Pick Your Own, MarketMaker provides the benefit of displaying the information about food producers/retail outlets on a map. Moreover, MarketMaker provides the ability to map consumer data related to several demographic characteristics. Thus, for farmers, it provides information to help better target markets and identify potential businesses with which to collaborate. For consumers—households, processors, handlers, retailers, and wholesalers—MarketMaker provides information to help make decisions about where to purchase products or to identify business-to-business opportunities all along the supply chain.

Previous studies that looked at several aspects of MarketMaker performance reported that 63 per cent of Ohio registered users including producers, farmers' markets, and wineries believed that the MarketMaker site was helping keep more food dollars in the regional economy. Cho and Tobias (2009) found that the average increase in annual sales attributed to MarketMaker among 374 New York farmers was between $225 and $790. Additionally, 12 per cent of the respondents reported receiving marketing contacts through MarketMaker and using the MarketMaker directory to contact other food industry business partners.

EVALUATION FRAMEWORK

In order to more clearly understand how an e-commerce platform such as MarketMaker can produce useful results, one must consider more than the platform itself. A useful way to analyze the components of a complex programme such as MarketMaker is to develop logic models which demonstrate the links between inputs, activities, outputs and outcomes of a programme. Logic models also facilitate the identification of relevant evaluation measures and are frequently used as project planning and evaluation tools (W.K. Kellogg Foundation, 2004).

Therefore, logic models were developed for each of the major identified MarketMaker user groups (Lamie et al., 2011): producers, consumers, food retailers, food wholesalers, restaurants/chefs, and farmers markets. However, given funding constraints, only producers and farmers' markets were included in the current evaluation effort. Results from a survey of MarketMaker administrators and partners identified producers and farmers' markets among the primary users of the web site. Farmers' markets were also included due to their growing importance as an alternative food distribution system connecting producers and consumers.

The logic model for producers. The inputs on the national and state levels of MarketMaker include human resources; adequate technological expertise to support programme requirements; funds to support planned activities—training, promotion, and networking; and availability of related public and private data such as the National Census and survey data from independent studies. These inputs are used to conduct a series of activities such as developing, updating and improving content, and usability and functionality of the platform. MarketMaker purchases, gathers, manages, and distributes relevant existing data such as consumers' socio-demographic characteristics. MarketMaker also conducts training and promotional sessions at national, state, and regional levels in order to create awareness and prepare producers to successfully participate in MarketMaker. The adequate combination of inputs and activities leads to accomplishing the desired outputs, which include the complete MarketMaker web site as well as the registration and participation of new producers in the MarketMaker programme.

The outcomes of the platform, which we believe should be the main focus of the evaluation efforts, can be classified as short-, intermediate-, and long-term. In the case of MarketMaker, short-term outcomes comprise the creation of an initial web presence for some producers, additional web presence for others, and active participation in the site. In the intermediate-term, producers should be easily identified by wholesalers, retailers, and other consumers who choose to use MarketMaker.

In the long-term, MarketMaker portends to assist producers to increase profitability as a result of reduced marketing transaction costs and increased revenues via increased purchases from new and existing customers. The tested logic model outcomes can be used to identify quantifiable metrics. For example, the time and other resources a business devotes to the management of MarketMaker can be used as measures of short-term outcomes. The number of new contacts and new customers can be used as metrics of intermediate-term outcomes. The changes in total sales and marketing costs or changes in profits can be used as metrics for long-term outcomes of the platform. We followed a similar approach for the development of the logic model for the farmers' markets segment of users and the subsequent identification of quantifiable evaluation metrics.

MARKETMAKER IMPACT ON PRODUCERS AND FARMERS' MARKETS

The data on the metrics describing the impact of MarketMaker on producers was collected through a survey that was distributed by e-mail and postal mail to 4,264 producers registered on MarketMaker between April 2011 and March 2012. The farmers' market managers' survey was distributed by e-mail to 1,295 managers registered on MarketMaker web sites (May–June 2011).

Both surveys were distributed in all 15 participating (as of 2010) states: Arkansas, Colorado, Florida, Georgia, Illinois, Indiana, Iowa, Louisiana, Michigan, Mississippi, Nebraska, New York, Ohio, South Carolina, and Washington D.C. The response rates were 18 per cent for producer survey and 10.2 per cent for farmers' market managers' survey.

Survey results (discussed in more detail in Zapata et al., 2011; and Carpio et al., 2013) indicate that the perceived impact of MarketMaker on producers and farmers' markets outcomes are presently relatively modest. As a result of their participation with MarketMaker, producers have received an average of 2.9 new marketing contacts, and have gained an average of 1.6 new customers. Additionally, MarketMaker has assisted producers in increasing their annual sales by an average of $152. Nearly 87 per cent of producers registered in MarketMaker participate in direct marketing to individual consumers and wholesale buyers. MarketMaker has helped these producers receive an average of 2.9 new marketing contacts and increase their annual direct sales to individual consumers by 1.1 per cent and to wholesale buyers by 0.8 per cent on average. Our findings for farmers' markets indicate that, as a result of their participation with MarketMaker, managers have been contacted, on average, about 1.6 times by customers and vendors, and obtained an average of 0.8 new vendors and 1.9 new customers. The average annual increase in sales due to participation in MarketMaker was estimated at about 3.6 per cent, or $4,889 per farmers' market.

These reported averages, however, mask substantial variability across survey respondents. We found that the perceived impacts of MarketMaker on producers tend to be positively related to self-registration in MarketMaker, the amount of time since registering on the site, and the amount of time users spend on MarketMaker activities. In fact, producers who registered themselves on the MarketMaker web site (83 per cent of respondents) have received, on average, almost twice as many additional contacts and customers than those who were registered by someone else or do not know how they were enrolled in MarketMaker. Registration by others can occur if an existing list of producers—usually the one maintained by a state department of agriculture—is used to populate the MarketMaker database. Self-registered users are very likely to be more aware of their business presence on the site which facilitates the attribution of additional contacts and sales to it. It is also possible that the quality of the information provided by self-registered producers is more accurate and up to date.

Producers who spend between 30 and 60 minutes per month (12 per cent of sample) on the MarketMaker web site have an average annual sales increase of $242 compared to only $32 for those users who spend less than 30 minutes a month (83 per cent of the sample) on MarketMaker-related activities. The most used site features include "logging on to check or update profile," "searching for products," and "searching for buyers and sales opportunities." Less commonly used features include "searching for business partnerships," "using of the buy/sell Forum," and "finding target market for your products." These findings about farmers' registration and use intensity suggest that more education and promotion of MarketMaker is needed to encourage self-registration and more active use of MarketMaker to achieve the desired benefits from participation.

Our analysis of factors that affect the increase in farmers' markets sales due to MarketMaker revealed the components needed for more successful use of MarketMaker by the farmers' markets, namely, the established MarketMaker programme, the established farmers' market, and the active user-manager. Thus, the track record in the states with a longer presence on MarketMaker demonstrates the program's potential for new users. The fact that more established farmers' markets are able to achieve higher increases in sales than the new ones suggests that MarketMaker is possibly more effective in expanding existing markets rather than helping create new capacity. Finally, higher sales among more active users indicates that, in order to see the impact of MarketMaker on their operations, users have to invest time and effort in making the programme work for them.

WILLINGNESS TO PAY FOR MARKETMAKER

Since the long-term MarketMaker outcome for producers is an increase in profitability, we also asked producers the following hypothetical question regarding their willingness to pay for the services provided by MarketMaker: "If MarketMaker becomes privately funded, while retaining all the features and services it currently provides, would you be willing to pay an annual participation fee of $X for the services you receive from MarketMaker?" (Carpio et al., 2013).

The willingness to pay (WTP) measure derived from the question is directly related to the increase in profits obtained from using the site (Zapata, 2012; and Hudson and Hite, 2003). The survey results indicate that, on average, producers are willing to pay $47.02 annually for the services they receive from MarketMaker. Willingness to pay analysis for the subsample of farmers' markets revealed that managers are willing to pay an average of $41.19 annually for the services provided by MarketMaker. Theoretically, this value reflects the value users assign to the entire basket of MarketMaker services. The estimated mean WTP for farmers' markets comprised about 1 per cent of their perceived increase in sales estimated in this study.

Our findings indicate that registration type, time registered on MarketMaker, time devoted to the web site, type of user, the number of marketing contacts received, and firm total annual sales have a significant effect on producers' WTP for the services provided by MarketMaker. Thus, the effectiveness of MarketMaker is strongly linked with how it is used by producers after registration. For example, WTP is positively related to the time devoted to MarketMaker activities after registration. The positive relationship between the time producers have been registered on the site and the stated WTP implies that the benefits associated with MarketMaker tend to increase over time. Results of this research also indicate that additional marketing contacts increase producer WTP. Hence, with the aim to increase the number of marketing contacts received, MarketMaker web site development should focus on encouraging producers to frequently update their site profiles, specifically their contact information—phone number, e-mail, web site URL—and products' attributes and availability. Ultimately, these findings provide valuable information for current and potential users trying to better understand the expected costs and returns from a wide range of marketing, promotion, and other competing e-commerce activities.

FUTURE E-COMMERCE DEVELOPMENT AND EVALUATION

The systemic approach to MarketMaker evaluation undertaken in this study offers several important lessons for future development and evaluation of e-commerce in agriculture.

1. E-commerce offers an alternative venue of promoting and marketing agricultural products that has a benefit of reaching extensive geographical populations and providing detailed product information at a relatively low cost.
2. The costs of an e-commerce platform are not limited to user participation costs and include web site development, support, and training and promotion costs and, therefore, tend to be frontloaded. On the other hand, the benefits tend to increase over time as more producers and consumers become more familiar and active users of the platform.
3. The benefits of an e-commerce platform include new or additional web site presence in the short term, new contacts and new customers in the intermediate term, and ultimately higher profitability in the long term.
4. Active user participation is critical to achieving the desired benefits from participation. Due to the pattern of front-loaded costs and back-loaded benefits, e-commerce platforms are likely to show negative net returns in the early stages, but the track record of the more established programmes shows the potential for new ones.
5. Attribution effects—credit given to MarketMaker for additional contacts and sales—may mask the benefits of e-commerce; therefore, every effort should be taken to improve the visibility of MarketMaker effects.
6. Electronic collection of information about web site users as well as

the application of homogeneous evaluation approaches, such as the one developed in this study, will facilitate future evaluation efforts.

MarketMaker and similar sites offer a way for businesses to gain inexpensive initial web presence, allowing them to be "known" to the rest of the world. MarketMaker differentiates itself form other web sites because of its ability to map consumer and producer-related data. The platform has been supported by the land grant Cooperative Extension programmes in participating states, as well as other state public and private agricultural organizations. The long-term potential benefits of MarketMaker to producers, farmers' markets, and other user groups are now beginning to become evident. But, it is too soon to tell just how extensive these impacts will be. Much depends on the continued commitment of the programme partners—and the users themselves—to turn this latent potential into realized net benefits. Perhaps the frameworks and initial analyses created under the scope of this study will help facilitate the execution of appropriate actions necessary by all parties, working in unison, to derive these benefits.

Since its creation in 2000, MarketMaker has offered its electronic infrastructure and resources to registered users at no cost. Federal and state governments have provided most of the funding for the initial platform development and maintenance. Public investment in this "cyber-infrastructure" project can be justified if some benefits also accrue to the public through increased access to a more efficient, transparent, and robust food supply chain. However, in this era of fiscal austerity, this financing model may not be sustainable.

If this project were fully privately funded by an individual food supply business, it is possible that consumer acceptance and producer participation would be thwarted. Funding through a collection of associations, and public-private or non-profit public benefit-oriented groups that represent private interests is a possible alternative, so long as these groups represent a broad cross-section of food supply chain businesses and possibly even consumers. For certain, potential funders should fully recognize that substantial in-kind financial and other support has come from state-level institutions including (*e.g.*, land grant universities, departments of agriculture). Involvement of these organizations has facilitated the connection of the platform with producers and consumers which might be harder to achieve for a private organization.

FARMER EMPOWERMENT THROUGH E-CHOUPALS

ITC's unique strength in this business is the extensive backward linkages it has established with the farmers. This networking with the farming community has enabled ITC to build a highly cost effective procurement system. ITC has made significant investments in web-enabling the Indian farmer. Christened 'e-Choupal', ITC's empowerment plan for the farmer centres around providing Internet kiosks in villages. Farmers use this technology infrastructure to access on-line information from ITC's farmer-friendly web site www.echoupal.com. Data accessed by the farmers relate to the weather, crop conditions, best practices in farming, ruling international prices and a host of other relevant information. e-Choupal today is the world's largest rural digital infrastructure.

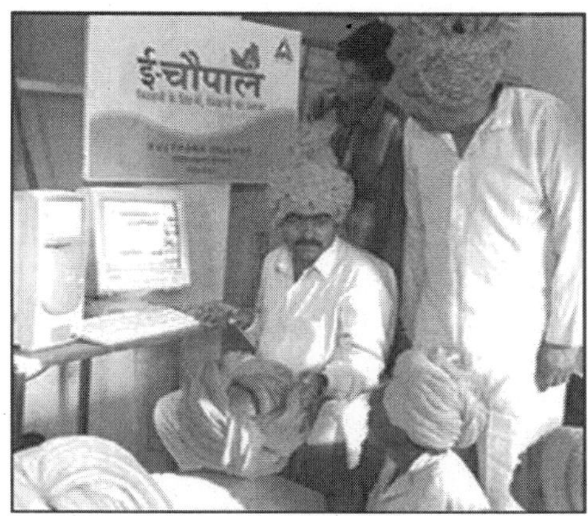

The unique e-Choupal model creates a significant two-way multi-dimensional channel which can efficiently carry products and services into and out of rural India, while recovering the associated costs through agri-sourcing led efficiencies. This initiative now comprises about 6500 installations covering nearly 40,000 villages and serving over 4 million farmers. Currently, the 'e-Choupal' web site provides information to farmers across the 10 States of Madhya Pradesh, Haryana, Uttarakhand, Uttar Pradesh, Rajasthan, Karnataka, Maharashtra, Andhra Pradesh and Tamil Nadu. Over the next 5 years it is ITC's Vision to create a network of 20,000 e-Choupals, thereby extending coverage to 100,000 villages representing one sixth of rural India.

Supporting the e-Choupal network are ITC's procurement teams, handling agents and contemporary warehousing facilities across India, enabling its Agri Business to source identity-preserved merchandise even at short notice. ITC's processors are handpicked, reliable high quality outfits who ensure hygienic processing and modern packaging. Strict quality control is exercised at each stage to preserve the natural flavour, taste and aroma of the various agri products.

ICT APPLICATIONS FOR AGRIBUSINESS SUPPLY CHAINS

The global food industry has undergone significant structural changes in recent years that have created opportunities for smallholder farmers in developing nations. The inclusion of these smallholders in agribusiness supply chains offers significant opportunities as well as challenges. ICTs can aid smallholders in taking advantage of opportunities and mitigating some of the challenges, as discussed in this module. Smallholders in the Global Food Industry: A Complex Relationship

The global food industry, with over US$ 4 trillion in annual retail sales, comprises agribusinesses of varying sizes. The largest are multinational corporations that operate internationally. In this module, "agribusiness" refers to a wide range of private companies:

- Retailers such as supermarkets or convenience stores (Walmart, Carrefour, ITC Choupal Fresh)
- Food processors and manufacturers (Nestlé, Kraft, Unilever)
- Input suppliers who produce fertilizer, seed, pesticide, irrigation equipment, and farm machinery (Bayer, Syngenta, Monsanto).
- Producers that may be smallholder farmers, cooperatives, or corporate farming entities.
- Wholesalers, traders, and other intermediaries in a supply chain that connect retailers or processors to producers (ADM, Bungee, Cargill, Olam).

In recent years, following deregulation of the food industry in many developing nations (Reardon et al. 2009:5) and the lowering of trade barriers in developed ones, private, market-driven agribusinesses have replaced state-supported entities. On the demand side, an increasingly urban population worldwide requires food to be delivered farther and farther from the farm; with rising incomes and changing preferences, this population also demands higher levels of food safety, quality, and traceability.

Changes in the informal supply environment have accompanied changes in the broader industry. The entry of large, private, often international agribusinesses from the formal economy has caused fragmented, informal suppliers to consolidate and formalize. To meet supply requirements arising from changing demands, agribusinesses often prefer to source through lead farmers or farmer cooperatives or directly from individual farmers. The alternative—purchasing from wholesale markets—can pose difficulties, be inefficient, and most important, cost more. When sourcing from wholesale markets, agribusinesses have little quality control, face uncertainties in supply and price, and lose the ability to trace products (which consumers increasingly demand).

When farmers are insulated by layers of intermediaries, it is difficult to communicate to farmers what items or quality levels the market demands. Reducing the number of intermediaries ("disintermediation") allows companies to reduce, deploy their market power more directly to garner lower prices, and improve quality control.

Direct procurement and improvements in production, transport, and supply-chain technologies make it possible to source competitively from vast numbers of suppliers and increase the relative importance of factor costs such as labour and raw materials. Companies looking to economize move production to places where factor costs are lower, which presents an enormous opportunity for farmers in developing countries (World Economic Forum 2009).

"Commercial supply chain"refers to a supply chain in which a private agribusiness is sourcing agricultural produce from farmers or selling products to farmers in accordance with a profit-seeking business model. "Supply chain" typically refers to the set of buy-sell interactions as goods flow from raw materials through production to the final retailer where consumers can buy them. "Value chain" generally refers to the whole ecosystem of players involved moving from the retailer backward to the producer. These terms are often used interchangeably, and a special distinction is not made in this module. Such chains can be of various types.

Although participation in commercial supply chains presents an opportunity for smallholders to attain higher incomes (between 10 and 100 per cent) and reduce poverty, these outcomes are not certain unless other important factors are addressed. For example, actual income changes depend on the crop, the time needed for farmers to learn to produce the crop more efficiently, and the quality and other standards required. Changes in income may not be sustainable unless accompanied by improved practices such as postharvest handling or risk management.

Market forces do not in and of themselves guarantee smallholders' inclusion in modern supply chains. When possible, companies might seek to source from larger producers, who can deliver economies of scale, often are better educated, and typically also have better access to finance. Including smallholders can present significant challenges for both the agribusiness and smallholder, but a strong business case can be made for both sides to work together.

For agribusiness:

- Smallholders can have distinct competitive advantages in certain situations. Compared to smallholders, large suppliers have greater market reach and multiple options to sell produce, so it can be riskier to source from them. It may also be less risky to source from numerous producers distributed across a wider geographic area, which can reduce systemic vulnerability to floods, droughts, and pests. Uncertainty about prices and quantities is reduced in the short term.
- Smallholders might simply have access to better land or other resources, and they are often more likely to follow the labour-intensive management practices required for higher-quality outputs.
- With smallholders under contract farming, production can adapt more rapidly to market demand.
- Government or donors may offer incentives to include smallholders.

On their side, smallholders can earn higher incomes. Participation also reduces their uncertainty as to who will buy at harvest and how much they will pay. Linking to a modern supply chain might be an important mechanism to address smallholders' lack of credit, inputs, extension services, and marketing resources (Reardon et al. 2009).

Smallholder inclusion is challenging for agribusiness because interacting with a large group of small suppliers implies high administrative and transaction costs (for example, in controlling quality, maintaining traceability, or ensuring adherence to certification standards). The agribusiness may have to supply physical assets (land, machinery, inputs) and information (on management practices and postharvest practices) for smallholders to produce the quantities and qualities required. Farmers may have to learn how to grow new crops or obtain more costly inputs. Farmers may not honour standing agreements with agribusiness at harvest, especially if the spot market offers higher prices than the company; the agribusiness may not honour its commitment to purchase from farmers.

In short, a supply-chain relationship between agribusinesses and smallholders is a complicated partnership with difficult requirements on both sides. Procurers need an agreement that will provide them with the right product mix, items that meet safety standards and are traceable, items of the right quality, timely delivery, and a cost-effective arrangement. Farmers require market information, extension services, risk-management capacity, financial services, and supporting physical infrastructure services (such as roads, storage, power, and telecommunications).

ICTS AND SMALLHOLDER INCLUSION IN COMMERCIAL SUPPLY CHAINS

Modern ICTs and their applications significantly affect smallholders' inclusion in commercial supply chains. ICT applications (hardware and software), guided by business logic, can foster smallholders' inclusion by making the following interventions in the supply chain: reducing costs of coordination (collection of production, distribution of inputs, and so on); increasing transparency in decision making between partners; reducing transaction costs; disseminating market demand and price information; disseminating weather, pest, and risk-management information; disseminating best practices to meet quality and certification standards; collecting management data from the field; and ensuring traceability.

Such interventions have been driven by the private and public sector. Their slightly different focus and resource base influence the kinds and the sustainability of ICT applications they propound.

The private sector views its supply-chain relationships as a competitive advantage. The ICT applications it develops to engage with the supply chain and provide information services are typically exclusive to its suppliers. Larger agribusinesses are also likely to have the scale and resources to deploy more expensive, commercially available ICT solutions within their supply chain. These interventions, if supported by a viable business model, are likely to be sustainable, but agribusiness-driven interventions may not necessarily focus on smallholders.

The public sector, donors, and civil society typically see the inclusion of smallholders as a public good. Their intervention in the supply chain is therefore focused on inclusion. The applications they create and develop are less likely to be exclusive. Because they have usually been designed to be specific to particular projects and used only once, public-sector interventions are unlikely to be easily generalizable to other contexts. Sustainability of donor-supported efforts has therefore been more uncertain, though future designs are expected to incorporate learning from previous experiences to enhance sustainability.

ICT FOR SUPPLY-CHAIN MANAGEMENT

In conditions of poor information flows supply chains are highly fragmented. Otherwise information technology driven innovations make it easier to acquire, manage, and process information and allow closer integration between adjacent steps in the value chain. There is therefore greater integration of supply chains based on information availability.

Organizations have understood for some time that logistics and supply-chain management (SCM) applications could reduce the transaction costs of procuring from smallholders. Indeed, any sizeable company in the developed world uses SCM systems to handle procurement and other tasks.

The lack of context-appropriate software, the prohibitive cost of hardware, and the lack of supporting infrastructure once made it quite difficult to use SCM systems in developing countries. The diffusion of ICT devices (especially mobile phones) and infrastructure has eased these constraints by making it possible to aggregate smallholders virtually. A secondary-source survey of ongoing or recent efforts towards smallholder inclusion using ICTs and their applications suggests that these technologies can solve many supply-chain problems associated with transactions (ordering, invoicing, payment); logistics (collection, storage, transport), quality assurance (safety, traceability); process management (production oversight, input distribution, extension support); and product differentiation (specialization in organic, fair trade, or regional labels).

The development of ICT applications for SCM can be driven by a wide variety of agents in the private and public sector, but collaborative partnerships appear to yield more effective applications. For example, agribusiness companies, mobile network operators, third-party service providers, and software firms as well as development institutions and research institutes may participate. It is rare for applications to be developed independently by any one party; collaborative partnerships focused on smallholder inclusion or value-chain competiveness are much more common.

No single ICT application is ideally suited for all procurement contexts or types of producers and actors along the chain. Organizations vary in size, budget, and operations. Some source perishables; others source staple grains. Supply chains encompass larger and smaller ranges of regions and producers (whose languages and education levels also vary). Not surprisingly, the varying degree of sophistication in ICT applications reflects this diversity. Bigger firms can extend their SCM solutions; other, smaller firms, turn to the off-the-shelf software or applications for mobile phones that are increasingly available; still others rely on spreadsheets. Some applications handle everything from transactions to logistics and quality control. Others focus on a smaller subset of areas. They rely on different combinations of software and hardware, but a combination of mobile phones, PDAs, networked computers, and centralized databases figure prominently in the architecture of most applications. (Module 2 discusses how the accessibility and affordability of ICT devices and infrastructure influence their use.)

Finally, the applications differ in their commercial approach. Some are public goods that do not have a revenue-generating model, while others adopt a one-time turnkey installation fee. Still others take a fee-per-transaction approach, while many follow an embedded service model in which revenues are generated from commercial trading (buying-selling) transactions and a fee for ICT services is not charged to farmers.

There is a sense that ICT applications can be the glue that holds together complex supply-chain partnerships. The rapid flow of information between buyers and producers that such applications allow minimizes misunderstandings, allows for risk management, provides higher levels of transparency, and ultimately fosters trust.

A particular area of concern on both sides is the possibility that one side or another will not uphold the pre-existing agreement. As mentioned, when prices are high, farmers have an incentive to sell to the spot market (side-sell) instead of to agribusinesses. Similarly when market demand for certain products changes or is lower than expected, procurers have an incentive to buy less than promised or at a lower price (finding produce to be of insufficient quality is a common tactic). Better communication between farmers and procurers, and systems that allow farmers to be paid faster, can reduce such myopic behaviour and help relationships endure. If farmers know that side-selling this season will have repercussions in the next because the company keeps electronic records, they might be less likely to engage in this behaviour. On the other hand, automated processes in the collection center make it more difficult for buyers to reject products arbitrarily or pay less.

ICTs will not sustain linkages that are fundamentally flawed, however. If the supply chain is not competitive or the business environment or trade laws prove restrictive, software to manage sourcing will not reverse the situation.

If market price fluctuations are sufficiently severe, ICT applications may not prevent farmers from side-selling or procurers from reneging.

Finally, the impact of these ICT applications on smallholders' inclusion in commercial value chains is not yet known. There is a general consensus that participation has a positive effect, but to what extent ICTs enhance or dilute that effect is unknown and requires research. The application of ICTs can be expensive from the perspective of software development or purchase, implementation, training, and so forth. The costs may not be justified in all cases. Better information on potential impact can help to make this determination.

KEY CHALLENGES AND ENABLERS

ICTs may create opportunities to incorporate smallholders more effectively into supply chains, but their impact will be limited without the requisite supporting infrastructure, policy, and culture of collaboration. This chapter describes the challenges and enabling factors associated with using ICT to manage supply chains and integrate smallholders.

Infrastructure is particularly critical for ICTs, which often require reliable electrical power and telecommunications networks. The presence of complementary infrastructure also has much to do with the success of ICT interventions for smallholder inclusion (roads, storage facilities, transportation, and financial infrastructure, among other types).

Commercial value chains prosper in an enabling business environment; policies that support such an environment are indirectly quite important to the effectiveness of ICT applications in supply chains. Policies can also discourage or encourage smallholder inclusion. In India, for example, limits on the size of landholdings make it difficult for agribusinesses to avoid smallholders in favour of larger producers. Until quite recently, policy barriers made it difficult to source directly from farmers at all.

Public-private partnerships have proven critical in developing ICT applications targeted towards smallholder inclusion. Public organizations lack the technical capacity, agribusinesses alone may not have sufficient incentive to reach out to smallholders, and technology companies are reluctant to absorb the risk of producing products unless they are assured of markets. Public institutions can lead such collaborative efforts if they are willing to share rights to outputs of the joint activities.

Public intervention in the private sector's use of ICTs in supply chains should focus specifically on improvements in the policy environment and the competiveness of smallholders. An important role of the public sector might be to incentivize smallholder inclusion and provide guidance on technologies that can be used to do so. The public sector might also work to organize farmers into groups and spread financial literacy.

Finally, the public sector should rigorously evaluate current ICT applications to determine their impact on smallholder inclusion and incomes. Quantitative and qualitative evaluation can include a variety of indicators to document outcomes.Key quantifiable indicators that are relevant to smallholders and can measure impact throughout the chain can include production volumes; product quality; net income; distribution of income among smallholders, within households, and along the supply chain; and the distribution of costs associated with risk mitigation and management.

These indicators can be complemented by additional quantitative measures that assess the overall viability of the supply chain, such as market position and penetration, profitability as compared to similar chains, and trends in volume and prices. Wherever possible, disaggregate data by gender.

Key qualitative or skills-based indicators that have an impact on farmers' incomes can include key skills related to: (1) the nature and quality of the relationship between farmers and trading intermediaries; (2) improvement in bargaining power; and (3) the governance functions of the chain itself. For chains linked to high-value markets, pay additional attention to issues related to product and process upgrading and collective innovation as the chain adapts to increasingly demanding market conditions. While this process does not occur fully at the farmer level, the existence of this skill set is critical for the entire system's continuing competitiveness. Unlocking innovation and opportunities for smallholders is a critical element of impact, because it leads to benefits that help drive farmers' incentives for inclusion (K. Kumar, personal communication).

When beginning an intervention, ascertain whether the barriers to smallholder inclusion are best addressed by an ICT application. Care should be taken to ensure the presence of key enablers—special attention is required to include women and other vulnerable groups. It is also important to consider the full cost of ownership beyond any one-time software and hardware fees. Installation charges, maintenance, upgrades, and the cost of training users must also be included.

After diligent consideration, if an ICT application is deemed appropriate, consider existing commercial products before attempting to develop new products. If the development of a new product cannot be avoided, sustainability should be a made a priority, and local partners must be included. A focus on developing standards for ICT applications and systems will allow interoperability between technologies and make it easier to develop new applications when necessary. Finally, human capacity is critical for the development and uptake of ICTs in supply chains. Farmers or farmer associations may find ICT tools challenging to use (illiteracy, a lack of training, or simply a lack of comfort with modern ICTs are typical barriers). Nor can ICTs be developed or deployed well if a technical talent pool with an entrepreneurial spirit is lacking.

TECHNOLOGICAL NEEDS AND FUTURE AGRICULTURE

It is apparent that the tasks of meeting the consumption needs of the projected population are going to be more difficult given the higher productivity base than in 1960s. There is also a growing realization that previous strategies of generating and promoting technologies have contributed to serious and widespread problems of environmental and natural resource degradation. This implies that in future the technologies that are developed and promoted must result not only in increased productivity level but also ensure that the quality of natural resource base is preserved and enhanced. In short, they lead to sustainable improvements in agricultural production.

Productivity gains during the 'Green Revolution' era were largely confined to relatively well endowed areas. Given the wide range of agroecological setting and producers, Indian agriculture is faced with a great diversity of needs, opportunities and prospects. Future growth needs to be more rapid, more widely distributed and better targeted. Responding to these challenges will call for more efficient and sustainable use of increasingly scarce land water and germplasm resources.

Technical solutions required to solve problems will be increasingly location-specific and matched to the huge agroecological/climatic diversity. Detailed indigenous knowledge and greater skills in blending modern and traditional technologies to enhance productive efficiency will be more than ever before, key to the farming success and sectoral growth. Most technological solutions will have to be generated and adapted locally to make them compatible with socio-economic conditions of farming community.

New technologies are needed to push the yield frontiers further, utilize inputs more efficiently and diversify to more sustainable and higher value cropping patterns. These are all knowledge intensive technologies that require both a strong research and extension system and skilled farmers but also a reinvigorated interface where the emphasis is on mutual exchange of information bringing advantages to all. At the same time potential of less favoured areas must be better exploited to meet the targets of growth and poverty alleviation. These challenges have profound implications for products of agricultural research.

The way they are transferred to the farmers and indeed the way research is organized and conducted. One thing is, however, clear – the new generation of technologies will have to be much more site specific, based on high quality science and a heightened opportunity for end user participation in the identification of targets. These must be not only aimed at increasing farmers' technical knowledge and understanding of science based agriculture but also taking advantage of opportunities for full integration with indigenous knowledge. It will also need to take on the challenges of incorporating the socio-economic context and role of markets.

With the passage of time and accelerated by macro-economic reforms undertaken in recent years, the Institutional arrangements as well as the mode of functions of bodies responsible for providing technical underpinning to agricultural growth are proving increasingly inadequate. Changes are needed urgently to respond to new demands for agricultural technologies from several directions. Increasing pressure to maintain and enhance the integrity of degrading natural resources, changes in demands and opportunities arising from economic liberalization, unprecedented opportunities arising from advances in biotechnology, information revolution and most importantly the need and urgency to reach the poor and disadvantaged who have been by passed by the green revolution technologies.

Another important implication of increasing globalization relates to the need for greater attention to the quality of produce and products both for the domestic and the foreign markets. This would imply that production must be tuned to actual rapidly changing product demand. Such adaptation to global markets would require state of the art research, which can be achieved only by setting global standards of research, focus on well defined priorities and mechanisms which permit close interaction of farmers with researchers, the private sector and markets.

CHALLENGE OF INDIAN AGRICULTURE

There are several concerns about Indian agriculture that we need to address. Even in the green revolution crops, rice and wheat, the Indian average yield ranks around 50th in the world. A weak agro-based industry, the problems of weak marketing, storage, transportation, credit support, etc, haunt us. Only 0.3 per cent of agricultural GDP is spent on agricultural R &D. The population is growing at an alarming rate of 1.8 per cent per annum. It will be around 1.3 billion by 2020 and would require 50 per cent additional foodgrains. A total of 200 million people are still below the poverty line and have insufficient access to food. Declining resource base, degeneration of natural resources, including soil fertility, water environment, etc, are some other concerns.

Food production systems centre around a few states only. Out of 17 big states, only 7 are self-sufficient and surplus in foodgrains. There was a significant decline in per capita intake of calories and protein between 1975 and 1990.

Inspite of all these problems, there is plenty of good news. India harbours a vast diversity of basic natural resources. Tremendous bio-diversity of agriculturally important micro-organisms exists, which are not yet been harnessed. We have a variety of crops, including quality fruits, vegetables, foodgrains, and a number of plantation and commercial crops. Indian fisheries resource has spread over 8,100 km of coastline on the east and west, and provides access to marine resources and transportation.

Very large proportion of area of various food crops is under low productivity category. It can be increased with inputs of high class science. There is a scope to cultivate sizeable portion of waste lands through soil amendment and introduction of agroforestry. Several million hectares of saturated soil can be potentially exploited from rainfed lowlands of eastern India.

The India Exclusive Economic Zone (EEZ) of the marine sector, of about 2 million sq.km., has high potential harvestable yields. Only about 20 per cent of the cultivated area is covered under tractor cultivation, great scope of making quantism quantum jump in production and productivity exists, changing consumption patterns offer scope for diversified agriculture. Trade liberalization provides opportunities to reach for outside markets, which were not accessible otherwise. Great gains are possible through agro-based industrialization and overall rural development.

The cultivators and farmers of India have a rich heritage of traditional agricultural wisdom on crop and livestock husbandry and fisheries; based on the principles of the conservation of natural resources and environment. India has an extensive network of extension channels for dissemination of technologies generated by research institutes. India has a strong manufacturing capability and a large network of about 20,000 manufacturers and nearly one million village artisans for indigeneous production of all types of agricultural machinery. Therefore, inspite of several disadvantages, we have a great chance to leap forward, if we enforce the right policy initiatives and implement them.

BIOTECHNOLOGY AND INDIA AGRICULTURE

We need to recongnise that, just as the food reequirements of today's population of nearly 6 billion people could not have been met by the technologies of the 1940's, we cannot assume that current practices will feed the population of 8 billion expected by 2020. New approaches are needed in addition to the continued improvement of existing methods of crop and animal husbandry and food processing.

The strategic integration of biotechnology tools into India agricultural systems can revolutionise Indian farming and usher in a new era. Genetic engineering is clearly the most revolutionary tool to impact agricultural research since the discovery of genetics by Mendel. Prior to genetic engineering, the exchange of DNA material was possible only between individual organisms of the same species.

With the advent of genetic engineering in 1972, scientists have been able to identify specific genes associated with desirable traits in one organism. For example, a gene from bacteria, virus or animals may be transferred into plants to produce genetically modified plants having changed characteristics. This method therefore allows mixing of the genetic material from species that cannot otherwise breed naturally.

Genetic engineering can be vital for an agrarian country like India. It can help in minimising the crop damage through disease and pest resistant varieties, reducing the use of chemicals, enhancing stress tolerance in crop plants, thus permitting productive farming on unproductive lands, etc. One could even extend the growing season of crops and minimise losses due to environmental factors on the one hand and increase the shelf life of fruits and vegetables, on the other, thus minimising losses due to food spoilage. We can expand the market vista and improve food quality. Biotechnology can also produce plants that possess healthy fats and oils, possess increased nutritive value, and create a whole range of higher value feeds. Biotechnology has even the power of producing biodegradable plastics, edible vaccines, etc.

It is only through the blending of the 'gene revolution' with our experience in the 'geen revolution' that we can reach our goal of 'evergreen revolution' and also 'nutritional revolution' The advantage of the gene revolution is that it is relatively scale neutral, benefiting big and small farmers alike, It is also environment friendly. Thus, it can be of great help to the smallest farmer with limited resources in increasing farm productivity through the availability of improved but powerful seed. It can also reduce a farmer's dependency on chemical inputs such as pesticides and fertilizers.

Modern biotechnology offers unlimited opportunity for enhancing genetic potential of crops and other commodities, management of biotic and abiotic stresses, bio-remediation and organic recycling. India is one of the main centres of agricultural biodiversity and its gene richness can greatly complement the developments in modern biotechnology. Likewise, new developments in GIS, remote sensing, and crop modelling, provide new opportunities for integrated management of natural resources. The revolution in informatics provides opportunities for sharing latest information for research and planning in a highly organised and efficient manner. With the globalisation of economy, the opportunities for value addition and post harvest management are also immense. With clever blending of technologies, the India farmer can usher in new era of confidence and performance. Sir Francis Bacon had once said: *'It would be an unsound fancy to expect that things which have never yet been done, can be done except by methods which have never been tried'* In the new context, we have to review the new role of biotechnology. Also, it is in this context that we have to view the rapid advances that modern biotechnology is making, including the area of transgenics. Let us focus on some of the developments in this area, since considerable interest and controversy has been created around the world today in this exciting area.

10

Agribusiness Environment and IT in Agribusiness

AGRIBUSINESS

From precision farming methods to complex environmental fate projects, Battelle is solving the problems that matter most to companies that feed the world. We deliver innovative agribusiness solutions that lead to new products and boost existing product productivity while maintaining safety and sustainability within the global food system.

Expert scientists at our facilities in the U.S. and Europe perform an array of testing and analysis (GLP and non-GLP compliant), develop new products and provide regulatory and risk assessments.

- Agricultural Formulation Development: Gain access to markets worldwide with Battelle's integrated development and analysis services. Our scientists provide novel technologies and product development, regulatory compliance assessment, and customized research studies that meet the demands of our clients worldwide.
- Biotechnology: Accelerate development timelines, control costs and support faster product registration with Battelle's agricultural biotechnology services. We are advancing agrobioscience with existing methodologies and novel technologies to support the development needs of our clients. We provide an integrated approach to research, development and regulatory compliance to help you move to market quickly while controlling costs and risks.
- Environmental Fate Studies: Meet regulatory requirements and get to market faster with accurate, defensible data. At our labs, Battelle scientists conduct single studies and manage complex, whole programmes. Studies include aerobic and anaerobic soil degradation, adsorption and desorption, leaching, soil dissipation, lysimeter and hydrolysis. We also provide computer modeling to support the registration of new and existing chemicals.

- Plant and Animal Metabolism Studies: Isolate, identify and characterize known and unknown metabolites in tissues and excreta. Our studies can be conducted in the greenhouse under environmentally controlled conditions or outside where typical farming and climatic conditions can be more closely simulated. Battelle's 14C-labeled animal metabolism include rat, hen, goat and swine.
- Residue Studies: Move your product to market quickly with coordinated, customized residue studies. Our more than 25 years of analytical experience gives Battelle scientists the competitive edge. We can develop, adapt, and validate new methodologies quickly and reliably. Our team can coordinate the entire residue process for both crop and animal feeding studies, from study planning to field/in-life phase management, and sample processing through analysis.
- Risk Assessment and Regulatory Support: Get the data you need to make effective risk management decisions and minimize remedial footprints while achieving project goals and objectives. Battelle has world renowned competency in regulatory affairs, both in the U.S. and Europe. We provide full-scale registration and re-registration support as well as provide expert advice in dossier defence.
- Toxicology: Ramp up your development timelines with Battelle's cutting-edge, multi-disciplinary toxicology services. For more than 30 years, Battelle has provided industry-leading general toxicology services for the pharmaceutical, biotechnology, medical device, veterinary health and nutrition, and agrochemical industries, as well as government agencies. We can execute both routine and non-traditional toxicology studies and customize our service to meet your unique needs.

AGRIBUSINESS AND RURAL ENTREPRENEURSHIP DEVELOPMENT

CHALLENGE

Many developing countries and economies in transition, particularly those with large rural communities, suffer from inadequate access to food and lack of employment. The problem is compounded by the dependence on outdated and inefficient technologies leading to poor productivity and slow economic growth. Agriculture-based industrial products account for half of all exports from developing countries, yet only 30 per cent of those exports involve processed goods compared to a figure of 98 per cent in the developed world.

ASSISTANCE

UNIDO provides a variety of technical cooperation activities to assist developing countries in adding value to the output of their agricultural sector and generate increased employment opportunities for rural communities, thereby increasing food security and sustainably reducing poverty.

Through its technical assistance, UNIDO links resources and markets in the agribusiness value chains and strengthens forward and backward industrial linkages in order to leg up the economic transformation of countries, improve employment and income opportunities, and reinforce sustainable livelihoods.

Activities benefit a number of groups, including poor and marginalized rural populations, urban agro-industries and communities facing human security challenges or requiring urgent supplies of agricultural equipment and the rehabilitation of food industries.

Technical cooperation and capacity-building services are provided to agro-based and agro-related businesses and industries, inter alia, in the food, leather, textiles, wood and agricultural equipment sectors.

UNIDO promotes investment in agribusiness and value chain development; builds partnerships and linkages with strategic financing institutions; organizes various global forums and expert group meetings in related fields; and publishes specialized training manuals, guides and electronic media.

To carry out its mandate in this area, UNIDO mobilizes expert services such as cluster development, conformity with quality and standards, rural energy, environmental management and cleaner production.

THE AGRIBUSINESS THREAT TO FARMERS, FOOD, AND THE ENVIRONMENT

The agribusiness/food sector is the second most profitable industry in the United States — following pharmaceuticals — with annual sales over $400 billion. Contributing to its profitability are the breathtaking strides in biotechnology coupled with the growing concentration of ownership and control by food's largest corporations. Everything, from decisions on which foods are produced, to how they are processed, distributed, and marketed is, remarkably, dictated by a select few giants wielding enormous power. More and more farmers are forced to adopt new technologies and strategies with consequences potentially harmful to the environment, our health, and the quality of our lives. The role played by trade institutions like the World Trade Organization, serves only to make matters worse.

Through it all, the paradox of capitalist agriculture persists: ever-greater numbers remain hungry and malnourished despite an increase in world food supplies and the perpetuation of food overproduction.

Hungry for Profit presents a historical analysis and an incisive overview of the issues and debates surrounding the global commodification of agriculture. Contributors address the growing public concern over food safety and controversial developments in agricultural biotechnology including genetically engineered foods. *Hungry for Profit* also examines the extent to which our environmental, social, and economic problems are intertwined with the structure of global agriculture as it now exists.

EFFECTS OF ENVIRONMENTAL CHANGES ON FOOD AND AGRIBUSINESS

The objective of this article is to share with China Daily readers some important changes in the macro-environmental variables that are affecting and may affect companies operating in food and agribusiness.

The traditional PEST analysis to summarize the major thoughts. Just remembering, the PEST (or STEP) analysis is a traditional tool to understand macro-environmental changes. The "P" represents the political-legal environment (institutional environment). "E" is the economic and natural environments. "S" is for the sociocultural environment and finally, "T" is for the technological environment. These four environments help to organize the variables and are a first and very important "step" in strategic planning processes.

Starting with the political-legal environment, we may point that instability in Iran, North Korea and other Middle East and northern African countries are affecting oil prices, together with the growth in oil consumption coming from emerging economies. This will lend more leverage to the biofuel industry, since high oil prices mean fewer the economic benefits and environmental benefits, which will boost investments and also stimulate governments' efforts to blend biofuels into gasoline.

Lower interest rates in Europe are driving enormous flow of resources to emerging economies, and these are suffering the valuation of their currencies, eroding their competitiveness. Also some expected reforms are not progressing in important food/agribusiness countries, which bring increases in cost in several commodities.

Confusing tax management policies in developing countries are also happening, with new protections and market access limitations. We face increasing risks of interference (regulation), for examples the limitation in food advertisement to kids and regulation towards international investments in land. In developed economies, we see insufficient federal budgets as an argument to remove support for some less efficient agribusiness industries.

The economic and natural environment shows that this year and probably in this decade, economic growth will be coming mostly from emerging countries (5,5 per cent average of GDP growth in 2012, and 1,5 per cent growth in developed economies) due to a larger pace of income distribution in populated emerging economies, the faster than expected recovery of the USA and Europe still not growing. Since there is an increasing influence of weathercoupled by land prices and labour factors, in some regions, we see production regions switching. New agricultural frontiers are being developed by local or international companies, following the governmental incentives for value capturing in producing regions (more processing and other). Also in the economic environment, bigger environmental pressure will increase production costs and we see more initiatives of buyers increasing coordination over suppliers (farmers).

The sociocultural environment shows some interesting changes. Migration and urbanization leveraging the growth of processed food, the protest and mobilization movements are increasing pressure over inclusion, thus signalizing companies that this can be an opportunity within the supply chain. Risks of consumer movements are getting bigger for companies. The demographic trends of reduction on family size and people living alone continue to boost foodservice and ready to eat markets. There is also an increasing concern about food waste, and we see growing debate in this field.

Consumers are also demanding more information about the story behind production (link to farmers), more direct trade and to value what is "local". Natural and healthy movements continue strong, increasing demand for certifications of products, companies and food chains and there is a larger acceptance of biotechnology, with focus over genetically modified products.

Another important point is an increasing pressure done by society and buyers towards protection of some industries. As an example, several sugar buyers in the USA are protesting against high import taxes and other support programmes for local sugar industry, with higher costs than international markets.

Finally, within the technological environment there is an increasing pressure over natural resources and we are entering the era of commodities. A lot of investments and the development of biotech and nanotechnologies are happening, and in the communication side, we feel the rapid transformation of society with the digital world and new media improving the speed of communication. The development of systems speeding up information availability is facilitating the tracing process, helping to identify products sources and other relevant information. Technologies that allow to recycle and re-use have higher value than before.

These were some of the changes coming from recent discussions with business managers and executives. These are facts that will bring specific impacts to the industries and desiring strategies, or acts of food chain participants in their planning processes.

AGRIBUSINESS BOOM THREATENS KEY AFRICAN WILDLIFE MIGRATION

Unreported, an environmental tragedy is unfolding in a remote corner of Africa, on the borders of the newly-designated state of South Sudan, that could imperil the second-largest mammal migration on the African continent. Most of us know about Africa's largest migration, the millions of wildebeest and their attendant predators who race across the Serengeti plains of east Africa in search of water each year. It is the stuff of countless photographs and hundreds of TV natural history programmes. But how many have heard of the second-largest migration? I certainly hadn't until I stumbled on it last month in the Ethiopian region of Gambella.

As we drove into the bush, the track ahead was alive with large animals. From the far distance they looked like cattle, but it soon became clear they were antelope. As we drew closer, their numbers grew, and they began running. I could see a dense column stretching in all directions. They numbered many thousands, with warthogs in among them, darting through the tall wet grass between a series of ponds and heading towards the Baro River, a tributary of the Nile.

But as the antelope ran I saw, not far away on the horizon, bulldozers and plumes of smoke. Someone else wanted this rich grassland and its water. This bush would soon be transformed — and the future of the great migration in grave doubt.

My guides said the antelope were white-eared kob. Along with the Nile lechwe, another endangered antelope, and the giant shoebill stork, they were the main reason for the creation back in 1974 of the Gambella National Park.

Geographically and ethnically, the swampy lowlands of Gambella look as if they should be in the neighbouring new state of South Sudan, which was formed in January when Sudan was formally divided in two. Gambella's tall, jet-black skinned Anuak and Nuer tribes are very different from the lighter-skinned peoples of highland Ethiopia. And the vicissitudes of war in Sudan and pogroms in Gambella have seen a constant flow of refugees across the border.

South Sudan is where most of the white-eared kob I saw came from, traveling across the open woodland bush at the end of the dry season in search of Gambella's open water and wetlands. More than a million of them are estimated to come this way each year, along with a scattering of elephants and giraffes.

But the park that is supposed to protect them is little more than a mark on the map. Two years ago, the Ethiopian ministry of agriculture declared that, whatever its wildlife credentials, the park had "a huge agricultural investment potential." And now the ministry is seeking to realize that potential through a series of major leases to foreign agribusinesses. Some

TWO YEARS AGO, ETHIOPIA'S AGRICULTURE MINISTRY DECLARED THAT GAMBELLA NATIONAL PARK HAD 'HUGE INVESTMENT POTENTIAL.'

400,000 hectares, an area 80 times the size of Manhattan, much of it within the 1974 boundary of the park, has been promised so far. Drive west from Gambella town, the capital of the region, and for most of the two hours it takes to reach Nyininyang near the Sudanese border, you are in the concession of the Indian firm Karuturi Global Limited. Take the road south and the bush suddenly gives way to the vast compound of the other large investor, Saudi oil billionaire Sheik Mohammed Hussein Ali Al Amoudi.

Both concessions are being developed at a breakneck pace with heavy machinery everywhere, clearing forests, draining swamps, and installing irrigation systems. Bush along the roads is burned and, as the dry season ended in late February, smoke plumes dotted the horizon.

Al Ahmoudi's company, Saudi Star, is constructing a 30-kilometer canal from an old unused state reservoir to irrigate tens of thousands of hectares of rice paddy. But so far the lead actor is Karuturi. The Indian company is the world's largest grower of roses, claiming 10 per cent of the global market. But when we met in Karuturi's compound by the road an hour west of Gambella town at Iliya, its local project director, Karmjeet Singh Sekhon, resplendent in turban and long twirling moustache, told me the world rose market was now saturated. So Karuturi is moving into other products. And that is bad news for the white-eared kob and much else.

Fig. Workers clear earth for irrigated rice farming on Ethiopian land leased by a Saudi firm.

Until recently, there were no roads around Iliya, except the main dirt road west. But the Karuturi concession will soon have 600 kilometers of roads. Within a couple of years, Sekhon expects to have 100,000 hectares under cultivation, with another 200,000 hectares awaiting the go-ahead. Every square meter of bush has been surveyed. His 15 huge John Deere tractors are ploughing 500 hectares a day. "This May we will plant 35,000 hectares of rice, 10,000 hectares of maize, and 10,000 hectares of sorghum," he told me. Some 20,000 hectares of oil palm and sugar cane will be added soon. This is land clearance on a gigantic scale.

"The soil is excellent," he said. "It's virgin land. You can grow anything here. We have no land like this in India. There you are lucky to get 1 per cent of organic matter in the soil. Here it is more than 5 per cent. We don't need fertilizer or herbicides."

Within five years, Sekhon expects to have 50,000 people living within the concession area, working its fields and operating processing mills in three townships. But locals expect most of the jobs to go to highlanders, which they contend routinely happens when businesses come into this region.

There is also growing concern about what is happening to the land. One former ranger at the park told me that Karuturi's engineers were draining the large Duma swamp 30 kilometers south of the highway and deep inside the Gambella National Park. It is one of the last refuges of the endangered Nile lechwe. And there is the kob migration, which appears to run right through Karuturi's concession.

I asked Sekhon about the wildlife. Yes, he said, the animals on his land were a "problem." What about the park? On my map all the land for the 60-kilometer stretch from Baro River in the north to the Gilo River in the

What is happening in Gambella is probably just the start for this forgotten corner of Africa south is part of the national park. At least part of that is on the concession. But he said he knew of no rules that prevented Karuturi from cultivating its concession.

He may be right. It turns out that the park, though marked on maps, has never been formally gazetted. In any case, according to Cherie Enawgaw of the Ethiopian Wildlife Conservation Authority, a government agency, the park may have a handful of rangers but it has "no management plan and no clear indicated boundary."

What angers Ethiopian environmentalists is that alternative economic uses such as tourism have not been explored. Sanne van Aarst of the Horn of Africa Regional Environment Centre at Addis Ababa University says Gambella has the same potential as the Serengeti and Maasai Mara tourist "hotpots" in Kenya and Tanzania.

Instead, the government has asked the conservation authority to "re-demarcate" the park's boundaries. There are three options, according to Enawgaw. Each involves moving the park boundaries south and west by several tens of kilometers to make way for the new concessions. But his own maps of sightings of wildlife, produced to help with the demarcation, show that the plan will allow migration paths to be blocked and "wildlife core areas" ploughed up.

Many of the local Anuak people, who farm small areas and hunt in the park, are not happy. I talked to a small group that had refused to move from their village right next to the Saudi Star canal and a road that ran right through their old fields.

"We used to hunt with dogs, but after the farm came the animals here disappeared," said Omot Ochan, sitting by an open fire on an old waterbuck skin and eating corn from a bowl. "Two years ago they began chopping down the forest and the bees went away. We used to sell honey. Now we only have fish."

He insisted the company had no right to be there. "Everything for two days' walk from here is ours," he said. But nobody is listening. Another truck rumbled past his straw hut, shattering the silence of the bush and creating a cloud of dust. After it had gone, I noticed a large, dead stork in the road.

Park rangers sporadically chase Anuak hunters through the swamp grasses, which can grow up to three meters high. On the road to Nyininyang, I spotted a small gang with dogs, rifles and a couple of chestnut-coloured kob slung over their shoulders. Elsewhere, we drove through dense smoke and flames that darted across the road, set by hunters trying to corral their prey.

Back in Gambella town in the evening, as power gave out in the old government hotel, I spoke to a park official. He took note of my report on the hunters. He would send his people out in the morning to check if they were still there. But on the subject of the land grabbers — the real threat to wildlife in the park — he could only shrug his shoulders. It wasn't his business.

And what is happening in Gambella is probably just the start for this forgotten corner of Africa. Across the border, the emerging government of the soon-to-be official nation of South Sudan is entertaining would-be agricultural investors in its capital, Juba. As in Gambella, the land of South Sudan is fertile, well watered and, by modern standards, hugely under-populated. It is ripe for land grab.

Environmentalists fear that the crown jewel of the upper basin of the White Nile could be under threat — the vast swampland known as the Sudd. Investors are eyeing its water to irrigate huge plantations, says Jane Madgwick, the head of Wetlands International, a Dutch-based NGO that wants a global campaign to resist the move.

In recent decades, few outsiders have visited these areas. They have become almost mythical gardens of Eden, surrounded by civil war. But, now that peace has broken out in southern Sudan, the economic dividend may be wholesale ecological destruction.

CONCEPTUALIZATION OF TECHNOLOGICAL CHANGE IN AGRICULTURE

Technological change is among a small handful of topics in what is now referred to as the sociology of agriculture — along with part-time farming and analysis of the farm family — that have received attention during all three major phases in rural sociology. From the 1930s through the 1950s there were several major studies by sociologists of the impacts of technological change, most of which centered on the rapid mechanization of southern agriculture that began during the Great Depression.

The concern of many of the early rural sociologists with the socioeconomic consequences of technological change, however, largely lapsed during the succeeding phase of rural sociology in which the diffusion-adoption perspective was dominant. In the era of the "new sociology of agriculture," we again see a rekindling of concern with the social consequences of technological change. Moreover, interest in the consequences of technological change has become expressed in more social structural terms and, as demonstrated later, has become more closely integrated with analyses of the genesis of new agricultural technologies.

Several other important observations can be made on the rural sociological literature on technological change in agriculture over the past decade. First, mechanization has received the lion's share of attention; technological change in the forms of improved varieties of crops and breeds of livestock or of agricultural chemicals has been the focus of only a few major ex post studies. Second, as Berardi has observed, there are many problems in decomposing the effects of technological change on farm structure from the effects of other variables such as public policy or changing wage rates; therefore, rural sociologists have often relied heavily on the primary research of others (*e.g.*, agricultural economists and geographers with complementary expertise in assessing techno logical impacts) when developing comprehensive assessments of the socioeconomic impacts of technological change.

Third, the conceptualization of technological change in agriculture by rural sociologists has been heavily influenced by the work of the agricultural economist, Willard Cochrane, who developed a theory of the "treadmill of technology" in the late 1950s. Briefly, Cochrane's theory of technological change builds on a key observation from rural sociological research on the diffusion-adoption of innovations-that large farmers who are entrepreneurial and non-risk-averse tend to be the first to adopt new agricultural technologies.

By adopting early, these farmers enjoy "innovators' rents" because of their ability to reduce their average per unit production costs. Cochrane posits that after these innovators and early adopters have employed the new agricultural technology, aggregate output of the particular commodity involved will begin to increase and prices will decline disproportionately (because agricultural commodities tend to have a low price elasticity of demand).

Declining prices then begin to force non-adopters to utilize the technology. These subsequent adopters gain very little from the technology; they adopt it merely to be able to stay in business. Slow adopters or non-adopters will tend to be forced out of agriculture because they can no longer compete due to their high average costs of production. Cochrane considers the technological change process to be a "treadmill" in that the majority of farmers, who receive little or no benefit from the technology, are nonetheless forced to continue to adopt new agricultural technologies in order to stay in business.

Cochrane has also observed that the benefits of technological change tend over the long term to be realized in the form of rising land values, particularly if there are commodity programmes in effect to cushion a commodity sector from declining crop prices caused by increases in output. The benefits of technological change (innovators' rents from early adoption) tend to be reinvested in farm land, causing increases in land prices and benefiting large landowners over smaller property owners and tenants. These effects, in Cochrane's view, are heightened when commodity programmes support crop prices, thereby increasing the level of innovators' rents and intensifying the extent to which these benefits are capitalized in asset values.

Cochrane's theory of the treadmill of technology has been widely accepted in recent rural sociological work as an orienting perspective because it has effectively linked knowledge from diverse origins (rural sociological research on diffusion-adoption, agricultural economics research on technological change, and so on). Cochrane's theory, however, has its limits — in particular, its inapplicability to industrial agriculture, as noted by LeVeen.

Moreover, the notion of the treadmill of technology gives little attention to the origins of technology, that is, to why technologies that have displaced labour or benefited larger operators over smaller ones have been developed, a topic that, has recently come to be a major research issue for those working in the sociology of agricultural science. Nonetheless, with the exception of sociologists of agriculture who have focused on industrial agriculture, rural sociologists have yet to develop a distinctly sociological theory of technological change and its socioeconomic consequences.

While rural sociologists have not yet developed a comprehensive and distinctly sociological theory of technological change and its consequences applicable to both family farming and industrial agriculture, the new sociology of agriculture as applied to technological change has made some significant advances.

Among the most significant research papers on the topic of technological change in agriculture were two published in the late 1970s that dealt with limitations of the classical diffusion-adoption model. The first, by Pampel and van Es, was an otherwise traditional adoption study that included, in addition to "commercial-profitable" innovations that were the major focus of previous diffusion-adoption studies, a set of "environmental innovations" such as contour farming, terracing, sod waterways, and use of rotation cover and reduced tillage. Pampel and van Es discovered that the correlates of adoption of environmental innovations were far different than those of commercial innovations. For example, while large-scale farmers were observed to be the earliest adopters of commercial technologies, there was no significant relationship between farm size and the adoption of environmental innovations.

They concluded their study by cautioning rural sociologists that the diffusion-adoption tradition, because it had ignored dimensions of innovative behaviour such as the adoption of environmental-conservation technologies, "may have provided within rural sociology a field of knowledge with a narrow empirical foundation on which to base its generalizations". They also emphasized that given the tendency of most environmental innovations to be unprofitable, it was not clear that voluntary strategies to achieve soil conservation and related environmental goals would be successful.

Writing two years after the Pampel and van Es study, Goss focused more systematically on the limitations of diffusion-adoption research. The principal failings of diffusion-adoption research, in Goss' view, were (1) that it was uncritical, if not promotional, towards patterns of technological change that may be detrimental to the interests of society and/or farmers, and (2) that the focusing of research effort on the adoption component of the technology change process had caused rural sociologists to ignore the consequences of technological change for various groups in society.

Goss summarized the early 1970s literature critical of the Green Revolution and pointed out that the Green Revolution had been initiated with the same promotional orientation towards technology that had predominated during the heyday of the diffusion adoption tradition in U.S. rural sociology. He noted that many social scientists who had once worked within the diffusion-adoption tradition in international agricultural activities had begun to recognize the limits of the adoption paradigm. He argued, however, that this point had not been well recognized in domestic rural sociology and that rural sociology must begin to develop a more comprehensive programme of research on the consequences of technological change in agriculture.

In a manner similar to Goss' exegesis of the Green Revolution literature in terms of the impacts of technological change on social groups and classes in agriculture, Stockdale, in an influential article, raised comparable issues in terms of the ecological aspects of modern agricultural technology. Noting that the contribution of technological change to American agricultural productivity had often been held up as a model for other nations to emulate, Stockdale argued that social scientists had tended to ignore the adverse ecological impacts of energy-intensive technology in the United States and elsewhere. Like Goss, Stockdale called for an expanded programme of research on technological change in agriculture that would emphasize the social and ecological consequences of new agricultural technologies.

Among the most innovative thrusts in sociological research on technological change in agriculture have been the series of books, monographs, and articles published by Friedland and associates. Friedland and his colleagues have focused on technological change in industrial agriculture in California and have identified a number of components of the technological change process that have tended to be ignored in the earlier rural sociological literature. As noted earlier, Friedland has focused on several commodity systems (chiefly tomatoes, lettuce, and grapes) and has found that the social organization of each commodity system — and hence the structural context of technological change — is distinctive.

Nonetheless, Friedland and his associates have demonstrated that the technological change process in industrial agriculture generally is heavily influenced by considerations of the cost of, access to, and control over labour and by the relationships of grower organizations with public and private components of the agricultural research system. They have also demonstrated that the nature of the technological change process and its consequences depend upon the structure of the commodity system. For example, in the aftermath of the termination of the bracero programme in 1964, a mechanical tomato harvester technology was widely adopted in California, displacing three-quarters of California tomato producers and the bulk of their agricultural wage labour forces. In lettuce, however, grower-shippers were able to utilize the threat of lettuce harvest mechanization to discipline the labour force and to avoid the costly process of harvest mechanization.

A further component of the recent rural sociological literature on technological change, one to which Friedland and Friedland and associates have made major contributions, is social impact assessment of new agricultural technologies. Social impact assessments of new agricultural technologies are a distinctive type of research in that they often tend to rely primarily on secondary data or on the results from the published literature rather than on primary research and data collection; this is particularly the case if the assessment must be prepared over a short period of time.

Other distinctive aspects of social impact assessment of new agricultural technologies are, first, that the assessment is ex ante (rather than ex post, as in most previous rural sociological research on technological change) and, second, that the orientation towards these emerging technologies tends to be questioning and skeptical, rather than accepting and promotional. Rural sociologists who have conducted social impact assessments of new technologies have approached their subject matter from the vantage point that most new technologies tend to involve an unequal distribution of costs and benefits and that advance projections of these patterns of the distribution of costs and benefits can enable social scientists and affected groups to redress these distributional impacts more effectively.

An example of the type of knowledge base that has been developed, in part, as a building block for social-impact assessment research is Berardi's comprehensive review of the impacts of agricultural mechanization. Berardi reviewed the re sults of over 180 studies of agricultural mechanization in the United States and elsewhere and summarized the major empirical regularities in studies of a variety of types of mechanization in a diversity of commodity systems. She concluded by pointing out that the rich literature on agricultural mechanization — most of it prepared by non-sociologists — is sufficient so that sociologists may contemplate conducting rigorous ex ante studies of the impacts of new mechanical technologies in agriculture.

Unfortunately, as noted earlier, sociological knowledge on the socioeconomic impacts of mechanization is far greater than that on "biochemical" technologies (plant varieties, fertilizers, insecticides, herbicides, and so on). There have only recently been sociological studies that have delivered on Stockdale's and Berardi's admonitions in the area of biochemical technologies in plant and animal agriculture. It is likely, however, that the great attention that has already been paid by rural sociologists to emerging biotechnologies in agriculture will result in a sustained research programme into the socioeconomic consequences of these new technologies.

It should be noted in closing, however, that while there is great merit in inventorying the results of existing studies on various facets of technological change — as, for example, Berardi has done — sociologists should exercise caution, as technological change should not be assumed to have invariant consequences across time and space. In particular, as noted above, there have been some major recent changes in the structure of U.S. agriculture that one might expect will have significant effects on the technological change process and its consequences.

For example, instead of technological change tending to result in the decline of small farms, as it did during the bulk of the post-World War II period, it may be quite likely that new agricultural technologies such as biotechnology-derived plant varieties, growth regulators, and bovine growth hormone may tend to result in the disproportionate demise of medium-sized, full-time family farmers. Thus future rural sociological research on technological change in agriculture must be cognizant of ongoing changes in the structure of agriculture and how these changes affect the types of agricultural technologies that are employed, the processes of technological change, and their consequences.

11

Entrepreneurship in Agribusiness

GOVERNMENT PROGRAMMES FOR AGRIBUSINESS

Both the central government and state governments have devised programmes to woo the entrepreneurs for setting up agribusiness in India. They may be classified as programmes falling under small-scale industries, Khadi village industries, small and medium industries, large industries based on the finance required for investment. Apart from these to harness the export market after signing the WTO agreement. They include conversion of Export Processing Zones (EPZ) into Special Economic Zones (SEZ), establishment of Agri-export zones. For which government is attracting private investors to make investment in infrastructure development like cold storage chains, improvement in road, rail, sea and air transport systems.

The Ministry of Small Scale Industries and Agro and Rural Industries designs and implements the policies through its field organizations for promotion and growth of small and tiny enterprises, including the coir industries. The Ministry also coordinates with other Ministries/ Departments on behalf the Small Scale Industries (SSI) sector.

The implementation of policies and various programmes/schemes for providing infrastructure and support services to small enterprises is undertaken through its attached office, namely the Small Industry Development Organization (SIDO), statutory bodies/other organizations likely Khadi and Village Industries Commission (KVIC) and Coir Board a Public Sector Undertaking - National Small Industries Corporation (NSIC) and three training institutes - National Institute of Small Industry Extension Training (NISIET), Hyderabad, National Institute for Entrepreneurship and Small Business Development (NIESBUD), New Delhi and Indian Institute of Entrepreneurship (IIE), Guwahati.

ENTREPRENEURIAL TRAINING INSTITUTES

There are three institutes engaged in training of small-scale entrepreneurs. These are Indian Institute of Entrepreneurship (IIE), Guwahati, National

Institute of Small Industry Extension Training (NISIET), Hyderabad, National Institute for Entrepreneurship and Small Business Development (NIESBUD), New Delhi.

FRAGRANCE AND FLAVOUR DEVELOPMENT CENTRE (FFDC), KANNAUJ

Fragrance and Flavour Development Centre (FFDC) has been set up as an autonomous body in the year 1991 by Govt. of India with the assistance of UNDP/UNIDO and Govt. of U.P/UNDP/UNIDO has provided technical expertise and imported equipments. Govt. of U.P has provided land and building while Govt. of India has been contributing for indigenous equipments and recurring expenditure. Main objectives of the Centre is to serve, sustain and upgrade the status of farmers and industry engaged in the aromatic cultivation and its processing, so as to make them competitive both in the local and global market.

EXPORT PROCESSING ZONES

The export zones (EPZ) set up as enclaves, separated from the Domestic Tariffs Areas by fiscal barriers, are intended to provide a competitive duty free environment for export production. There are four EPZs set up by the Government at Noida(Uttar Pradesh), Chennai (Tamil Nadu), Palta (West Bengal) and Vishakapatnam (Andhra Pradesh).

SPECIAL ECONOMIC ZONES

A new scheme for setting up of Special Economic Zones (SEZs) in the country to promote exports was announced by the Government in the Export and Import Policy on 31st March, 2000. The policy provided for setting up of SEZs in the public, private, joint sectors or by State Governments. It was also announced that some of the existing Export Processing Zones would be converted into SEZs. Accordingly, the Government has issued notification on 1-11-2000 for conversion of the existing Export Processing Zones at Kandla (Gujarat), Santa Cruz(Maharashtra) and Cochin (Kerala) into SEZs. Notification has also been issued for conversion of the private sector EPZ at Surat (Gujarat) into the Special Economic Zone at the request of the promoters.

EXPORT ORIENTED UNITS (EOU)

The export Oriented units (EOU) scheme introduced in the early 1981, is complementary to the EPZ scheme. It adopts the same production regime but offers a wide option in locations with reference to factors like source of materials, ports of export, hinter land facilities, and availability of technological skills, existence of industrial base and the need for a large area of land for the project. And 1,536 units are in operation under the EPU scheme as on March, 2001. Product range: EOUs are mainly concentrated in textiles and yarn, food processing, electronics, chemicals, plastics, granites and minerals/ores. Majority of units are located in Tamil Nadu, Andhra Pradesh, Karnataka, Maharashtra and Gujarat.

Export Promotion Industrial Park (EPIP) Scheme: A centrally sponsored " Export Promotion Industrial Park (EPIP) Scheme has been introduced in 1993-94 with a view to involving the State Governments in the creation of infrastructure facilities for export oriented production. The scheme provides that 75 per cent of the capital expenditure incurred towards creation of such facilities, ordinarily limited to ₹ 10 crores in each case, will be met from a central grant to the State Governments.

New Anna Marumalarchi Thittam: The State Government of Tamil Nadu has introduced the New Anna Marumalarchi Thittam in April 2002 to set up agribusiness units with a minimum investment of ₹ 35 lakhs at the rate of one unit in each block. This scheme provides scope for setting 385 agribusiness units in Tamil Nadu.

Agri-Clinic and Agribusiness Centres: Small Farmers Agribusiness Consortium in co-operation with MANAGE has drawn plans to provide training on management capacity building for those willing to set up Agriclinics and Agribusiness Centres either as individual or a group five (four agricultural and allied graduates and one management graduate) with a maximum loan assistance of ₹ 10 lakhs for individuals and ₹ 50 lakhs for a group of five entrepreneurs.

AGRIBUSINESS, AGRI CLINICS AND AGRI ENTREPRENEURS

Agricultural development over years has been the result of continuous agri skill generation and its popularization. The earliest agriculture was animal domestication over thousands of years ahead, man domesticated wild fowl, dog, goat and smaller animals, whom he could overpower easily and subjugate to his sub-ordination. Agriculture thus since beginning has been the results of trails, experiments and experiences over years, learned first though behavioral changes, psychic reoccurrences, memories passed through parents to children and later on through doing and learning and now through sharing experiences and writing them or dotting them as an Entrepreneurship concern.

ENTREPRENEURSHIP CONCEPT

The Entrepreneurship adds economic profits and cost-benefit ratios to Agricultural Output. Entrepreneurship is dominated by four factors like:

a. Social systemic changes
b. Support system availability and use
c. Resource base and its utilization
d. Self confidence, exploration work capacity and intellectual potency.

An entrepreneur has to have a thinking of his own, a capacity building interest in acquiring needed technique. An explorative and analytic faculties to judge the way of procuring cheap raw material. He must be equipped with "knowledge" and mindset to use and benefit out of it.

FARM BUSINESS

A potential entrepreneurship must strive from getting maximum output. Decades back agricultural development and industrial setups was a public sponsored and heavily subsidized but over time "knowledge" explosion in India Agriculture, have brought us on threshold of a system, where wide distances exist between industry and farm business. Where huge subsidies are benefiting Agro-Industrialists. The Farmers who use fertilizers or agro-chemical are crushed under economic pressures. The gaps between technology generated and technology use at farmers door is increasing day after day.

The farm technology adoption rates are not more than 20-30 per cent by any higher prospective. The use of information and communication technology (I&CT) for reducing the gaps and increasing productivity is the need of the hour. The modern technology and knowledge flow is fast expanding and bringing change. It demands more educated and trained farmers. Our education system has produced more literates but not educationally trained youth to earn their own bread.

They after attaining graduation in agriculture and allied sectors, beg for job. The system has to be corrected to make these graduates as employers and not employees. I wrote a treatise as back as 1992, emphasizing a system. Germans are smart to have Farmers school, Farmer business training institutes, practical agri-farmers training centres and like, where way farmer or animal husbandry man is essentially a trained fellow. The banking system is so organized that they are on the door of convocation hall to sell their agri-business and agri-clinics to graduates, without any personal investments. Banks are so smart, that they have surveyed the villages who need vets or agri-graduates or have attained land and all facilitation, so that agricultural or veterinary or even other medico-biological graduates are used as bank investment. This is what is envisaged in India under agriclinic, Agribusiness venture. We have trainings not in the hands of banks but universities.

SUCCESS OF AGRIBUSINESS

A systematic liaison and support system between Govt. banking and University culture has made this otherwise an remunerative and lucerative programmes into a failure inspite of its personal monitoring of PMO. The success of Agri-business and Agri-clinics success rates. The universities involve and their success stories. Both these details are distressing inspite of huge moral, financial support from Govt.

Agricultural professionals are getting converted into Agribusiness and agri-clinical experts. More than 14,000 applicants and 615 agriclinics came to existence in Indian 12 states. The agri clinic trained persons in J&K many number in hundreds. Among them 34 have registered agri clinics earning a handsome profit annually.

We visited Bandipora district and unregistered Agriclinics were earning a handsome salary, more than the Rahbar-e-zerat or Agriculture Asstt. A visit documentary is enclosed and shall be shown. It consisted of Agribusiness viz sale of pesticides, cattle feed, poultry feed and agri-extension services. At a small village in Papchan, one agri graduate Mr. Iqbal Shah earns ₹ 10,000/- per month by selling the services and input. At a distance of few kms. In same district one Mr. Khyatlani owns a big poultry farm and earns around ₹ 20,000/ - per month. Both these entrepreneurs employ 2-3 persons at present. Similarly, the success shown by one Mr. Shah at Malangam in Agri products and pesticide sale and one Mr. Bhat in Dairy production and milk product sale earn a handsome income besides generating employment for poor.

FARMER AS ENTREPRENEUR

Indian Farming and farmer has to change if proper WTO recommendation and GATT agreements are to be followed.

The present day poultry scenario has emerging high profile agri-business prospects in India.

The conversion of poultry farmer's into poultry entrepreneurs shall make the present day 6 per cent contribution of poultry products to 25 per cent share of Global market from India and China. This when translated into action shall increase employment generation by manifolds. The introduction of rural based Vanraja, Gramapriya, Giriraja, Cari Gold and vast other locally grown varieties of poultry have adopted well to our agri-rural base. The market acceptability is higher than exotic poultry concerns. Research to farmers doors in generating free-rang-poultry is like BT cotton hybrid spreading through villages of India and assuring high returns and exports.

POULTRY AS AGRI-BUSINESS

Dr. Gordon Butland, president of Global poultry strategies presents "Backyard poultry production" as a tool of alleviating poverty and malnutrition. We have tried to distribute "birds" under free-rang system in all our KVK's our results were excellent and income generation was totally in favour of the Agri-business and agri-clinics as will be shown in case histories and success stories.

A grand show of using poultry, rabbit meat processing introduction at SKUAST-K have innovated white meat usage.

This all will need the involvement of Agri-Veterinary and food processing technocrats to develop rural-based establishments so as to faster export and fast returns.

HOLISTIC VISION FOR LIVESTOCK ENTERPRISE

Improving income, employment and self-reliance are among educated graduates and un-employed youth especially women needs fostering community development, women empowerment, environmental protection. Rural-based backyard poultry subscribes to all these norms and could be a rich resource for developing agri-entrepreneurship. Govt. of India is liberally financing such agri-business ventures and a proposed infrastructure cost set-up. A vast and finance assured schemes are available for agri-graduates for establishing poultry ventures. An initial allocation of 107 crores for initiating nucleus breeding farms. Further more provision of hatcheries to provide chicks to more than 2 lac farmers and farm women will need many agri-business centres for providing basic germplasm, medicine and above all training. Some of the success stories in animal husbandry section can be reproduced as follows:

Backyard poultry and incubation

Though the Vanraja are the most suitable for back yard poultry, they do not have habit of broodiness. There is a problem among the farmer to get a broody hen in all season. KVK solve this problem of hatching by installing small unit of hatchery. Every month 15-20 farmers are benefited by purchasing chicks for backyard poultry. There 200 back yard poultry units of Vanraja. Each farmer is rearing 10 to 25 in the backyard. There is a good demand and response for the chicks and eggs of Vanraja. KVKs are now planning to expand this unit.

Semi-stall-fed Goat Rearing

KVK's made an intervention to improve this enterprise by conducting short durational training programmes for rural youth. Similarly exposure visit were organized on goat feed, breed and health management. More emphasis was given on O)smanabadi goat and up-gradation in selected non-descript goat breed by osmanabadi pure buck and given the knowledge about semi stallfed goat rearing concept.

Broiler Production

KVK has conducted many durational training programmes for 165 trainees. Due to training and demonstrations awareness was increased about contract farming in broiler production with private sector which provide chicks, feed and medicine and after 40 days purchases ₹ 3 to 3.50 per kg on live weight and FCR basis and changed their attitude. They acquired skills through learning by doing at KVK demonstration unit.

The technology has been adopted by 10 per cent of youths now in the radius of 20 km there are 27 poultry units having capacity of 5000-10000 poultry birds on contract farming basis. These self employed rural youth earning ₹ 10000-15000 per lot.

Recently a seminar-cum-farmer's meet was arranged at SKUAST-K on 26-27th of Oct.2007. The knowledge -sharing and use for making agricultural graduate and scientists was emphasized by our worthy Chancellor. A vision of poverty alleviation through backyard poultry intervention was the theme of the seminar. Many belts in Gurez, Tangdar, Telail and Zanskar are rearing native livestock species. Who are better suited and need improvement and identification. The cooking methods will need more expansion and scientific intervention for export.

AGRIPRENEURSHIP EDUCATION AND DEVELOPMENT

Indian economy is basically agrarian economy. On 2.4 per cent of world land India is managing 16 per cent of world population. At the time of independence, more than half of the national income was contributed by agriculture. At the same time more than 70 per cent of total population was dependent on agriculture.

The first five year plan has emphasis on agriculture development. Also the green revolution strategies adopted during 60s has contributed a lot in making India self sufficient in food production. With advent of new economic policy adopted since 1991, the picture has changed drastically. The contribution of agriculture in national income has declined to 26 per cent and that of service sector has increased to more than half of the total national income.

Service sector is emerging as driver of economic growth. Service sector, though growing at faster rate in term of income generation, has contributed little in terms of employment generation. Employment opportunities in the service sector are basically for educated and skilled manpower and are centered in urban centers. Uneducated and unskilled mass of India population, living in rural areas are not fitting in to the employment market created by service sector growth. They are therefore depends in agriculture for their livelihood. With employment of more than 50 per cent of labour force agriculture sector is major employment provider even today. The seasonal nature of agriculture and lack of irrigation facilities creates problem of seasonal and cyclical unemployment. Large numbers of persons employed in agriculture are of disguised nature. They seem to be employed but their marginal productivity is zero. Withdrawing of some of the persons from agriculture will not affect agriculture production at all. Disguised nature of agriculture forced the people to migrate from rural to urban areas creating pressure on cities in terms of additional facilities for housing, sanitation, water and also employment. The situation of an uneducated, unskilled rural labour migrating from rural to urban areas is like second class citizens without their own identity.

This situation can be changed by generating employment opportunities for them in rural areas itself. Agro entrepreneurship can be used as best medicine for the solution of this problem. Developing entrepreneurs in agriculture will solve all the problem viz. (a) reduce the burden of agriculture (b) generate employment opportunities for rural youth (c) control migration form rural to urban areas (d) increase national income (e) support industrial development in rural areas (f) reduces the pressure on urban cities etc.

The aim of present paper is to highlight the type of self-employment opportunities which can be generated in agriculture through the development of entrepreneurship skills. Following areas offer to appear a good hope for future;

(1) Food processing an packaging
(2) Preservation of seasonal vegetables
(3) Preservation of seasonal fruits
(4) Seed processing and preservation
(5) Use of solar community dehydration centers for seeds
(6) Selective farming for vegetables and flowers seedlings

This is also not an exhaustive list. I the filed of preservations, the role of women and cooperatives can be explored to best extent. Agro based industries to be planned in away that major products may be manufactured in a medium and small enterprises while middle level services can be provided by groups or co-operatives at the rural level and repair and maintenance services could be taken up be the artisans at individual level.

Basic objectives of new entrepreneur in agro based

1. Selection of market
2. Licensing and control
3. Location
4. Applicability of labour legislations
5. Technical know-how
6. Finance

Important Steps in the development of Agro based EDP programme;

1. Identification of and location of perspective self-employees
2. Selection of potential self-employed/entrepreneurs from amongst prospective candidates
3. Agro based entrepreneurship development training
4. Providing help/guidance in selection of product ad preparation of project report
5. Mobilizing different resources
6. Organisational support in setting a enterprise
7. Follow up

Agro based EDP requires expert handing process which must be carried out to meet various good entrepreneurs;

1. Good promotional activities
2. Proper selection (every one can not be made entrepreneur)
3. Good opportunity guide and developing a systematic business plan
4. Developing motivation and competencies

5. Developing managerial capabilities
6. Providing all information/counsellng and follow up

Good Agro based EDP should cover the following areas;

1. Information sources
2. Product guidance
3. Market Survey
4. Achievement motivation
5. Project report preparation
6. Managerial inputs
7. Counselling
8. Follow up

Trainer conducting agro based EDP should be competent as facilitators, and motivator possessing certain personalities and to act as a leader/ councellor and motivators and should also have adequate knowledge about agro-based sources of information and support system and skill to pensive entrepreneur potentials opportunities in the region and capabilities to mould the raw materials in to owners of agro enterprise. Agro EDPs are important for human resource development and enlarging the number of enterprises.

12

Rural Technology in Agriculture

CURRENT INFRASTRUCTURE FOR RURAL TECHNOLOGY DISSEMINATION

India currently has one of the largest technical manpower pool in the world (approx.38 lakh as per 1991 census). About 25 per centof this manpower is engaged in technology/ research activities, 21 per cent in administrative activities and 10 to 11 per cent in design and development. Fundamental research accounts for only 2.6 per cent of this total manpower.

Only about 6 per cent of the total technical manpower is employed in agricultural and rural activities. Besides, most of the manpower at the grass root level responsible for technology dissemination are not fully qualified and these positions are also understaffed.

There has been a substantial achievement in education but the illiteracy in actual numbers have gone up. Almost 68 per cent of the adults in rural areas are currently illiterate. This is despite the fact that 85 per cent of all the educational institutes are located in rural areas. Non-formal education is being increasingly advocated. National Literacy Mission has ambitious programs of adult education and of covering 80 million people by 1995.

There are however, very few linkages for application of science and technology to rural development. One of the experiments being tried is in the form of Vigyan Ashrams wherein the youth are involved in scientific endeavours.

In training, there are several institutions in the country offering a wide range of training coursers, orientation work shops, refresher courses, diplomas, degrees and post graduate courses. There are advanced training centers set up in select State Agricultural Universities for subject matter areas. National Centre for Management of Agriculture Extension caters to training requirement of senior level managers. Regional rural technology centers have been set up in some of the areas for technology transfer to rural areas. Many States training institutions and community polytechnics exist for specialized training. There, however, exist very few training courses oriented towards attitudinal and behavioural changes among beneficiaries and functionaries.

In extension and demonstration, there is a fairly good infrastructure in the country for agricultural programs through T & V systems ad other research based extension systems. For technologies, other than agriculture, the extension and demonstration system need upgradation as well as specialized innovative thinking.

The current expenditure on S & T is ₹ 3,771/- crore (1988-89). In relation to the GNP it is only about 1 per cent which is considered to be quite low compared to even other moderately developed countries. Regarding R & D infrastructure, there are two premier councils – ICAR and CSIR engaged in several R&D activities. This is followed by several central government institutions, universities and colleges, district rural development agencies, district industry centre, State S & T councils, and private industries who are also actively involved in R&D activities.

Voluntary agencies/NGOs at the grass root level also pay a key role in rural technology dissemination. By one estimate, there are about 10,000 NGOs in the country. The work of only a handful of them is, however known. CAPART is the apex agency coordinating the activities of voluntary agencies.

CHANGING RURAL INDIA

Given that 70 per cent of the Indian population lives in rural areas, economic growth must expand beyond city centers to include the entire country. In the modern world, connectivity is not just about roads. An opportunity exists today for villages to be connected to the national mainstream by broadband and internet telephony (also known as Voice over Internet Protocol – VoIP) and virtual markets. Though India is starting late in comparison to other countries, we can leapfrog ahead to implement cutting edge technologies. Companies that have recognized and tapped this opportunity have prospered in rural India.

Rural India today faces a severe technology deficit. While there are serious physical and social infrastructure shortages in power, water, health facilities, roads, etc., the role of technology in solving these problems is barely acknowledged. The actual availability of technology in rural areas is, at best, marginal. Science and technology are often hyphenated and spoken of in the same breath. One would, however, like to differentiate the two. Technology generally (though not always) derives and draws from science, and often manifests itself in physical form for example, as a piece of hardware. Science, on the other hand, is knowledge. In rural India, there is a dire inadequacy of both. Crop yields are therefore far lower than they are in demonstration farms where science and technology are more fully applied. The scope for applying technology to both farm and non-farm activities in rural areas is huge, as are the potential benefits.

Any introduction of technology in the rural space should address the following criteria, not only to be successful but to be widely used.

Four criteria for introducing technology-led interventions are:

1. How is it useful for the community in which you are introducing it?
2. How does it improve productivity?
3. Does it enhance or undermine employment?
4. Is the control of the technology and decision making process democratic?

That is, technology must be in the control of and answerable to the people it directly affects.

Post the first green revolution, there hasn't been any steady break-through in changing production and productivity, and as a result, both have stagnated. Agriculture, once the backbone of rural economy, is losing its share in rural GDP. This is a pressing problem and an exciting challenge. It represents a unique three dimensional convergence of technological capability, economic opportunity and societal need for technology.

Here are some examples:

Irrigation and Water Technologies: Managing the release and distribution of water is critical for maximizing production. Sophisticated power transmission systems use information and communication technologies to optimize and monitor the distribution of electricity. Despite many similarities, there is hardly any use of Information and Communication Technology (ICT) in water distribution.

In rain-fed areas, the construction of check-dams is vital. Pilot projects have already demonstrated that choosing the right location for such water-harvesting structures can be greatly facilitated by using satellite remote sensing data. Where irrigation comes from wells, the simple technology for the pump to be automatically switched on when power is available (and timed to then switch-off) – so common in cities – is still rare in villages. As a result, the farmer has to manually switch on the pump (generally in the middle of the night) when power becomes available.

While many technologies exist for water-purification, there is need for developing context-specific technologies (ideally, low-cost, reliable and not power-dependent) for providing safe drinking water. Recently launched water purification systems that have taken such considerations into account have been successful in the rural space.

Transportation and Distribution: Cold storage and cold-chains for transportation to market is of great importance for many agricultural products particularly, fruits and vegetables, but they remain insufficient. These are clearly technologies with an immediate return on investment and benefits for all: the farmer, the end-consumer, the technology provider. However, regulatory and structural barriers are holding back investments.

Information: Commodity prices, agricultural practices, weather, etc. are crucial for the farmer. Technology can now provide this information easily and instantaneously either at a village computer kiosk or on a mobile handset. While such information is delivered and used "passively", we need to look at making them transaction oriented. Transactions, including the purchase of agricultural inputs as well as other goods and services, can be handled on kiosks, wireless cable TV or a mobile phone. All these technologies are proven; the challenge now is to ensure connectivity and scale them for wider usage and application.

The growth of India's technological base has made it possible to meet these needs. Further, economic growth, however skewed and iniquitous, has made rural India an attractive market. What's now needed is to exploit this opportunity in socially relevant ways. Understanding the consumers' needs from a socio-cultural standpoint is not easy but essential.

Technological Interventions in rural space: Many decades ago, India's program for practical application of space technology had a large team of social scientists, dedicated to understanding the true needs of rural India and acting as a bridge between villagers and technologists. Such an effort is needed today for new and emerging technologies in fields such as communication, irrigation, energy, and others. This could be the next big thing for techno-entrepreneurs and enterprising corporates.

The following examples demonstrate the use of technology in the rural space and how it has impacted the life of farmers and non-farmers in our villages.

Finance:

- Swabhiman is a nationwide program on financial inclusion in which five crore households will receive access to banking services in unbanked areas. The banking will be taken door to door through business correspondents who called 'Bank Saathi' in Hindi. The scheme will also promote the micro-insurance and micro-pension plans for the villagers. This is made possible by the same V-sat technology that has been powering the spread of ATM's and other services for more than a decade in other parts of the country.
- SBI and Airtel have formed an alliance for financial inclusion for unbanked India using mobile telephony. Developers like Bangalore based Integra Micro Systems have created micro banking and branchless banking solutions based on these technologies, bringing banking to thousands of villages in India. More recently 3i InfoTech has launched a range of value-added services for rural India under the ISERV brand in over 12,000 locations.

Knowledge and Information Delivery:

- Hariyali Kisaan Bazaar (HKB), the rural retail venture of DCM Shriram Consolidated Limited (DSCL), is planning to tie up with mobile telephone companies to provide farm and commodity advisory services.
- Reuters Market Light (RML), a mobile phone-based service offers

up-to-date local and customized commodity pricing information, news and weather updates.

- Nokia India and Tata DOCOMO partnered to offer Nokia Ovi Life Tools to plug information gaps in the areas of agriculture, education and entertainment.
- Infibeam's 2,000 crore rural e-commerce platform in Gujarat will allow access to FMCG and other products currently not easily accessible to rural households at transparent prices.

These initiatives aim to pass the power of expert knowledge and information to individual farmers, increasing their negotiating power and enabling them to realize better prices.

Non-Conventional Energy and Telecom:

- Electronic solar rickshaws for postmen in India.
- Ashden award winner D. Light and UK based Christian Aid have joined forces to deliver a micro finance program designed to bring solar lighting to 4400 rural households in India.
- The Ericsson sponsored Gram Jyoti Project (which literally means 'Village Light' in Hindi), a pilot project, has brought the benefits of mobile broadband connectivity to a few selected villages and small towns in Tamil Nadu. It uses the Wireless CDMA/ HSPA (High Speed Packet Access) technology. This will enable a wide range of services like telemedicine, e-education, e-governance, online local information like local rates for agricultural products, video conferencing and live TV and entertainment.
- Radio-Frequency Identification (RFID) and Near Field Communication (NFC) based smart cards, an evolving technology that is used for toll ticketing and goods tracking, enable traceability of farm produce and smart card based financial transactions. These technologies can mark something as small as a needle or as large as a missile. They have the precision to overcome issues like line of sight and and the memory to incorporate numerous product details and history.

We can also look at adopting ideas from other countries and continents:

- Australia: the Rural Transaction Centers (RTC) Program helped establish locally run units that introduce new services or brings back services into rural towns.
- Finland: Telemedicine services have allowed a specialist doctor in Helsinki to provide diagnosis on X-rays taken in sparsely populated regions thousands of kilometers away.
- Africa: The Solar Electrification Light Fund (SELF) uses solar arrays to power irrigation pumps for growing high-valued crops in the dry season (solar market gardens).

With such initiatives, quality of life as well as income growth can be vastly and quickly improved. Access through technology can provide a new momentum in our efforts to promote inclusive growth. The technology exists. It is up to us to evolve cost effective and socially relevant ways to bring it to and use it in rural settings.

AGRICULTURAL RESEARCH AND TECHNOLOGY TRANSFER TO RURAL COMMUNITIES

The role of agriculture research and technology transfer to rural communities in Central and Eastern Europe as well as in Countries in Transition (CT) derives from the critical importance of agriculture in their overall economy, and large resource base. The overall agro-industrial complex is large and accounts for about 25-50 per cent of GDP and for about 30-40 per cent of employment in these countries. Population in all CT is over 400 million, of which the rural population accounts for about 33 per cent and the agro-industrial sector employs over 32 million people. Only the Czech Republic, Slovakia and Slovenia have parameters close to those of the EU, where its share of GDP is 1.7 per cent and employment 5.1 per cent.

Cereals are the largest crops in most of the countries, so they are a good indication of plant production productivity, occupying 42-60 per cent of arable land, with wheat the most important crop, grown on 25-50 per cent of the total cereal area.

In general, 16 countries are net importers of agricultural products and the other 8 are net exporters. However, agriculture on these areas has a large export potential and considerable importance for import substitution. At present, agricultural resources in most of these countries are not used efficiently. Agriculture research is vital to increase productivity and efficacy in the sector and convert the CT to a market economy. It is long-term in nature, but should not be used as a justification for inaction or low priority. Agricultural research capacity is a strategic resource that can be justified from a number of perspectives.

Extensive literature on agricultural economics convincingly demonstrates that investment in agricultural research fields yields a high pay off. For public sector agricultural research, average returns were 48 per cent for developed countries and 80 per cent for developing ones. However, the size of the rate of return varies from one crop to another, between sectors (crops vs livestock) or aggregate agricultural production, and from one country to another. The rate of return to investments in research varies from 22-42 per cent for potatoes in Peru and is 19 per cent for maize in South America. The nature of agricultural technology, the level of agricultural productivity and appropriateness of agricultural policies greatly influences returns on investment in agricultural research. Careful and informed research management and public investment are essential and must take place within a set of constraints defined by national resources and policies.

Positive and higher rates of return mean that the stream of societal benefits from research outweigh the cost over a planning horizon of several years. The costs of these investments are repaid because the economy grows as a consequence of the reduced food and fibre costs, with benefit to both the consumer and the producers, the relocation of physical and human capital into higher and better uses, and increased economic activity, including trade.

Public support for agricultural research can also be an important action to increase the competitiveness of the national agricultural sector, whether through direct public investment or through public action to foster private research. Operationally, increased competitiveness means that the agricultural sector is better able to sell products abroad or to produce substitutes for products being imported.

Food security is relevant to CT in two important ways. First, CT can go a long way forward in realizing waste potential in agricultural production and in meeting future food needs, in increasing efficiency and quality in agricultural processing and distribution. Second, there is an important domestic need for food security. The transition to a democratic and civil society is significantly influenced by the price of sausages and bread. Agricultural research place an important role in improving productivity, increasing competitiveness and also standards of living for rural as well as urban people and supporting democratization, so it should be a priority investment.

Table. Rate of return to agricultural research in OECD and some developing countries

Country	Study	Commodity	Period	Rate of return
Australia	Duncan (1972)	Pasture Improvement	1948-69	58-68
Finland	Sumelius (1987)	Aggregate	1950-84	21-62
Germany	Burian (1992)	Aggregate	1950-87	21-56
Ireland	Boyle (1986)	Aggregate	1963-83	26
Japan	Hayami & Akiro (1977)	Rice	1932-61	73-75
United kingdom	Thirtle & Bottomley (1988)	Aggregate	1950-81	70
Mexico	Ruvalcaba (1986)	Maize	-	78-91
South america	Evenson (1989)	Maize	-	19
Indonesia	Parday (1993)	Rice	-	60-65
India	Evenson (1990)	Rice	-	65
Pakistan	Nagy (1983)	Wheat	-	58
Brazil	Ayers (1985)	Soybean	-	46-69
Philippines	Libero (1987)	Sugar cane	-	51-71
Peru	Norton (1987)	Potato	-	22-42
Senegal	Schuartz (1987)	Cowpea	-	60-80

AGRICULTURAL RESEARCH AND DEVELOPMENT

The size of investments in research and development (R&D) in CT is much lower than the average in the EU; in Poland it is 7 times lower and in Czech

Republic 3 times less. Investment in research as a share of GDP is significantly lower than that observed in the EU (1.84 per cent), reaching 0.76 per cent in Poland, 0.66 per cent in Hungary and 1.07 per cent in the Czech Republic. It is primarily financed from state budgets, which is typical for less-developed countries, in contrast to industrial countries, where this is financed through non-treasury resources. Most costs in R&D are for biological and engineering sciences, including agricultural disciplines. Hungary allocated 82.6 per cent; Poland, 94.0 per cent; and Czech Republic, 95.8 per cent of R&D budgets to this sector. In Poland and Hungary, most is spent on basic research, with only 66.7 per cent on applied R&D, which is the lowest rate among all OECD countries. Czech Republic allocated as much as 83 per cent to this sector, which is similar to the USA (83.5 per cent). Applied R&D in Poland is mainly through higher education units (43 per cent) and specialized R&D (86 per cent) units (Wanke-Jakubowska and Wanke-Jerie, 1999).

Effectiveness of research in agronomy in the period 1995-1999, assessed by the number of publications and citations in Science Citation Index, was quite large in Czech Republic and in Hungary (369 and 359 respectively), and lower in Poland and Slovakia (139 and 132 respectively). This represented 1.36-1.32 per cent and 0.51-0.48 per cent respectively of all such publications worldwide. The other countries of the region published only a few papers or none. In this period, on average, EU countries published 465 papers each, which was 1.71 per cent of all publications. The relative impact factor - indicating the quality of these papers - appeared to be higher for Polish papers (0.73), and lower for Czech, Hungarian and Slovakian (0.51, 0.43 and 0.36, respectively); in the EU it was 1.34.

A very important reason for delays in agriculture research in CT is the anachronistic structure of the agricultural science sector, usually divided into three independent organizations: agriculture education institutions, with a large number of faculties; scientific units of academies of sciences; and specialized R&D units. This leads to difficulties in efficient use of existing research potential, and creation of large and multidisciplinary research groups, and results in waste of the limited research funds.

Agricultural education institutions, especially in the FSU, mainly prepared people for work in the academies of sciences, R&D units, and agriculture entities. Unfortunately, the relationship between agricultural science and education was very weak. The most important reason was that they were separate structures, with weak linkage between their leadership, even though the objectives of the three organizations overlapped. The education institutions prepared mainly MSc-equivalent specialists. All three organizations prepared PhD-equivalent specialists; in FSU, mostly through research institutions. Normally, the bright MSc Students would be kept after graduation in the same education institutions, as assistant professors. However, a PhD degree was required for promotion. The choice was either to enter a 3-year PhD study programme in one of the research institution, or to obtain it by distance learning and continue teaching. So, some teachers had been exposed to a research environment in the research institutions, while others were not.

Table. Papers on agronomy published in the CEEC, CIS and CT

Country	No. of publications	Publication as % of total world publications*	Impact factor**	Relative impact factor
Albania	2	0.01	1.50	1.30
Armenia	2	0.01	0	0
Azerbaijan	0	0	0	0
Belarus	28	0.1	0.07	0.06
Bosnia	0	0	0	0
Bulgaria	45	0.17	0.27	0.23
Croatia	33	0.12	0.59	0.51
Czech rep.	369	1.36	0.59	0.51
Estonia	11	0.04	0.82	0.71
Hungary	359	1.32	0.49	0.43
Kazakhstan	4	0.01	0.5	0.43
Kyrgyzstan	0	0	0	0
Latvia	3	0.01	0	0
Lithuania	10	0.04	0	0
Macedonia (fyr)	0	0	0	0
Moldova	6	0.02	0	0
Poland	139	0.51	0.84	0.73
Romania	22	0.08	1.09	0.95
Russia	1 051	3.86	0.09	0.08
Slovakia	132	0.48	0.41	0.36
Slovenia	18	0.07	0.89	0.77
Tajikistan	0	0	0	0
Turkmenistan	0	0	0	0
Ukraine	50	0.18	0.1	0.09
Uzbekistan	4	0.01	0	0
Yugoslavia	30	0.11	0.73	0.63
European union	6 969	25.59	1.54	1.34

Note:

* Total number of publications reported worldwide (sample size) = 27 229. ** Impact base 1.15

Most of the teaching positions allowed only 10-20 per cent of time to be spent on research. Since the facilities for research in educational institutions were inferior to those in the research units of the academies of sciences and more specialized R&D units, the scientific level of the teachers was far lower than that of scientists in research institutions. Only elite universities, like those in Moscow or Leningrad (now St Petersburg), maintained a high level of research for the benefit of the students. Scientists were very rarely invited to give lectures to students or to take part-time teaching positions. Furthermore, there was a low level of rotation within educational institutions and of interchange of staff between the academies of sciences and higher education institutions.

The scientific societies played a minor role in CT, in contrast to the situation in western countries. The activity, for example, of the All-Union Society of Geneticists and Breeders in FSU was limited to the organization of meetings every 4-5 years and the publication of their proceedings. There was no regular publication and communication between members and no formal linkages between these societies and the academies.

In the international arena, the highest priority was cooperation with the other socialist countries. There were several mechanisms for cooperation. First, one of the most important mechanisms was establishment of joint research programmes addressing common problems. For example, a successful programme that focused on winter wheat breeding united plant breeders from Russia, Ukraine, Hungary, Romania, Bulgaria and other countries. This programme resulted in a number of advanced varieties. Second, cooperation at the research- and the R&D-unit level was encouraged. This involved both joint research activities and exchange of scientists. It seems that cooperation and coordination of agricultural research at this international level was much better than at the national level. Third, reciprocal membership of the academies was one of the mechanisms to maintain communication.

The relationship with the scientific community of the Western countries could be characterized as one of isolation. This isolation was determined by two major factors: the lack of language skills and very limited exchange of people, for political reasons. Very few scientists were trained abroad. The physical isolation from the West did not undermine information exchanges. The important scientific journals published globally were received by central libraries and were available to scientists. They were also abstracted - mainly in Russian - and published on a monthly basis. Journal such as *Agriculture Abroad* and others published monthly or quarterly consisted of literature reviews and trip reports, and could be easily obtained on subscription.

Priority setting, as understood in the west, i.e. consultation with the participation of stakeholders, was unknown within the agricultural scientific communities of CT. The process of priority setting was highly politicized, as one might expect, and it was top-down. The communist parties gave the general directions for societal development. Whether they were based on scientific knowledge or the personal ambitions of the leaderships is open to question. The scientific communities were often consulted, but were not always listened to. The mechanism of transformation of the overall priorities set by the government into the priorities of particular research programmes is difficult to describe. Sometimes government regulations set up the priorities. In the case of plant breeding, for example, if the Ministry of Agriculture decided not to accept varieties susceptible to a certain diseases, the breeders had to emphasize breeding for resistance to the pathogen. However, the major factor changing priorities was selective funding for the institutions and programmes involved.

Customer identification and participation as known in western management culture was missing in CT. The uniformity of agriculture enterprises in the countries did not allow a differentiated approach. They were similar in their farming methods and level of agriculture. Currently, things have clearly changed. Each research programme had to compile a detailed plan describing what would be done, what output was expected, what would be the impact on the output for producers, and what funding would be needed to implement the objectives. However, there were no uniform criteria to judge the results of the work of the research programme and often no criteria at all. For example, a crop breeding programme that did not release a variety for a number of years would still be funded, equally to another, more successful, programme. Some authorities used the production figures in the region to evaluate the impact of a research programme, but administrations were satisfied with merely proper formal reporting. There was no formal system of research programme evaluation that could be applied to research institutions.

After the collapse of the SU, new systems of research started to develop in newly-independent countries. In most of them, they remain essentially the same as at the time of establishment. The separation between agricultural research and education has changed only a little. As a result of reduced funding, some institutions have suffered up to 50-70 per cent staff reduction in 1997 compared to 1990. The result of the first few years of activities of new scientific organizations can be summarized as the weakest point being application of research results in practice. The extension services do not have resources to introduce new products. At the same time, producers are weak economically and are not able to pay extra for development of new technology.

Many countries have now drafted the framework for improvement of the system of research and extension to support the development of agro-industrial complexes. The concepts define the research priorities, the mechanisms of research planning, funding, management, and implementation of the results. The necessity of re-structuring and optimizing the research network is also mentioned. At the time of writing, available information indicated that in the countries of FSU, there were 363 research institutions employing in total 30 913 scientists. The highest number of research institutions are located in Russia (203, employing some 17 000 scientists), in Ukraine (51 with 7 000 scientists) and in Kazakhstan (29 units with 1 900 staff). The number of scientists per 1

000 ha of arable land ranges from 0.05 in Kazakhstan, 0.13 in Russia, 0.22 in Ukraine to as much as 1.03 in Armenia and 1.13 in Georgia (Morgounov and Zuiderma, 1998).

Poland has 37 higher education and research institutions, employing some 4 000 scientists, equivalent to 0.29 scientists per 1 000 ha of arable land. Assessing cooperation of a few Polish agriculture research units in 1990-1997, measured by number of publications, showed that, out of 259 papers authored from more than one organization, only 4 were done in cooperation with an industrial organization; none involved extension centres or grower associations. This should be seen as a lack of flow of results from research into agricultural practice. Agricultural research in most cases appeared duplicative or adaptive work, so this obscures real achievements, i.e. what might be considered original work with significance on a global scale.

GENETICS RESEARCH AND PLANT BREEDING PROGRAMMES

In almost all CT, genetic and breeding research was conducted mainly by academies of sciences, and sometimes in higher education units. In contrast, applied research and breeding programmes were done mostly by R&D units. However, in many countries, including Poland, Czechoslovakia and Hungary, large breeding programmes were executed by state breeding establishments. Those programmes were financed directly from state budget resources. At present, in many countries, R&D units have abandoned breeding programmes, and simultaneously growing interest is observed among breeding companies, which originated from the former state breeding establishments, which are gradually being privatized.

Nevertheless, the larger part of their expenses is covered still by state budgets. Those subsidies in Poland and Czech Republic reached US$ 40 million/year on average in the period 1998-2000. There are a great number of such breeding companies, but according to European criteria they fall into the category of small and very small units.

On the seed markets of CT, only those varieties can be introduced that are included in the National List of Varieties. The variety can be included into the National List after at least two years of tests for distinctness, uniformity and stability (DUS), and only after tests for cultivation and use as well for crops of greater industrial importance. Previously, this required three to four years of variety testing, mainly for cultivation and use. In the National List up to 1989, there were mainly domestic varieties (54-99 per cent of all varieties). In the last decade, the number of varieties in these lists almost doubled and the proportion of domestic varieties dropped to 50-60 per cent of the total. A similarly trend was observed among cereals - the main crop category - in the same period, with a drop in the proportion of domestic varieties in the official lists.

Table. Market for crop cultivars in CEEC, CIS and CT

Country	1989				1999			
	Total	of which domestic (%)	Cereals	of which domestic (%)	Total	of which domestic (%)	Cereals	of which domestic (%)
Czech rep.	370*	86	85	80	909*	40	273	34
Hungary	1 262	72	49	57	3 833	44	216	49
Lithuania	185	30	31	32	3 373	25	74	28
Poland	1 061	99	75	93	2 271	60	260	65
Russia	5 577	54	816	36	7 620	62	906	83
Slovakia	1 102	17	131	28	2 233	11	264	24
Slovenia	-	-	-	-	1 885	4	153	6
Ukraine	1 352	88	183	96	3 144	79	644	76
Yugoslavia	6 291	90	563	77	7 165	92	983	80

Note:

* = field crops only

Improved wheat varieties, which can be treated as an example, typify the quality of new tested varieties, introducing with them important yield potential. Yields of those varieties in trials in 1999 ranged from 3.41 t/ha in Russia to 6.60 t/ha in Slovakia, 7.01 t/ha in Poland and 7.0-9.0 t/ha in Slovenia. They were double the average yields obtained previously in commercial growing in those countries, and could increase the productivity and profitability of farms. Unachieved potential reached 40-50 per cent of yields of improved varieties. It should be borne in mind that average yearly rise in yields as a result of introduction of the new wheat varieties in the UK and France reaches 100-120 kg/ha. In Central and Eastern Europe it is lower, and in Poland reaches only 30-35 kg/ha. An interesting fact should be noted that, in 1999 in Hungary, Russia and Ukraine, new improved wheat varieties yielded less than those tested 10 years earlier.

Table. Wheat (winter) yield in CEEC, CIS and CT

Country	Yield of new varieties in official trials (t/ha)		National average yield (t/ha)		Unachieved potential (%)	
	1989	1999	1989	1999	1989	1999
CZECH REP.	6.99	7.79	4.94	4.65	29	40
HUNGARY	6.43	5.75	5.24	3.29	19	43
LITHUANIA	5.80	6.20	3.61	2.67	38	57
POLAND	6.76	7.01*	3.44	3.46	45	51
RUSSIA	4.01	3.47	2.62**	1.79	35	48
SLOVAKIA	5.85	6.60	5.53	4.13	5	37
SLOVENIA	-	70-90	-	3.94	-	-
UKRAINE	4.68	4.43	2.85	2.32	39	48
YUGOSLAVIA	-	-	2.80 (4.20***)	2.58 (4.50***)	-	-

Notes:

* = 1997-1999;

** = 1952;

*** according to information from M. Milosevic

SEED MARKET

There are no precise data on commercial markets and international seed trade of CT. According to International Seed Federation (FIS) estimations, internal commercial markets of seed and plantlets in some CT are: CIS, US$ 2 000 million; Poland, US$ 400 million; Hungary, US$ 200 million; Czech Republic, US$ 150 million; and Slovakia, US$ 90 million. CT with significant international seed trade exports are: Hungary, US$ 36 million; Czech Republic, US$ 18 million; and Russia, US$ 13 million.

FORMAL SEED SUPPLY

Among CT, 10 are members of OECD or affiliated to OECD seed certification schemes, namely Bulgaria, Croatia, Czech Republic, Estonia, Hungary, Lithuania (from end of 1999), Poland, Romania, Slovakia and Slovenia. Sowing material for the market in some countries has an obligatory certification requirement; in others it is only laboratory tested and has to meet national standard requirements.

Production and seed trade is mostly the domain of specialized seed companies, which originated from state seed establishments. In a few countries, the production and trade of sowing material of owned varieties is done by state R&D units. Almost all seed companies in Czech Republic, Slovakia, Hungary and Poland have been privatized. In the other countries, the privatization process is still in progress. In many countries, the emergence of new, private seed companies can be seen. In each country, there are from a few to a few thousand seed companies, mainly small or very small. Production of sowing material is done on privately owned farms, on a contract basis, and basic seed delivered for reproduction. In many countries, like Czech Republic and Poland, use of fresh seed of cereals in command economy times was obligatory and its production allowed exchange of the entire sowing material every two to four years. In the other countries it was 20-60 per cent. Adaptation to a market economy has led to abandonment of mandatory seed exchange, and in 1995-1999 the demand for cereal seed decreased noticeably. Exchange of cereal seed at present is only 20-40 per cent in many countries; only in Slovenia is it 80 per cent.

Some cereal sowing material, mostly wheat, is not used for sowing purposes. This is due to overproduction, and supply exceeds demand. It has been estimated that, frequently, 20-30 per cent of cereal seed has to be used for human consumption and as fodder. It seems that accessibility to new varieties with high yielding potential to a large degree satisfies the needs of agriculture in the majority of the countries. It is difficult to estimate the degree of satisfaction of the specific needs of the end users, as well as farmers operating in specific agro-climatic conditions, because the flow of information from them to plant breeders and decision-making bodies is weak, if it exists at all. The quality of seed very frequently is poor, including varietal purity; and this is seen even in the countries using the OECD certification schemes.

Table. Production of Seed Material in CEEC, CIS and CT

Country	Total ('000 t)		Wheat ('000 t)		Rate of exchange of wheat (%)	
	1989	1999	1989	1999	1984-1989	1995-1999
Czech rep.	430	115	218	62	98	30
Hungary	570	340	240	142	63	51
Lithuania	225	212	60	50	20-25	17-20
Poland	562	384	187	197	90-100	25
Slovakia	150	120	73	72	100	30-40
Slovenia	-	10	-	8	-	80
Ukraine	1 025	430	26	16	35	28
Yugoslavia	300	260	169	143	2.2*	2.2*

Note:

* = through commercial channels.

Falling seed sales (mainly cereals) is to a large extent a result of lack of management abilities and bad marketing by breeding and seed trade companies, but is also due to poor productivity and efficiency in the seed sector. The common feeling is that state-run organizations and companies are less effective than private ones can be confirmed on the basis of many observations. Reasons include unclear breeding objectives, having to deal with many minor crops, reacting slowly to the rapid changes of a free market economy, and having high costs due to excessive employment levels and high price of inputs.

Establishing private breeding companies is hard, mainly because of financing. There is a lack of money for financing breeding programmes because very often plant breeders' rights (PBR) have not been implemented and there is no organized royalty collection system. Privatization of state-run breeding companies is difficult because of their high nominal value (in terms of money) and for political reasons. Sometimes it is felt that they are strategic companies because they support national food security by producing varieties well adapted to regional agro-climatic conditions and introducing them more cheaply than foreign cultivars.

INFORMAL SEED SUPPLY

All CT have large informal seed markets, with farmers managing seed production and general crop production activities, based on indigenous knowledge and local diffusion mechanisms. It includes methods such as retaining seed on-farm from a previous harvest to plant them the following season, and farm-to-farm seed exchange networks (including local market places). There has been little or no attention paid to this informal seed supply sector, and little is known about its operation in the region. As a result, documentation is scarce. At present, probably from 20-30 per cent up to sometimes 90 per cent of seed used by farmers is supplied through this informal arrangement. Major reasons for the existence of informal seed markets can be summarized as:

- It is difficult for the formal seed sector to reach every farmer, especially in the eastern part of FSU,
- Most of the improved varieties produced by the formal seed supply sector are targeted at the better-endowed farmers, where rainfall, irrigation and other means of production are easily available, and
- The very poor financial situation of many small farmers, inefficient seed company marketing strategies, and ineffective agriculture.

TRANSFER OF TECHNOLOGY TO RURAL COMMUNITIES

In a command economy, with the existence of mostly large and state-owned or quasi-cooperative farms, there was no need for organizing systems for transfer of technology (TOT) from agricultural research centres to rural communities, taken to mean farmers. This function was fulfilled by investment sections of those establishments and their management bodies. When the large establishments were disbanded and privatized, even that remnant TOT system vanished.

Only in a few countries, mainly in Poland, where private farmers still owned 80 per cent of agricultural land, did the system of linkage between science, extension and farmers survive. The flow of information in that system was in one direction only.

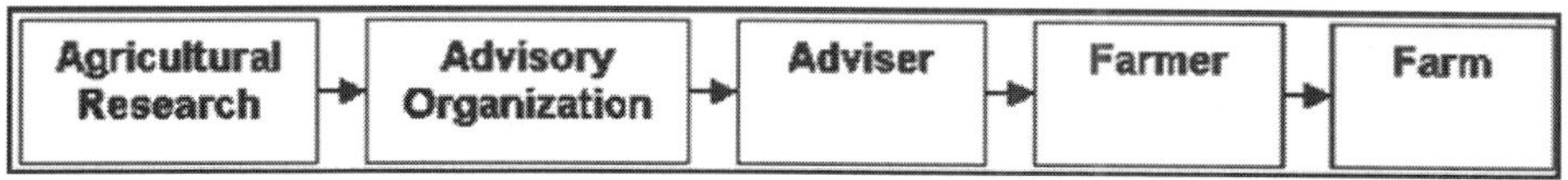

Fig. The information flow in the agricultural technology system

Extension organizations obtained their information from agriculture research, which was then tried on a single farm and then disseminated to others. Now farmers are using many different sources of information, because knowledge is not only the domain of research centres, but also spreads from many other sources, including:

- Extension centres,
- Agricultural chambers,
- Results of research from different disciplines of applied science, distributed by scientific centres in the form of innovation, i.e. techniques, technologies, systems of farming, etc.,
- Private extension organizations and services,
- Breeding and seed companies,
- Companies selling fertilizers, pesticides, etc.,
- Companies buying agricultural raw materials,
- Banks and other agencies,
- Ministries of Agriculture, of Finance and of Environmental Protection,

- NGOs, and
- Other farmers.

At present the system is undergoing change, and in the future it will develop into an Agricultural Knowledge and Information System.

It is difficult to imagine that the TOT in CT could exist without creating such systems in each of them and could take place without the involvement of public agriculture extension services. Institutions of that type played a key role in the period of modernization of the agriculture sector and change in rural environment in Western Europe and the USA. As the development of agriculture proceeds, the role of public extension should diminish, and competition among other providers of services and information should increase and private extension services should appear on the scene.

Among the strengths of public extension services should be an emphasis on rural communities, whose membership extends beyond just farmers; consideration of the growing degree of important social and environmental problems, not only purely agricultural; a wide range of services; and introduction and dissemination of sustainable farming and environmental protection programmes and stimulating an ecological consciousness among farmers and rural area inhabitants. Such extension is costly for taxpayers, not always convergent with actual farmer needs and open to manipulation by political decisions of government.

Autonomous extension services are more directed to the needs and expectations of farmers and quicker to react changes in their economic conditions; always know their requirements; specialize in services satisfying needs of mainly commodity farms; and is less costly for tax-payers. Weaknesses of such extension services are high costs of extension services for farmers themselves, and exclusion of a high percentage of rural area inhabitants, who are not farmers. Another weakness is the low number of employees in these services.

In all CT, modern TOT into rural communities is currently heavily supported by external aid agencies, such as EU, USA and World Bank support in Albania, Czech Republic, Estonia, Hungary, Poland, Slovakia, Russia and Ukraine. Taking into consideration the low effectiveness of agriculture and the weakness of newly established farmer associations and other NGOs, the extension services have to be based on funds from state budgets and built initially as a public system. The system should then be increasingly supported and displaced by breeding and seed companies, traders selling fertilizers and buying raw materials, banks and NGOs.

Poland has begun to organize a country network of post-registration variety trials (for the National List of Varieties). It should be seen as a system of variety trials, directed to providing information of value for cultivation and use, looking at the same time at cultivar reaction to environmental and agro-climatic conditions. On the basis of such variety trials, a Recommended List of Varieties will be created for the most important crops. In variety research, cooperation is expected to build up with governmental and local administrations, agriculture councils, extension services, breeding and seed companies, research units, food processing industry and other organizations and institutions connected to agriculture. Responsibility for creation of the basic framework for such post-registration variety research has been delegated to the Centre of Variety Testing Office (COBORU). In the future, extension should be transformed into a network of local agriculture research systems, managed by agricultural councils and grower associations, and financed by variety users.

Existing Polish governmental and regional extension centres and extension departments of agricultural research and higher education units demonstrate, at indoor meetings during the winter, the advantage of improved varieties, simplifying the results of variety trials to make them understood more easily. Advisory officers of extension services conduct some demonstration trials on growers' own land, sowing improved varieties alongside the varieties usually grown locally.

International seed companies, such as Pioneer, that have entered CT markets have well-organized TOT systems. Some of them established their own research stations and processing plants, but in most cases the seed companies limit their operation to production of hybrid seeds of maize, oil seed crops or vegetables. They have well-organized programmes for farmer education and for dissemination of information in relation to improved crop varieties and usage of quality seed. They also have more credibility with the growers compared to the public sector organizations and are considerably more effective than state-run entities.

In some cases, multinationals dealing with maize have contracted farmer growers to use improved seeds, and supplied necessary inputs, and have guaranteed to buy back products at an agreed price for further processing. Several other agro-processing industries have adopted this mechanism to promote improved technologies with their contract growers.

There is considerable interplay between the public and private sector in the area of seed improvement and distribution. Various public institutions and private organizations do field testing of varieties. Breeders and agronomists in both public and private sectors share knowledge on varieties and production of good seeds to encourage adoption of improved technologies and for further research.

Local variety demonstration trials are organized mostly by seed companies. They also cover improved cultural practices, and attract great interest among the rural community and advisory officers. Seed company representatives invite farmers in the area for a field day, where the trials are demonstrated and explained.

Seed exhibitions are held occasionally, or in some cases at regular intervals. This creates great interest and is given wide publicity, both in the daily and agricultural press or radio and TV. Another, and perhaps more direct, way of bringing home to the farmer the value of good seed is a so-called drill box survey. The normal procedure in such cases is to compare the quality of the farmer's seed - determined by the seed testing station - with the quality standards for certified seed or other kind of high quality seeds. If a farmer has saved seeds from his own previous crop or he has obtained seed, likewise homegrown, from his neighbour, the comparison with standard certified seed often gives a real shock and he becomes converted to the use of only the best possible seed for the rest of his life. Such a farmer is often good local support for the advisory officer in efforts to make all farmers in the area "seed minded."

Existing general producer or seed grower associations in some countries create interest in the production of improved varieties and wider use of good seeds by direct information to their members and their friends, through lectures, meetings, etc. Such TOT is well accepted by farmers for two reasons, namely:

- Farmers can see with their own eyes the benefits that can be obtained by use of improved and good seed, and
- The use of good seed provides assurance of credit and sale of produce at pre-agreed prices.

Evaluation of improved varieties by rural farming in most countries is in an early stage, and must be extended to become an important factor in increasing productivity and effectiveness of agriculture. Only an organized system of TOT, along with local agriculture extension services and wider spread of present ways of promoting improved varieties and good seed, can ensure the benefit of the full yield potential of improved varieties and good quality seed, which was lost in the past. In the end, we expect to see growth in effectiveness in agriculture.

HOW DOES THE RESEARCH SYSTEM MEASURE UP?

Public research systems in many industrial countries are becoming more demand driven, efficient and closely coordinated with the private sector. If these characteristics can serve as a benchmark, it is clear that most CT have a long way to go. Transformation strategies need to be based on a clear analysis of the current system's strengths and weaknesses, and the opportunities and threats it faces at present and in the future.

STRENGTHS AND OPPORTUNITIES

The difficult challenges faced by most transition-economy agricultural research systems should not obscure their many strengths, assets and opportunities (Mudahar *et al.*, 1998).

- National Agricultural Research Systems (NARS) are extensive and have comparatively high levels of investment in human and physical capital.

- Research systems are distributed throughout major agro-climatic zones, so they can be decentralized and closer to stakeholders.
- The basic training of the agricultural scientists is good, with well developed fundamental skills.
- Agricultural researchers in most institutions have a long tradition of entering into joint projects or consultancies with agricultural enterprises. This can form a foundation for more effective problem recognition and technology transfer.
- Despite low salaries, dwindling operating funds and isolation from the world scientific community, many dedicated agricultural scientists continue to work their trade as best they can.
- Many national - especially Russia - stocks of agricultural research products have not been widely shared with the rest of the world. The potential for collaboration and mutual benefit in research with other scientists, public or private, around the world has scarcely been explored. This stock includes knowledge, expertise and new technologies, as well as data.
- There are reform-minded institutions and individuals in agricultural research systems, and many positive examples of institutional innovations and reforms can be seen. Many of the reforms, such as increasing contract research, are driven by financial necessity. Others, such as re-directing research away from minor crops and irrigation, toward work on improved tillage systems and erosion control practices, are driven by scientists responding to client demands. Many agricultural researchers and institutions are seeking collaborative arrangements with foreign universities, government agencies and private firms. There are new activities for many scientists, despite the novelty of these relationships, and an entrepreneurial drive to create opportunities in research is clearly evident.
- Many agricultural research institutes, especially in Russia, control extensive land holdings. These assets might form the basis for real "land grant" institutions. Well-managed farms can provide the much-needed cash flows to support research activities. Furthermore, it may be possible to sell excess land and re-invest the proceeds in needed research facilities or as an endowment for financing research.

WEAKNESSES AND THREATS

The fundamental weakness of CT NARS is that they cannot be financially or politically supported and sustained as currently configured. Many of weaknesses are legacies of central planning that weakened the vitality of the agricultural research system.

- Centralized management is still the norm in most countries. Research managers and scientists are largely accountable to the centre, rather than to the end user. Only to the extent that top administrators correctly anticipate end-user needs can the current system be viewed as demand driven.
- The organization of NARS is extremely complex. Scientific councils, boards and committees abound. The effectiveness of those systems in establishing appropriate priorities, incentives and oversight is questionable.
- The system appears to have a great deal of overlap and duplication of responsibilities and a lack of coordination among R&D units, as well with institutions of higher education with research programmes. Duplication is difficult to pinpoint in any NARS, but the sheer size, isolation and lack of local accountability would suggest that duplicated effort is likely.
- Research has focused on increasing primary agricultural production. Research objectives have tended to be quota-driven, with little regard for economic efficiency, product quality, environmental consequences or the safety of agricultural workers. This orientation is still evident throughout the system.
- Little, if any, research capability exists in agricultural economics, agribusiness management and all related social sciences. This fact remains, despite the large number of economists working in separate research units. For ideological reasons, there had been little contact between all CT, but especially little between FSU and Western agricultural economists in the past. The two groups probably share knowledge of constraint optimization. Beyond that, however, little in the training or orientation of economists allows them to tackle the problems of market-based agriculture or the necessary transition.
- Agricultural scientists have had almost no exposure to concepts of agricultural economics or farm management. Limited economic literacy makes it difficult for them to understand incentives for farmers to adopt new technologies. Consequently, the design attributes of new production technologies do not reflect the realities of decentralized, profit-oriented farm management. Research capacity in utilization, food science, storage, transportation, logistics and marketing is rudimentary at best. It is found mainly in specialized research units with little contact with agriculture scientists and end users.
- The integration of production research with environmental disciplines

is extremely limited. To some degree this reflects the limited scientific development of agro-ecology. This concerns mostly the FSU. However, some environmentally relevant research capacity does exist, e.g. in soil conservation and land reclamation. Again, the vertical, discipline-based structure of agricultural research system limits opportunities for multidisciplinary research. This working environment is essential if integrated production systems are to be developed and successfully transferred.

- Laws governing Intellectual Property Rights (IPR) are being developed but will be extremely difficult to enforce in most CT. As a consequence, publicly funded intellectual property - often in the form of crop varieties - can end up in private hands without payment of royalties to the institutions that developed it. Along similar lines, the lack of strong and enforceable intellectual property laws inhibits the growth of privately-funded research and technology transfer.
- Central planning has determined the location and structure of many agricultural enterprises. These decisions have in turn influenced the structure and orientation of agricultural research. Economic reforms and price liberalization will change the scope, scale and location of many country's agriculture to some, largely unknown, extent. The current configuration of agriculture research systems in many countries does not necessarily reflect emerging changes in agriculture.

PRODUCTIVITY OF AGRICULTURAL RESEARCH SYSTEMS

The productivity of a public agricultural research system is determined by several interrelated factors. Among the most important are:

- The management of the research enterprise itself, including priority setting, problem focus, scientist training and motivation;
- The level of support and investment for scientists;
- The efficacy of public education and technology transfer systems;
- The ability and incentives for the private sector to commercialize research funding; and
- The efficiency and profitability of the agricultural sector.

In most CT, the public NARS appears unproductive because many of these conditions are not met. Private national agricultural research institutes are just emerging, and seem more efficient and customer driven.

AGRICULTURE AND RURAL DEVELOPMENTS

From a nation dependent on food imports to feed its population, India today is not only self—sufficient in grain production, but also has a substantial reserve.

The progress made by agriculture in the last four decades has been one of the biggest success stories of free India. Agriculture and allied activities constitute the single largest contributor to the Gross Domestic Product, almost 33 per cent of it. Agriculture is the means of livelihood of about two—thirds of the work force in the country. This increase in agricultural production has been brought about by bringing additional area under cultivation, extension of irrigation facilities, the use of improved high yielding variety of seeds, better techniques evolved through agricultural research, water management, and plant protection through judicious use of fertilizers, pesticides and cropping practices.

IRRIGATION

As efforts continued to increase the irrigation potential in the country, the last 40 years saw the gross irrigated area reach 8~ million hectares. Flood forecasting has become an important activity over the years. Over 500 hydrological stations collect and transmit data through 400 wireless stations for issuing forecasts for 157 sites. About 5000 forecasts are issued in a year with 94 per cent accuracy. The country also receives international support, with the World Bank as a primary source, for developing the water resources. International cooperation is also envisaged in setting up a National Centre for Information on Water and Power. As there is a broad seismic belt in the country, particularly along the Himalayan, the Kutch region and ports of Maharashtra, a scheme is being evolved to collect all data on seismic activity at various dam sites.

CROPS

The 1970s saw a multi-fold increase in wheat production that heralded the Green Revolution. In the next decade rice production rose significantly; in 1995-96, rice production was 79.6 million tons. Total grain production crossed 191 million tons in 1994-95, a big leap from 51 million tons in 1950-5 1. During the Seventh Plan, the average grain production was 155 million tons, 17 million tons more than the Sixth Plan average.

To carry improved technologies to farmers, a National Pulse Development Programme, covering 13 states, was launched in 1986. The Special Food Production Programme augmented efforts to boost pulse production further. In 1995-96, pulse production was 13.2 million tons. With some States offering more than the statutory minimum price, sugarcane production also received a boost, in 1995-96 a record 283.0 million tons was registered.

FERTILIZERS

The fertilizer industry in India has grown tremendously in the last 30 years. The Government is keen to see that fertilizer reaches the farmers in the remote and hilly areas. It has been decided to decontrol the prices, distribution and movement of phosphatic and potassic fertilizers. Steps have been taken to ensure an increase in the supply of non-chemical fertilizers at reasonable prices.

There are 53 fertilizer quality control laboratories in the country. Since bio—fertilisers are regarded as an effective, cheap and renewable supplement to chemical fertilizers, the Government is implementing a National Project on Development and Use of Bio-fertilisers. Under this scheme, one national and six regional centres for organizing training, demonstrating programmes and quality testing of bio-fertilisers has been taken up.

It was a challenging decision of the Government to take Bombay High gas through a 1,700-km pipeline to feed fertilizer plants located in the consumption centres of North India.

However, the major policy which has ensured the growth of the fertilizer industry is the thrust on accelerating fertilizer consumption by fixing, on the one hand, low and uniform price for fertilizers, and on the other hand providing the manufacturers adequate compensation through the retention price and subsidy scheme. As expected, fertilizer nutrient demand has gone up from 0.29 million tons in 1960-61 to 13.9 million tons at the end of 1995-96, compared to 12.15MT during 1992-93.

FOOD PROCESSING

A Ministry of Food Processing Industries was established in July 1988 to ensure better utilization of farmers' output by inducting modem technology into the processing of food products, thus augmenting the income of farmers and generating employment opportunities in rural areas. A new seeds policy has been adopted to provide access to high quality seeds and plant material for vegetables, fruit, flowers, oil-seeds and pulses, without in any way compromising quarantine conditions. Initiatives have been taken to encourage private sector investment in the food processing industry.

FISHERIES

Fish production achieved an all-time high of 4.9 million tons at the end of 1995-96. Programmes that have helped boost production include the National Programme of Developing Fish Seeds, Fish Farmers' Development Agencies and Brackish Water Fish Farmers' Development Agencies. The Central Institute of Fisheries Nautical and Engineering Training trains the necessary manpower. To diversify fishing methods and introduce processed fish products on a semi-commercial scale, an Integrated Fisheries Project has been launched. A National Fisheries Advisory Board has also been established.

AGRICULTURAL RESEARCH

The apex body for education, research and extension education in the field of agriculture is the Indian Council of Agricultural Research (ICAR), established in 1929. India's transformation from a food deficit to a food surplus country is largely due to ICAR's smooth and rapid transfer of farm technology from the laboratory to the land. ICAR discharges its responsibilities through 43 research institutes, four national research bureaus, 20 national research centres, nine project directorates, 70 all-India coordinated research projects, and 109 Krishi Vigyan Kendras (farm science centres). Besides, the programme of Agricultural Education is coordinated by ICAR with the curricula and other normative guidance given to the 26 Agricultural Universities and four National Research Institutes.

DRINKING WATER

A Technology Mission on Drinking Water and Related Water Management has been constituted to cover the residual problem villages and provide potable water at 40 liter per capita per day, and 70 liters per capita per day in desert areas inclusive of 30 liters for cattle. The Mission is tackling the problem through 55 mini-missions in project districts and countrywide problem oriented sub-missions. A Village Level Operation and Maintenance (VLOM) pump called India Mark-11 has been developed and is being exported to 40 countries. By March 31, 1993, over 79 per cent of the rural and about 85 per cent of the urban population was provided drinking water facilities.

OILSEEDS PRODUCTION

A Technology Mission on Oilseeds was launched in 1986 to increase production of oilseeds in the country and attain self-sufficiency. Pulses were brought under the Technology Mission in 1990. Before the Mission was launched in 1985-86, oilseed production was 10.83 million tons; during 1995-96, it was estimated at 22.42 million tons, which is a record. Soybean, rapeseed and mustard largely contributed the increase in production. Production of pulses has seen many ups and downs, which is expected to be checked under the Mission. The country grows mainly nine oilseeds, with groundnut, rapeseed and mustard accounting for 62 per cent of total production. Lately, soybean and sunflower have shown major growth potential.

Bibliography

Alexis Leon and Mathews Leon: *Database Management Systems*, Leon Vikas, Delhi, 2002.

Amandeep Kaur Hundal: *Financial Management*, Knowledge Book Distributors, Delhi, 2010.

Amit Kumar Arora: *Financial Management*, Global Vision Publishing House, Delhi, 2013.

Arun Bhargav: *Rural Marketing and Agribusiness in India*, Surendra Publication, Delhi, 2010.

G. Ramesh Babu: *Financial Management*, Concept Publication, Delhi, 2012.

Gagan Varshney: *Database Management System (DBMS)*, Global Vision, Delhi, 2010.

J. M. Talathi, V. G. Naik and V. N. Jalgaonkar: *Introduction to Agricultural Economics and Agribusiness Management*, Ane Books, Delhi, 2008.

L.L. Somani: *Dictionary of Agribusiness Management*, Agrotech Publishing Academy, Delhi, 2007.

Lokanadan, Kornam: *Innovations in Agribusiness Management*, New India Publishing Agency, Delhi, 2009.

Mukesh Pandey and Deepali Tewari: *The Agribusiness Book : A Marketing and Value-Chain Perspective*, IBDC Publication, Delhi, 2010.

P S Gill: *Database Management Systems*, I.K. International, Delhi, 2009.

P.C.Tulsian: *Financial Management*, S. Chand Publisher, Delhi, 2009.

P.N. Mishra: *Database Management System and Digital Libraries*, Alfa Publications, Delhi, 2010.

Pankaj K. Agarwal and Neeraj Agarwal: *Financial Management*, Word Press, Delhi, 2011.

Pranab Kumar Das Gupta: *Database Management System, Oracle SQL And PL/SQL*, PHI Learning, Delhi, 2003.

R. Shekhar: *Financial Management*, Arise Publishers, Delhi, 2011.

Rajesh Kumar and Santosh: *Financial Management*, Shree Publication, Delhi, 2011.

Rajesh Narang: *Database Management Systems*, PHI Learning, Delhi, 2001.

Rajiv Chopra: *Database Management System A Practical Approach*, S. Chand Publisher, Delhi, 2010.

Rajiv Srivastava and Anil Misra: *Financial Management*, Oxford University Press, Delhi, 2011.

Ravi M. Kishore: *Financial Management - Problems and Solutions*, Taxmann, Delhi, 2009.

Ravi M. Kishore: *Financial Management*, Taxmann, Delhi, 2009.

S.C. Gaur and D. Singh: *A Handbook of Agribusiness*, Agrobios Publication, Delhi, 2012.

S.C. Gaur: *Agribusiness : Market and Marketing*, Agrobios Publication, Delhi, 2014.

S.C. Gaur: *Agribusiness: Management Information System*, Agrobios Publication, Delhi, 2014.

Smita Diwase: *Indian Agriculture and Agribusiness Management*, Scientific Publication, Jaipur, 2014.

Sudhir Gupta: *Financial Management*, Centrum Press, Delhi, 2011.

Swarup K. Das: *A Textbook of Database Management Systems*, Wisdom Press, Delhi, 2011.

Wesley J. Obst, Rob Graham and Graham Christie: *Financial Management for Agribusiness*, Scientific Publication, Delhi, 2010.

William W. Sihler, Richard D. Crawford and Henry A. Davis: *Financial Management*, Jaico Publication, Delhi, 2003.

Y K Singh: *Database Management System*, Shree Publication, Delhi, 2006.

Index

G

I

L

M

O

P

R

S

T